Android 炫酷应用 300 例
实战篇

罗帅 罗斌 编著

清华大学出版社

北京

内 容 简 介

本书以"问题描述＋解决方案"的模式，以 Android 5.0 为核心例举了 300 个实用性极强的移动端应用开发案例，旨在帮助广大读者快速解决实际开发过程中面临的诸多问题，从而不断提高开发效率、拓展应用领域。全书根据实例功能将内容分为 UI 布局、常用控件、文字、图形和图像、动画、音频和视频、文件和数据、系统和设备、第三方 SDK 开发等 9 章，以所见即所得、所学即所用的速成思维方式展示了个性化布局、特效文字和图形、矢量图形动态绘制、颜色矩阵特效、PorterDuff 特效、路径特效、场景过渡、三维空间旋转、图像轮播、头像裁剪、网格图像动画、旋转 3D 地球、各种传感器应用、网络文件断点续传等超炫超酷实例的实现过程和代码。使用第三方 SDK 进行应用开发，如使用腾讯 SDK 实现将文本、图像、音乐、视频等分享到 QQ 好友及微信朋友圈等；使用百度 SDK 实现查询驾车和步行线路、查询指定城市的热力图、调用百度地图的导航功能、以俯视角度观察街道三维图、在百度地图上添加悬浮框和动画等；使用新浪 SDK 实现将图像发布到微博、根据微博简介内容生成二维码等。

为了突出实用性和简洁性，本书在演示或描述实例时，力求有针对性地解决问题，所有实例均配有插图。本书适于作为 Android 应用开发人员的案头参考书，无论是编程初学者，还是编程高手，本书都极具参考价值和收藏价值。

本书封面贴有清华大学出版社防伪标签，无标签者不得销售。
版权所有，侵权必究。侵权举报电话：010-62782989 13701121933

图书在版编目(CIP)数据

Android 炫酷应用 300 例·实战篇/罗帅,罗斌编著. ——北京：清华大学出版社,2019
ISBN 978-7-302-52627-8

Ⅰ. ①A… Ⅱ. ①罗… ②罗… Ⅲ. ①移动终端—应用程序—程序设计 Ⅳ. ①TN929.53

中国版本图书馆 CIP 数据核字(2019)第 046905 号

责任编辑：黄　芝
封面设计：刘　键
责任校对：胡伟民
责任印制：沈　露

出版发行：清华大学出版社
网　　址：http://www.tup.com.cn, http://www.wqbook.com
地　　址：北京清华大学学研大厦 A 座　　　　　　　　　　　邮　编：100084
社 总 机：010-62770175　　　　　　　　　　　　　　　　　邮　购：010-62786544
投稿与读者服务：010-62776969, c-service@tup.tsinghua.edu.cn
质量反馈：010-62772015, zhiliang@tup.tsinghua.edu.cn
课件下载：http://www.tup.com.cn, 010-62795954

印 装 者：清华大学印刷厂
经　　销：全国新华书店
开　　本：210mm×285mm　　印　张：29.75　　字　数：1006 千字
版　　次：2019 年 7 月第 1 版　　　　　　　　　　印　次：2019 年 7 月第 1 次印刷
印　　数：1～2500
定　　价：99.80 元

产品编号：081948-01

　　Android 是一个以 Linux 为基础的半开源操作系统，主要用于管理手机、手表、眼镜、电视等智能设备。Android 操作系统最初由 Andy Rubin 开发，2005 年 8 月由 Google 收购注资。2007 年 11 月，Google 与 84 家硬件制造商、软件开发商及电信营运商组建开放手机联盟，共同研发改进 Android 操作系统。第一部 Android 智能手机发布于 2008 年 10 月，在随后几年中，Android 开始了迅猛发展的历程，并成为全球最主要的移动端操作系统。

　　随着支持 Android 操作系统的智能设备的不断普及和推广，运行在 Android 操作系统上的智能应用项目也如雨后春笋般涌现。从 IT 发展潮流来看，越来越丰富的移动应用是大势所趋，手机支付、手机拍照、手机游戏、手机导航、物联网等不断改变着人们的生活方式和工作方式。一种优秀的 Android 应用即可造就一家 IT 公司，打造一个产业链、诞生十个富豪，这已不再是神话。如何将最新的技术、理念和创意融入到应用开发中，是每个 Android 程序员需要不断思考的问题，也是本书创作的初衷。

　　本书以"问题描述＋解决方案"的模式，以 Android 5.0 为核心例举了 300 个实用性极强的移动端应用开发案例，旨在帮助广大读者快速解决实际开发过程中面临的诸多问题，从而不断提高开发效率、拓展应用领域。全书根据实例功能将内容分为 UI 布局、常用控件、文字、图形和图像、动画、音频和视频、文件和数据、系统和设备、第三方 SDK 开发等 9 章，以所见即所得、所学即所用的速成思维方式展示了个性化布局、特效文字和图形、矢量图形动态绘制、颜色矩阵特效、PorterDuff 特效、路径特效、场景过渡、三维空间旋转、图像轮播、头像裁剪、网格图像动画、旋转 3D 地球、各种传感器应用、网络文件断点续传等超炫超酷实例的实现过程和代码。使用第三方 SDK 进行应用开发，如使用腾讯 SDK 实现将文本、图像、音乐、视频等分享到 QQ 好友及微信朋友圈等；使用百度 SDK 实现查询驾车和步行路线、查询指定城市的热力图、调用百度地图的导航功能、以俯视角度观察街道三维图、在百度地图上添加悬浮框和动画等；使用新浪 SDK 实现将图像发布到微博、根据微博简介内容生成二维码等。

　　本书所有实例均基于 Android 5.0，在 Android Studio 2.2 集成开发环境中使用 Java 和 XML 语言编写，因此测试手机或模拟器的 Android 版本不能低于 5.0。部分实例在模拟器上无法测试，建议在学习时使用屏幕分辨率为 1920×1080 像素，操作系统为 Android 5.0 及其以上版本的手机作为主要测试工具。

　　全书所有内容和思想并非一人之力所能及，而是凝聚了众多热心人士的智慧并经过充分的提炼和总结而成，在此对他们表示崇高的敬意和衷心的感谢！本书编写人员包括罗帅、罗斌、汪明云、曹勇、陈宁、邓承惠、邓小渝、范刚强、何守碧、洪亮、洪沛林、江素芳、蓝洋、雷国忠、雷惠、雷玲、雷平、雷治英、刘恭德、刘兴红、罗聘、唐静、唐兴忠、童缙嘉、汪兰、王彬、王伯芳、王年素、王正建、吴多、吴诗华、杨开平、杨琴、易伶、张志红、郑少文等，书稿由罗斌完成统稿。由于时间关系和作者水平原因，少量内容可能存在认识不全面或有偏颇的问题，以及一些疏漏和不当之处，敬请读者批评指正。

读者可将购书凭证发送至邮箱 huangzh@tup.tsinghua.edu.cn，索取本书源代码。

罗帅　罗斌
2019年于重庆渝北

第 1 章 UI 布局 ... 1

- 001 使用纯 Java 代码创建应用 UI 界面 ... 1
- 002 使用自定义 View 代替布局文件 ... 2
- 003 使用 TableLayout 布局多个输入框 ... 4
- 004 使用 TextInputLayout 管理输入框提示 ... 5
- 005 使用 GridLayout 创建计算器按键布局 ... 7
- 006 使用 RelativeLayout 按照相邻关系布局 ... 9
- 007 使用 ConstraintLayout 在右下角布局 ... 10
- 008 使用 TableLayout 拉伸控件填充容器 ... 12
- 009 使用 TableLayout 缩小控件适应容器 ... 12
- 010 使用 LinearLayout 纵向居中对齐控件 ... 14
- 011 使用 LinearLayout 按权重分配控件空间 ... 16
- 012 使用 ConstraintLayout 平分剩余空间 ... 17
- 013 使用 ConstraintLayout 无间隙布局控件 ... 18
- 014 使用 TabLayout 和适配器创建选项卡 ... 19
- 015 使用 TabLayout 和 Fragment 创建选项卡 ... 22
- 016 使用 FrameLayout 创建纵向选项卡 ... 25
- 017 使用 TabHost 创建横向选项卡 ... 28
- 018 使用 AbsoluteLayout 实现平移控件 ... 29
- 019 使用 FrameLayout 实现闪烁控件 ... 31
- 020 自定义 FrameLayout 创建翻页卷边动画 ... 34

第 2 章 常用控件 ... 36

- 021 在 TextView 中创建空心文字 ... 36
- 022 在 TextView 中实现上文下图的布局 ... 37
- 023 在 TextView 中为文本添加超链接 ... 38
- 024 在自定义 View 中实现垂直滚动文本 ... 39
- 025 在 EditText 中指定输入法的数字软键盘 ... 41
- 026 禁止在 EditText 中插入非字符表情符号 ... 41
- 027 使用 AutoCompleteTextView 实现自动提示 ... 43
- 028 使用 SearchView 和 ListView 实现过滤输入 ... 44
- 029 在 EditText 右端设置输入提示内容和图标 ... 45
- 030 通过自定义 Shape 创建不同的圆角按钮 ... 46

031　通过设置背景图像创建立体的质感按钮 ………………………… 49
032　使用 FloatingActionButton 创建悬浮按钮 ……………………… 50
033　以全屏效果显示在 ImageView 中的图像 ………………………… 51
034　在自定义 ImageView 中显示圆形图像 …………………………… 52
035　使用单指滑动拖曳 ImageView 的图像 …………………………… 54
036　使用 Gallery 实现滑动浏览多幅图像 ……………………………… 56
037　使用 SwipeRefreshLayout 切换图像 ……………………………… 57
038　使用 AdapterViewFlipper 自动播放图像 ………………………… 58
039　使用两幅图像定制 ToggleButton 开关状态 ……………………… 60
040　使用 GridView 创建网格显示多幅图像 …………………………… 62
041　使用 ViewPager 实现缩放轮播多幅图像 ………………………… 64
042　使用 Handler 实现自动轮播 ViewPager ………………………… 67
043　使用 ViewPager 实现苹果风格的 cover flow …………………… 69
044　使用 RecyclerView 创建水平瀑布流图像 ………………………… 72
045　以网格或列表显示 RecyclerView 列表项 ………………………… 74
046　使用 RecyclerView 仿表情包插入输入框 ………………………… 77
047　使用 CardView 显示 RecyclerView 列表项 ……………………… 79
048　在 ListView 中创建图文结合列表项 ……………………………… 82
049　使用 ListPopupWindow 实现下拉选择 …………………………… 84
050　使用 Elevation 创建阴影扩散的控件 ……………………………… 85
051　在单击 CheckBox 时显示波纹扩散效果 ………………………… 86
052　使用自定义形状定制 Switch 开关状态 …………………………… 88
053　自定义 selector 以渐变前景切换控件 …………………………… 90
054　使用 ViewSwitcher 平滑切换两个 View ………………………… 92
055　使用 SlidingDrawer 实现抽屉式滑动 …………………………… 93
056　自定义 ScrollView 实现下拉回弹动画 …………………………… 96
057　使用 CollapsingToolbarLayout 实现滚动折叠 ………………… 98
058　使用 BottomNavigationView 实现底部导航 …………………… 99
059　在 ProgressBar 上同时显示两种进度 …………………………… 101
060　使用 ViewOutlineProvider 创建圆角控件 ……………………… 103
061　使用 AnalogClock 创建自定义时钟 ……………………………… 104
062　在 TextClock 中定制日期格式 …………………………………… 105
063　使用 RatingBar 实现星级评分 …………………………………… 106
064　在登录窗口中使用 SeekBar 实现手动校验 ……………………… 107

第 3 章　文字 ……………………………………………………………… 110

065　使用 ScaleXSpan 创建扁平风格的文字 …………………………… 110
066　使用 MaskFilterSpan 实现文字边缘模糊 ………………………… 111
067　使用 MaskFilterSpan 实现文字中心镂空 ………………………… 112
068　使用 MaskFilterSpan 实现文字整体模糊 ………………………… 113
069　使用 MaskFilterSpan 模糊多个字符串 …………………………… 114
070　使用 BulletSpan 在文本首字前添加小圆点 ……………………… 115

071	使用 StrikethroughSpan 添加文字删除线	116
072	使用 URLSpan 为部分内容添加超链接	117
073	使用 ImageSpan 同时显示 QQ 表情和文字	119
074	使用 StyleSpan 实现以粗斜体显示文字	120
075	使用 SuperscriptSpan 绘制勾股定理公式	121
076	使用 SubscriptSpan 绘制硫酸亚铁分子式	122
077	使用 TypefaceSpan 定制文本的部分内容	123
078	使用 ForegroundColorSpan 创建光照文字	125
079	使用 BlurMaskFilter 创建阴影扩散文字	126
080	使用 EmbossMaskFilter 创建浮雕文字	127
081	通过自定义 View 在半圆弧上绘制文字	128
082	通过自定义 View 在圆弧上滚动文字	130
083	通过自定义 View 绘制渐变色的文字	132
084	通过自定义 View 绘制线条描边文字	133
085	通过自定义 View 绘制阴影扩散文字	134
086	加载字库文件显示自定义草书字体	135
087	加载字库文件显示自定义液晶字体	136
088	判断在一个字符串中是否包含汉字	137

第 4 章　图形和图像 ……… 138

089	在自定义 View 中绘制径向渐变的图形	138
090	在自定义 View 中实现图像波纹起伏效果	139
091	在自定义 View 中使用椭圆裁剪图像	141
092	通过 PorterDuff 模式增暗显示两幅图像	142
093	通过 PorterDuff 模式将图像裁剪成五角星	143
094	通过 PorterDuff 模式改变 tint 属性叠加效果	144
095	使用 Region 的 DIFFERENCE 实现抠图功能	145
096	使用 ShapeDrawable 裁剪三角形图像	146
097	使用 ClipDrawable 裁剪图像实现星级评分	147
098	使用自定义 Drawable 实现对图像进行圆角	149
099	使用 Matrix 实现按照指定方向倾斜图像	150
100	使用 ColorMatrix 为图像添加泛紫效果	151
101	使用 ColorMatrix 实现图像的加暗效果	152
102	通过自定义 ColorMatrix 调整图像蓝色色调	153
103	使用 RenderScript 实现高斯算法模糊图像	154
104	使用拉普拉斯模板实现图像的锐化特效	156
105	通过像素操作实现在图像上添加光照效果	158
106	通过像素操作使彩色图像呈现浮雕特效	159
107	使用 BitmapShader 实现文字线条图像化	161
108	使用 BlurMaskFilter 为图像添加轮廓线	162
109	使用 PathDashPathEffect 实现椭圆线条	163
110	使用 SumPathEffect 叠加多种路径特效	164

111	通过 BitmapShader 实现以图像填充椭圆	165
112	使用 ComposeShader 创建渐变图像	166
113	使用 ImageView 显示 XML 实现的矢量图形	168
114	使用 BitmapFactory 压缩图像的大小	169
115	在自定义类中使用 Movie 显示动态图像	170
116	通过使用图像作为画布创建带水印图像	172
117	通过操作根布局实现将屏幕内容保存为图像	174
118	通过手势变化实现平移旋转缩放图像	175
119	使用 ThumbnailUtils 提取大图像的缩略图	177
120	通过采用取模的方式实现轮流显示多幅图像	178

第 5 章 动画 ... 180

121	使用 ObjectAnimator 创建上下振动动画	180
122	使用 ObjectAnimator 实现沿弧线路径平移	181
123	使用 ObjectAnimator 滚动显示多幅图像	182
124	使用 ObjectAnimator 实现图形数字形变	184
125	使用 ObjectAnimator 改变图像的色相值	186
126	使用 AnimatorSet 组合多个 ObjectAnimator	188
127	使用 TypeEvaluator 实现颜色过渡动画	189
128	通过 trimPathEnd 实现动态生成手指图形	191
129	使用 ValueAnimator 动态改变扇形转角	193
130	使用 ValueAnimator 实现分段转圈动画	195
131	使用 ValueAnimator 在三维 Z 轴上平移图像	197
132	使用 ValueAnimator 实现起飞转平飞动画	198
133	自定义 TypeEvaluator 以 GIF 动画显示图像	199
134	使用 Animation 实现图像围绕自身中心旋转	201
135	自定义 Animation 实现旋转切换扑克牌正反面	202
136	使用 AnimationSet 实现组合多个不同的动画	204
137	使用 Animation 实现按照顺序显示网格 Item	205
138	使用 windowAnimations 实现缩放对话框窗口	207
139	使用 AnimationDrawable 播放多幅图像	209
140	使用 AnimationDrawable 创建爆炸动画	211
141	使用 RotateAnimation 实现围绕自身中心旋转	213
142	使用 AlphaAnimation 创建淡入淡出动画	214
143	使用 ScaleAnimation 创建缩放图像动画	215
144	在 ViewPager 中实现上下滑动的转场动画	216
145	通过下拉手指实现两个 Activity 的相互切换	217
146	在应用启动时使用进场动画启动 Activity	218
147	以左入右出的动画效果切换两个 Activity	220
148	以收缩扩张的动画效果切换两个 Activity	221
149	使用转场动画 Explode 切换两个 Activity	223
150	使用转场动画 Slide 切换两个 Activity	224

151	以指定位置的转场动画切换两个 Activity	225
152	在切换 Activity 时叠加缩放动画和转场动画	227
153	在切换 Activity 的转场动画中共享多对元素	229
154	使用 FragmentTransaction 自定义转场动画	230
155	使用 TransitionManager 实现上下滑动动画	232
156	使用 TransitionManager 实现围绕 Y 轴旋转	234
157	使用 TransitionManager 实现 Fade 动画效果	235
158	使用 TransitionManager 组合多个不同动画	238
159	使用 TransitionManager 实现单布局过渡动画	240
160	使用 TransitionManager 实现平移过渡动画	241
161	使用 TransitionManager 实现缩放部分图像	243
162	使用 TransitionManager 实现矢量路径动画	244
163	使用 TransitionManager 同时实现多种动画	247
164	使用 TransitionManager 实现 XML 定制动画	248
165	使用 TransitionManager 指定控件执行动画	250
166	使用 TransitionManager 实现列表项滑入动画	251
167	使用 TransitionManager 实现弧线路径动画	252
168	使用 TransitionManager 实现裁剪区域动画	254
169	通过设置和获取控件的 Tag 确定动画过渡行为	255
170	在 TransitionSet 中指定多个动画的执行顺序	256
171	使用 TransitionDrawable 透明切换两幅图像	258
172	使用 AnimatedVectorDrawable 实现转圈动画	259
173	创建 AnimatedVectorDrawableCompat 动画	261
174	使用 ViewPropertyAnimator 创建多个动画	263
175	自定义 selector 实现以动画形式改变阴影大小	264
176	使用 ripple 标签创建中心波纹扩散动画	265
177	使用 GLSurfaceView 实现 3D 地球的自转	266

第 6 章 音频和视频 268

178	使用 MediaPlayer 播放本地 mp3 音乐文件	268
179	使用 MediaPlayer 播放本地 mp4 视频文件	269
180	使用 MediaPlayer 播放指定网址的音乐文件	270
181	使用滑块同步 MediaPlayer 播放音频的进度	273
182	使用滑块同步 MediaPlayer 播放视频的进度	275
183	使用 MediaController 创建视频播放控制栏	278
184	使用 MediaMetadataRetriever 实现视频截图	280
185	使用 MediaMetadataRetriever 获取视频缩略图	281
186	使用 VideoView 播放本地 mp4 视频文件	282
187	使用 VideoView 播放指定网址的视频文件	283
188	使用 MediaRecorder 录制音频文件	284
189	使用 RemoteViews 在通知栏上创建播放器	286
190	在使用 SurfaceView 播放视频时实现横屏显示	288

191	在选择音乐曲目窗口中选择音乐文件并播放	291
192	在 RecyclerView 中加载音乐文件并播放	292
193	依次播放在 RecyclerView 中的音乐文件	296
194	在 ListView 上加载手机外存的音乐文件	298
195	使用 SoundPool 播放较短的声音片段	300
196	使用 AudioManager 增大或减小音量	301
197	使用 AudioManager 播放系统预置的声音	303
198	使用 AudioManager 获取和设置铃声模式	304

第 7 章 文件和数据 ············ 306

199	使用 JSONObject 解析 JSON 字符串	306
200	使用 JSONArray 解析 JSON 字符串	307
201	使用 JSONTokener 解析 JSON 字符串	308
202	使用 JsonReader 解析 JSON 字符串	309
203	使用 JSONStringer 创建 JSON 字符串	311
204	使用 JSONObject 根据 IP 显示所在城市	312
205	使用 Gson 将数组转换成 JSON 字符串	314
206	使用 Gson 解析 JSON 字符串	315
207	使用 XmlPullParser 解析城市天气数据	317
208	采用 SAX 方式解析 XML 文件内容	320
209	使用 Pattern 根据正则表达式校验手机号码	321
210	使用 SharedPreferences 保存账户和密码	323
211	使用 ListPreference 读写单选按钮值	324
212	在代码中获取 CheckBoxPreference 值	326
213	通过 PreferenceScreen 跳转到 Wifi 设置	327
214	使用 Intent 实现在 Activity 之间传递小图像	329
215	使用 Intent 在 Activity 之间传递图像和文本	330
216	使用 Intent 在 Activity 之间传递集合数据	332
217	在 Intent 传递数据时使用 Bundle 携带数组	333
218	使用 Intent 在 Service 和 Activity 之间传递数据	335
219	使用 FileInputStream 和 FileOutputStream 读取和保存文本文件	337
220	将浮雕风格的特效文字保存为图像文件	338
221	在 SD 卡上将 Bitmap 保存为 PNG 图像文件	339
222	从手机相册中选择图像文件并裁剪头像	341
223	在 ListView 上加载手机外存的图像文件	342
224	使用 DownloadManager 下载网络文件	344
225	使用 RandomAccessFile 实现断点续传下载	346
226	使用 HttpURLConnection 下载图像文件	348

第 8 章 系统和设备 ············ 351

| 227 | 使用 QuickContactBadge 访问联系人 | 351 |
| 228 | 使用 ContentProviderOperation 增加联系人 | 352 |

编号	标题	页码
229	使用 ContentProviderOperation 修改联系人	354
230	使用 ContentProviderOperation 删除联系人	356
231	使用 ContentResolver 检测飞行模式的状态	358
232	使用 ContentResolver 检测手机的时间格式	359
233	使用 ContentResolver 获取所有短信	359
234	使用 ContentResolver 获取通话记录	361
235	使用 ContentResolver 获取 SD 卡的文件	363
236	使用 ContentResolver 改变屏幕亮度值	365
237	使用 ContentResolver 设置屏幕亮度值	366
238	使用 ContentResolver 检测旋转屏幕功能	367
239	使用 BroadcastReceiver 监听来电电话号码	368
240	使用 BroadcastReceiver 判断手机电池是否正在充电	369
241	使用 BroadcastReceiver 监听屏幕开启或关闭	371
242	自定义 BroadcastReceiver 实现短信拦截	372
243	使用 RingtoneManager 设置手机闹钟铃声	373
244	使用 RingtoneManager 设置手机通知铃声	375
245	使用 AlarmManager 以指定时间执行操作	376
246	使用 AudioManager 获取和设置音量	377
247	使用 PowerManager 实现屏幕一直亮着	379
248	使用 WallpaperManager 设置壁纸	380
249	使用 PackageManager 获取支持分享的应用	381
250	使用 WifiManager 开启或关闭 WiFi 信号	382
251	使用 WifiManager 获取 IP 地址	384
252	使用 ConnectivityManager 判断网络状态	385
253	使用 BluetoothAdapter 打开或关闭蓝牙	386
254	使用 LocationListener 获取当前经纬度值	387
255	使用 SensorManager 获取传感器信息	390
256	使用传感器监测耳朵与手机听筒的距离	391
257	使用加速度传感器监听手机的三维变化	393
258	通过传感器实现自动进行横屏和竖屏切换	395
259	使用 setRequestedOrientation()实现横屏	397
260	根据手机是横屏或是竖屏进行控件布局	398
261	使用 FLAG_FULLSCREEN 标志实现全屏显示	399
262	使用 Display 获取屏幕宽度和高度	401
263	使用 StatFs 获取内部总空间和可用空间大小	401
264	使用 GestureDetector 实现纵向滑动切换	403
265	自定义手机振动器(Vibrator)的振动模式	405
266	使用 SurfaceView 实现照相机的预览功能	406
267	使用 Camera 实现缩小和放大预览画面	408
268	使用 Camera 实现预览时摄像头手动对焦	409
269	从相册中选择图像并设置为手机壁纸	412
270	使用 Runnable 间隔执行重复的任务	413

271　使用 Timer 实现促销活动的倒计时功能 ……………………………………………… 414
272　使用 Runtime 执行系统命令静默安装应用包 …………………………………… 415

第 9 章　第三方 SDK 开发 ……………………………………………………………… 419

273　使用腾讯 SDK 获取授权 QQ 账户的简介 ………………………………………… 419
274　使用腾讯 SDK 实现以第三方登录 QQ 账户 ……………………………………… 422
275　使用腾讯 SDK 将指定文本分享给 QQ 好友 ……………………………………… 423
276　使用腾讯 SDK 将本地图像发表到 QQ 空间 ……………………………………… 424
277　使用微信 SDK 将视频链接分享给微信好友 ……………………………………… 426
278　使用微信 SDK 将音乐链接分享到朋友圈 ………………………………………… 428
279　使用百度 SDK 根据起点和终点规划步行线路 …………………………………… 429
280　使用百度 SDK 实现将驾车线路分享给好友 ……………………………………… 432
281　使用百度 SDK 调用百度地图 App 的驾车导航 …………………………………… 434
282　使用百度 SDK 调用百度地图 App 的 POI 检索 …………………………………… 436
283　使用百度 SDK 实现在地图中定位手机位置 ……………………………………… 437
284　使用百度 SDK 获取在地图上点击位置的地名 …………………………………… 438
285　使用百度 SDK 在地图的城市之间绘制连线 ……………………………………… 440
286　使用百度 SDK 在地图上添加图文悬浮框 ………………………………………… 441
287　使用百度 SDK 在地图上添加淡入动画 …………………………………………… 442
288　使用百度 SDK 在地图上添加弹跳型动画 ………………………………………… 444
289　使用百度 SDK 在地图上查询指定城市兴趣点 …………………………………… 445
290　使用百度 SDK 在地图上为行政区添加边界线 …………………………………… 446
291　使用百度 SDK 在地图指定范围添加圆角矩形 …………………………………… 448
292　使用百度 SDK 查询指定地点的热力图 …………………………………………… 449
293　使用百度 SDK 实现隐藏或显示地名标注信息 …………………………………… 450
294　使用百度 SDK 实现以俯视角观察街道三维图 …………………………………… 452
295　使用百度 SDK 实现根据经纬度计算两地距离 …………………………………… 453
296　使用新浪 SDK 实现跳转到微博主页 ……………………………………………… 454
297　使用新浪 SDK 获取授权微博账户的简介 ………………………………………… 456
298　使用新浪 SDK 将微博账户简介生成二维码 ……………………………………… 458
299　使用新浪 SDK 实现搜索指定关键字的微博 ……………………………………… 460
300　使用新浪 SDK 实现发布图像至微博 ……………………………………………… 461

第 1 章

UI布局

001　使用纯 Java 代码创建应用 UI 界面

此实例主要通过在 LinearLayout 线性布局管理器中使用 addView()方法添加控件,从而以纯粹 Java 代码的方式实现 UI 界面。当实例运行之后,最初的界面效果如图 001.1 的左图所示。单击"刷新内容"按钮,则显示最新的日期信息,如图 001.1 的右图所示。

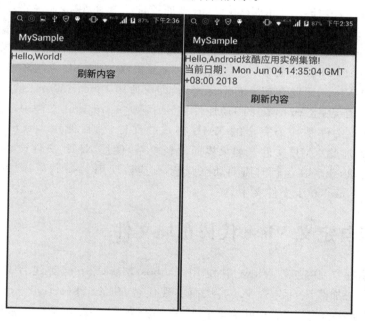

图　001.1

主要代码如下:

```java
public class MainActivity extends Activity {
    @Override
    protected void onCreate(Bundle savedInstanceState) {
        super.onCreate(savedInstanceState);
        // setContentView(R.layout.activity_main);
        LinearLayout myLinearLayout = new LinearLayout(this);      //创建线性布局管理器
        super.setContentView(myLinearLayout);                       //加载线性布局管理器
```

```java
//在线性布局管理器中垂直布局控件
myLinearLayout.setOrientation(LinearLayout.VERTICAL);
final TextView myTextView = new TextView(this);          //创建 TextView 控件
myTextView.setText("Hello,World!");                       //设置 TextView 控件的文本
myTextView.setTextSize(20);                               //设置 TextView 控件的字体大小
Button myButton = new Button(this);                       //创建 Button 控件
myButton.setText("刷新内容");                              //设置 Button 控件的文本
myButton.setTextSize(20);                                 //设置 Button 控件的字体大小
//为 Button 控件添加 Click 单击事件响应方法
myButton.setOnClickListener(new View.OnClickListener() {
    @Override
    public void onClick(View v) {
        myTextView.setText("Hello,Android 炫酷应用实例集锦!\n当前日期:"
                + new java.util.Date());
    } });
myLinearLayout.addView(myTextView);                       //在线性布局管理器中添加 TextView 控件
myLinearLayout.addView(myButton);                         //在线性布局管理器中添加 Button 控件
} }
```

上面这段代码在 MyCode\MySample001\app\src\main\java\com\bin\luo\mysample\MainActivity.java 文件中。在这段代码中,所有的控件都使用 new 关键字创建实例,然后通过 LinearLayout 线性布局管理器的 addView()方法添加控件。当使用 new 关键字创建控件时,在控件的实例化方法中都传入了一个 this 参数,这是由于在 Android 中创建 UI 控件时需要传入一个 Context 参数,Context 代表访问 Android 应用环境的全局信息的 API。如果 UI 控件带有一个 Context 参数,则可以让 UI 控件通过该 Context 参数来获取 Android 应用环境的信息。Android 的大多数控件都继承自 View,因此在控件实例化之后,即可使用 LinearLayout 的 addView()方法向 UI 中添加该控件。默认情况下,Android 的 UI 控件在 activity_main.xml 文件中以可视化的方式进行设计,activity_main.xml 文件是一个完全的 XML 格式的文件,然后使用 setContentView(R.layout.activity_main)方法添加此 XML 文件。如果界面比较复杂,使用 XML 文件设计 UI 控件应是最优选择;如果界面比较简单,或者在运行时需要动态调整,则应该使用 Java 代码添加控件。此实例的完整项目在 MyCode\MySample001 文件夹中。

002 使用自定义 View 代替布局文件

此实例主要通过在自定义 View 中使用 EmbossMaskFilter 创建浮雕文字,并直接使用 setContentView()方法加载该自定义 View,从而实现代替布局文件(activity_main.xml)的效果。当实例运行之后,将显示在自定义 View 中创建的浮雕文字,如图 002.1 所示。主要代码如下:

```java
public class MainActivity extends Activity {
    @Override
    protected void onCreate(Bundle savedInstanceState) {
        setContentView(new MyView(this));                 //加载自定义 View
        super.onCreate(savedInstanceState);
    }
    class MyView extends View {                           //创建自定义 View
        private Paint myPaint;
        EmbossMaskFilter myFilter;
        public MyView(Context context) {
```

```java
    super(context);
    myPaint = new Paint(Paint.ANTI_ALIAS_FLAG);
    myPaint.setColor(Color.RED);
    myPaint.setStyle(Paint.Style.STROKE);
    myPaint.setStrokeWidth(64);
    myPaint.setTextSize(800);
    float[] myDirection = new float[]{1, 1, 1};
    float myLight = 0.05f;                              //设置环境光亮度
    float mySpecular = 5;                               //设置反射等级
    float myBlur = 13;                                  //设置模糊级别
    myFilter = new EmbossMaskFilter(myDirection,
        myLight, mySpecular, myBlur);                   //自定义浮雕滤镜
    myPaint.setMaskFilter(myFilter);                    //设置浮雕滤镜的画笔
}
@Override
protected void onDraw(Canvas myCanvas) {
    super.onDraw(myCanvas);
    Display myDisplay = getWindowManager().getDefaultDisplay();
    int myWidth = myDisplay.getWidth();
    int myHeight = myDisplay.getHeight();
    myCanvas.drawText("炫", myWidth / 10,
        myHeight / 2 + 70, myPaint);                    //显示浮雕文本
}}}
```

图 002.1

上面这段代码在 MyCode\MySample086\app\src\main\java\com\bin\luo\mysample\MainActivity.java 文件中。在这段代码中，setContentView（new MyView（this））用于加载自定义 View，以显示浮雕文字。此外，在使用 EmbossMaskFilter 时，需要在 AndroidManifest.xml 文件中添加 android：hardwareAccelerated ＝"false"，即关闭硬件加速，否则不能显示浮雕效果。此实例的完整项目在 MyCode\MySample086 文件夹中。

003 使用TableLayout布局多个输入框

此实例主要通过使用TableLayout布局管理器,实现在其中以列对齐的方式布局多个TextView控件和EditText控件。当实例运行之后,使用TableLayout布局的多个TextView控件和EditText控件的效果如图003.1所示。

主要代码如下:

```xml
<TableLayout android:id="@+id/TableLayout1"
            android:layout_width="match_parent"
            android:layout_height="match_parent"
            android:paddingBottom="@dimen/activity_vertical_margin"
            android:paddingLeft="@dimen/activity_horizontal_margin"
            android:paddingRight="@dimen/activity_horizontal_margin"
            android:paddingTop="@dimen/activity_vertical_margin">
    <TableRow>
     <TextView android:text="用户名:"
            android:textSize="20dp"/>
     <EditText android:layout_width="300dp"
            android:hint="请输入用户名"
            android:inputType="textPersonName"
            android:textSize="20dp"/>
    </TableRow>
    <TableRow>
     <TextView android:text="密码:"
            android:textSize="20dp"/>
     <EditText android:hint="请输入密码"
            android:password="true"
            android:textSize="20dp"/>
    </TableRow>
    <TableRow>
     <TextView android:text="确认密码:"
      android:textSize="20dp"/>
     <EditText android:hint="请再次输入密码"
            android:inputType="textPassword"
            android:textSize="20dp"/>
    </TableRow>
    <TableRow>
     <TextView android:text="电话号码:"
            android:textSize="20dp"/>
     <EditText android:hint="请输入电话号码"
            android:inputType="phone"
            android:textSize="20dp"/>
    </TableRow>
    <TableRow>
     <TextView android:text="邮箱地址:"
            android:textSize="20dp"/>
     <EditText android:hint="请输入邮箱地址"
            android:inputType="textEmailAddress"
            android:textSize="20dp"/>
    </TableRow>
</TableLayout>
```

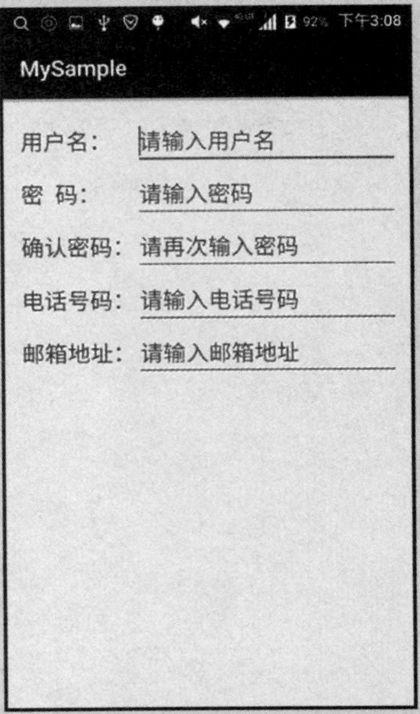

图 003.1

上面这段代码在 MyCode\MySample369\app\src\main\res\layout\activity_main.xml 文件中。

在这段代码中,每一行的 TextView 控件和 EditText 控件均使用 TableLayout 控件的 TableRow 标签来控制,如果试图删除 TableLayout 标签,仅用 TableRow 标签,则应用在运行时将立即崩溃。此实例的完整项目在 MyCode\MySample369 文件夹中。

004　使用 TextInputLayout 管理输入框提示

此实例主要通过使用 TextInputLayout 控件,实现管理输入框 EditText 的提示信息。在实例运行之后,当输入框获得焦点时,提示信息"请输入用户名称"将自动滑向左上角,从而在原始位置留出空白以输入新内容,如图 004.1 的左图所示;当"请输入用户密码"输入框获得焦点时,将实现类似的功能。如果输入框的内容为空白,则在单击"登录"按钮之后,在输入框的下面将显示红色的提示信息,如"用户密码不能是空白",如图 004.1 的右图所示。主要代码如下:

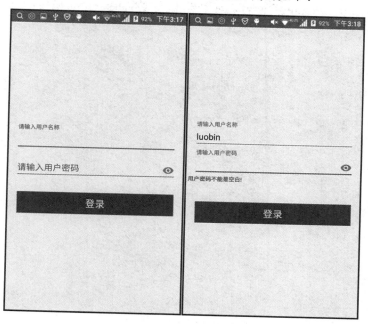

图　004.1

```java
public class MainActivity extends Activity {
  EditText myEditAccount, myEditPassword;
  TextInputLayout myAccountWrapper, myPasswordWrapper;
  @Override
  protected void onCreate(Bundle savedInstanceState) {
    super.onCreate(savedInstanceState);
    setContentView(R.layout.activity_main);
    myEditAccount = (EditText) findViewById(R.id.myAccount);
    myEditPassword = (EditText) findViewById(R.id.myPassword);
    myAccountWrapper = (TextInputLayout) findViewById(R.id.accountWrapper);
    myPasswordWrapper = (TextInputLayout) findViewById(R.id.passwordWrapper);
    myAccountWrapper.setHint("请输入用户名称");
    myPasswordWrapper.setHint("请输入用户密码");
  }
  public void onClickmyBtn1(View v) {                    //响应单击"登录"按钮
    String myAccount = myEditAccount.getText().toString();
```

```
        String myPassword = myEditPassword.getText().toString();
        if (myAccount.length()<1) { myAccountWrapper.setError("用户名称不能是空白！");}
        else if (myPassword.length()<1) {
        myPasswordWrapper.setError("用户密码不能是空白！");
        } else {
        myAccountWrapper.setErrorEnabled(false);
        myPasswordWrapper.setErrorEnabled(false);
        //doLogin();
} } }
```

上面这段代码在 MyCode \ MySample435 \ app \ src \ main \ java \ com \ bin \ luo \ mysample \ MainActivity.java 文件中。在这段代码中，myAccountWrapper.setHint("请输入用户名称")用于设置输入框的提示信息，myAccountWrapper.setError("用户名称不能是空白！")用于设置输入框的错误提示信息。从 TextInputLayout 控件名称可以看出，该控件是一个布局控件，输入框 EditText 控件则包裹在其中，主要代码如下：

```xml
<LinearLayout  android:layout_width = "match_parent"
               android:layout_height = "wrap_content"
               android:layout_centerInParent = "true"
               android:orientation = "vertical">
    <android.support.design.widget.TextInputLayout
        android:id = "@ + id/accountWrapper"
        android:layout_width = "match_parent"
        android:layout_height = "wrap_content"
        android:layout_marginTop = "4dp">
    <EditText  android:id = "@ + id/myAccount"
               android:layout_width = "match_parent"
               android:layout_height = "45dp"
               android:layout_marginLeft = "16dp"
               android:layout_marginRight = "16dp"
               android:layout_marginTop = "20dp"
               android:gravity = "center_vertical"
               android:paddingLeft = "12dp"/>
    </android.support.design.widget.TextInputLayout>
    <android.support.design.widget.TextInputLayout
        android:id = "@ + id/passwordWrapper"
        android:layout_width = "match_parent"
        android:layout_height = "wrap_content"
        android:layout_below = "@id/accountWrapper"
        android:layout_marginTop = "4dp">
    <EditText  android:id = "@ + id/myPassword"
               android:layout_width = "match_parent"
               android:layout_height = "45dp"
               android:layout_marginLeft = "16dp"
               android:layout_marginRight = "16dp"
               android:layout_marginTop = "20dp"
               android:gravity = "center_vertical"
               android:inputType = "textPassword"
               android:paddingLeft = "12dp"/>
    </android.support.design.widget.TextInputLayout>
    <Button  android:id = "@ + id/login"
             android:layout_width = "match_parent"
             android:layout_height = "45dp"
             android:layout_marginLeft = "20dp"
             android:layout_marginRight = "20dp"
```

```
            android:layout_marginTop = "30dp"
            android:background = "@color/colorPrimary"
            android:onClick = "onClickmyBtn1"
            android:text = "登录"
            android:textColor = "#fff"
            android:textSize = "20dp"/>
</LinearLayout>
```

　　上面这段代码在 MyCode\MySample435\app\src\main\res\layout\activity_main.xml 文件中。需要说明的是，在 Android Studio 中使用 TextInputLayout 控件需要在 gradle 中引入 compile 'com.android.support：design：24.2.0' 和 compile 'com.android.support：appcompat-v7：24.2.0' 依赖项，24 是当前开发环境的 SDK 版本号。如果应用的主题是 android：Theme.Material.Light.DarkActionBar，应用可能在运行时会崩溃，因此需要将 app\src\main\res\values\styles.xml 文件的主题 <style name="AppTheme" parent= "android：Theme.Material.Light.DarkActionBar">修改为 <style name="AppTheme" parent="Theme.AppCompat.Light.NoActionBar">或 <style name="AppTheme" parent="Theme.AppCompat.Light.DarkActionBar">。此实例的完整项目在 MyCode\MySample435 文件夹中。

005　使用 GridLayout 创建计算器按键布局

　　此实例主要通过使用 GridLayout 布局管理器，将控件放置在指定行和指定列的网格，从而创建包括不同大小按钮的计算器界面。当实例运行之后，包括不同大小按钮的计算器界面如图 005.1 所示。
　　主要代码如下：

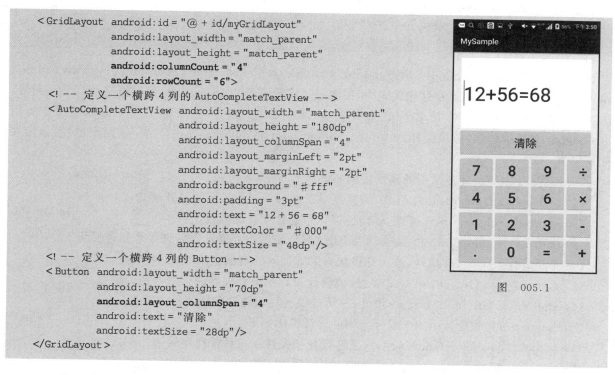

图　005.1

```
<GridLayout android:id = "@+id/myGridLayout"
            android:layout_width = "match_parent"
            android:layout_height = "match_parent"
            android:columnCount = "4"
            android:rowCount = "6">
    <!-- 定义一个横跨4列的 AutoCompleteTextView -->
    <AutoCompleteTextView android:layout_width = "match_parent"
            android:layout_height = "180dp"
            android:layout_columnSpan = "4"
            android:layout_marginLeft = "2pt"
            android:layout_marginRight = "2pt"
            android:background = "#fff"
            android:padding = "3pt"
            android:text = "12 + 56 = 68"
            android:textColor = "#000"
            android:textSize = "48dp"/>
    <!-- 定义一个横跨4列的 Button -->
    <Button android:layout_width = "match_parent"
            android:layout_height = "70dp"
            android:layout_columnSpan = "4"
            android:text = "清除"
            android:textSize = "28dp"/>
</GridLayout>
```

　　上面这段代码在 MyCode\MySample011\app\src\main\res\layout\activity_main.xml 文件中。在这段代码中，android：columnCount="4"表示 GridLayout 布局管理器有 4 列，android：rowCount="6"表示

GridLayout 布局管理器有 6 行，即此实例的 GridLayout 布局管理器是 4 列 6 行组成的网格。android：layout_columnSpan＝"4"表示 Button 控件占据 GridLayout 布局管理器 4 列，即第二行全部列（跨列显示）。如果使用 Java 代码向 GridLayout 布局管理器添加控件，则必须在 addView()方法的第二个参数中指明行号和列号。

主要代码如下：

```java
public class MainActivity extends Activity {
    GridLayout myGridLayout;
    //定义字符串数组，保存 16 个按钮的标题
    String[] myTitle = new String[]{"7", "8", "9", " ÷ ",
                                    "4", "5", "6", " × ",
                                    "1", "2", "3", " - ",
                                    ".", "0", " = ", " + "};
    @Override
    protected void onCreate(Bundle savedInstanceState) {
        super.onCreate(savedInstanceState);
        setContentView(R.layout.activity_main);
        myGridLayout = (GridLayout) findViewById(R.id.myGridLayout);
        for (int i = 0; i < myTitle.length; i++) {
            Button myButton = new Button(this);
            myButton.setText(myTitle[i]);
            myButton.setTextSize(36);                            //设置该按钮的字号大小
            //指定该按钮所在的行号，2 表示从第 3 行开始计数
            GridLayout.Spec myRow = GridLayout.spec(i / 4 + 2);
            GridLayout.Spec myColumn = GridLayout.spec(i % 4);   //指定该按钮所在的列号
            GridLayout.LayoutParams myParams =
                    new GridLayout.LayoutParams(myRow, myColumn);
            myGridLayout.addView(myButton, myParams);
        }
    }
}
```

上面这段代码在 MyCode\MySample011\app\src\main\java\com\bin\luo\mysample\MainActivity.java 文件中。在这段代码中，仅实现了将计算器的 16 个按钮，根据指定的行号和列号，放置在了 GridLayout 布局管理器的第 3 行至第 6 行的 16 个单元格中。一般情况下，使用 GridLayout 布局管理器，必须设置 android：columnCount 和 android：rowCount 这两个属性，以定义网格大小。

GridLayout 布局管理器常用的属性说明如下。

（1）android：alignmentMode 属性，该属性设置布局管理器采用的对齐模式。

（2）android：columnCount 属性，该属性设置布局管理器的列数。

（3）android：columnOrderPreserved 属性，该属性设置布局管理器的列序号是否保留。

（4）android：rowCount 属性，该属性设置布局管理器的行数。

（5）android：rowOrderPreserved 属性，该属性设置布局管理器的行序号是否保留。

（6）android：useDefaultMargins 属性，该属性设置是否使用默认的外边距。

GridLayout 布局管理器中的子控件常用的属性说明如下。

（1）android：layout_column 属性，该属性设置子控件在布局管理器的第几列。

（2）android：layout_columnSpan 属性，该属性设置子控件在布局管理器中跨几列。

（3）android：layout_gravity 属性，该属性设置子控件采用何种方式占据该布局管理器的空间。

（4）android：row 属性，该属性设置子控件在布局管理器的第几行。

（5）android：rowSpan 属性，该属性设置子控件在布局管理器中跨几行。

此实例的完整项目在 MyCode\MySample011 文件夹中。

006　使用 RelativeLayout 按照相邻关系布局

此实例主要通过使用 RelativeLayout 布局管理器以相对布局的方式，实现将 5 张扑克牌按照相邻关系摆放到指定位置。当实例运行之后，5 张扑克牌按照相邻关系的摆放效果如图 006.1 所示。

主要代码如下：

```xml
<RelativeLayout xmlns:android = "http://schemas.android.com/apk/res/android"
    xmlns:tools = "http://schemas.android.com/tools"
    android:id = "@ + id/activity_main"
    android:layout_width = "match_parent"
    android:layout_height = "match_parent"
    android:paddingBottom = "@dimen/activity_vertical_margin"
    android:paddingLeft = "@dimen/activity_horizontal_margin"
    android:paddingRight = "@dimen/activity_horizontal_margin"
    android:paddingTop = "@dimen/activity_vertical_margin"
    tools:context = "com.bin.luo.mysample.MainActivity">
    <!-- 设置该控件位于 RelativeLayout 布局管理器的中间 -->
    <TextView android:id = "@ + id/view01"
        android:layout_width = "wrap_content"
        android:layout_height = "wrap_content"
        android:layout_centerInParent = "true"
        android:background = "@mipmap/center"/>
    <!-- 设置该控件位于 view01 控件的上方 -->
    <TextView android:id = "@ + id/view02"
        android:layout_width = "wrap_content"
        android:layout_height = "wrap_content"
        android:layout_above = "@id/view01"
        android:layout_alignLeft = "@id/view01"
        android:background = "@mipmap/top"/>
    <!-- 设置该控件位于 view01 控件的下方 -->
    <TextView android:id = "@ + id/view03"
        android:layout_width = "wrap_content"
        android:layout_height = "wrap_content"
        android:layout_alignLeft = "@id/view01"
        android:layout_below = "@id/view01"
        android:background = "@mipmap/bottom"/>
    <!-- 设置该控件位于 view01 控件的左边 -->
    <TextView android:id = "@ + id/view04"
        android:layout_width = "wrap_content"
        android:layout_height = "wrap_content"
        android:layout_alignTop = "@id/view01"
        android:layout_toLeftOf = "@id/view01"
        android:background = "@mipmap/left"/>
    <!-- 设置该控件位于 view01 控件的右边 -->
    <TextView android:id = "@ + id/view05"
        android:layout_width = "wrap_content"
        android:layout_height = "wrap_content"
        android:layout_alignTop = "@id/view01"
        android:layout_toRightOf = "@id/view01"
        android:background = "@mipmap/right"/>
</RelativeLayout>
```

图　006.1

上面这段代码在 MyCode\MySample010\app\src\main\res\layout\activity_main.xml 文件中。

在这段代码中，@mipmap/right、@mipmap/left、@mipmap/bottom、@mipmap/top、@mipmap/center 是将 5 张 jpg 格式的扑克牌图像文件拖放到 res\mipmap 节点（目录）生成的图像资源，用于设置 5 个 TextView 控件的背景图像。5 个 TextView 控件的相邻关系则直接由 layout_alignTop、layout_toRightOf 等属性指定。下面是 RelativeLayout 布局管理器的子控件与相邻位置有关的属性说明。

(1) android：layout_centerHorizontal 属性，该属性表示是否水平居中。

(2) android：layout_centerVertical 属性，该属性表示是否垂直居中。

(3) android：layout_centerInParent 属性，该属性表示是否相对于父控件完全居中。

(4) android：layout_alignParentBottom 属性，该属性表示是否贴紧父控件的下边缘。

(5) android：layout_alignParentLeft 属性，该属性表示是否贴紧父控件的左边缘。

(6) android：layout_alignParentRight 属性，该属性表示是否贴紧父控件的右边缘。

(7) android：layout_alignParentTop 属性，该属性表示是否贴紧父控件的上边缘。

(8) android：layout_alignWithParentIfMissing 属性，该属性表示如果找不到对应的兄弟控件，则以父控件作参照物。

(9) android：layout_below 属性，该属性表示在某控件的下方。

(10) android：layout_above 属性，该属性表示在某控件的上方。

(11) android：layout_toLeftOf 属性，该属性表示在某控件的左边。

(12) android：layout_toRightOf 属性，该属性表示在某控件的右边。

(13) android：layout_alignTop 属性，该属性表示本控件的上边缘和某控件的上边缘对齐。

(14) android：layout_alignLeft 属性，该属性表示本控件的左边缘和某控件的左边缘对齐。

(15) android：layout_alignBottom 属性，该属性表示本控件的下边缘和某控件的下边缘对齐。

(16) android：layout_alignRight 属性，该属性表示本控件的右边缘和某控件的右边缘对齐。

(17) android：layout_marginBottom 属性，该属性表示离某控件底边缘的距离。

(18) android：layout_marginLeft 属性，该属性表示离某控件左边缘的距离。

(19) android：layout_marginRight 属性，该属性表示离某控件右边缘的距离。

(20) android：layout_marginTop 属性，该属性表示离某控件上边缘的距离。

此实例的完整项目在 MyCode\MySample010 文件夹中。

007 使用 ConstraintLayout 在右下角布局

此实例主要通过使用 android.support.constraint.ConstraintLayout 布局控件，实现将布局里面的一个子控件放置在另一个子控件的右下角。当实例运行之后，克林顿头像（ImageView 控件）放置在奥巴马头像（ImageView 控件）的右下角，如图 007.1 所示。主要代码如下：

```xml
<?xml version = "1.0" encoding = "utf-8"?>
<android.support.constraint.ConstraintLayout
        xmlns:android = "http://schemas.android.com/apk/res/android"
        xmlns:app = "http://schemas.android.com/apk/res-auto"
        android:id = "@+id/activity_main"
        android:layout_width = "match_parent"
        android:layout_height = "match_parent"
        android:paddingLeft = "20dp"
        android:paddingTop = "100dp">
    <ImageView  android:id = "@+id/myImage1"
```

```xml
            android:layout_width = "wrap_content"
            android:layout_height = "wrap_content"
            android:src = "@mipmap/myimage1"/>
    <!-- 实现居中效果 -->
    <!-- app:layout_constraintBottom_toBottomOf = "parent" -->
    <!-- app:layout_constraintLeft_toLeftOf = "parent" -->
    <!-- app:layout_constraintRight_toRightOf = "parent" -->
    <!-- app:layout_constraintTop_toTopOf = "parent" -->
    <ImageView  android:layout_width = "wrap_content"
            android:layout_height = "wrap_content"
            android:src = "@mipmap/myimage2"
            app:layout_constraintLeft_toRightOf = "@ + id/myImage1"
            app:layout_constraintTop_toBottomOf = "@ + id/myImage1"/>
</android.support.constraint.ConstraintLayout>
```

上面这段代码在 MyCode\MySample363\app\src\main\res\layout\activity_main.xml 文件中。在这段代码中，app：layout_constraintLeft_toRightOf ＝"＠＋id/myImage1"表示当前子控件 myImage2 在子控件 myImage1 的右边。app：layout_constraintTop_toBottomOf＝"＠＋id/myImage1"表示当前子控件 myImage2 在子控件 myImage1 的下边。ConstraintLayout 支持的具有相对定位的属性组合说明如下。

（1）layout_constraintLeft_toLeftOf 属性，该属性表示当前控件的左边在某控件的左边，即左对齐。

（2）layout_constraintLeft_toRightOf 属性，该属性表示当前控件的左边在某控件的右边。

（3）layout_constraintRight_toLeftOf 属性，该属性表示当前控件的右边在某控件的左边。

（4）layout_constraintRight_toRightOf 属性，该属性表示当前控件的右边在某控件的右边，即右对齐。

（5）layout_constraintTop_toTopOf 属性，该属性表示当前控件的上边和某控件的上边对齐，即上对齐。

图　007.1

（6）layout_constraintTop_toBottomOf 属性，该属性表示当前控件的上边在某控件的下边。

（7）layout_constraintBottom_toTopOf 属性，该属性表示当前控件的下边在某控件的上边。

（8）layout_constraintBottom_toBottomOf 属性，该属性表示当前控件的下边在某控件的下边，即下对齐。

需要说明的是，在 Android Studio 中使用 ConstraintLayout 之前，需要在 app 下的 gradle 文件中添加 ConstraintLayout 依赖项，如：

```
dependencies {
    compile 'com.android.support.constraint:constraint-layout:1.0.2'}
```

此实例的完整项目在 MyCode\MySample363 文件夹中。

008　使用 TableLayout 拉伸控件填充容器

此实例主要通过设置 TableLayout 布局管理器的 stretchColumns 属性，实现拉伸指定列中的控件以填充 TableLayout 布局管理器的剩余空间。当实例运行之后，由于指定了 TableLayout 布局管理器的 stretchColumns 属性是第二列，因此"B 图（拉伸）"占用的空间最大，如图 008.1 所示。

主要代码如下：

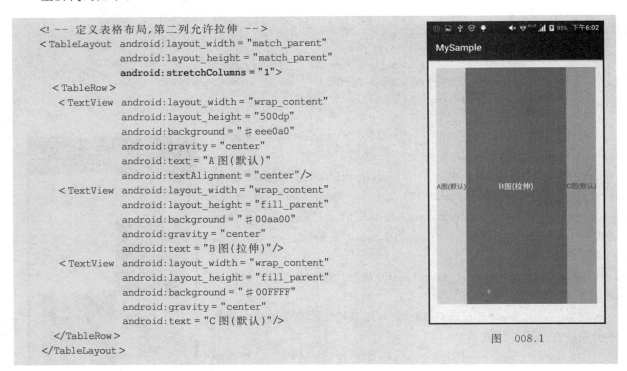

```xml
<!-- 定义表格布局,第二列允许拉伸 -->
<TableLayout android:layout_width = "match_parent"
             android:layout_height = "match_parent"
             android:stretchColumns = "1">
  <TableRow>
    <TextView android:layout_width = "wrap_content"
              android:layout_height = "500dp"
              android:background = "#eee0a0"
              android:gravity = "center"
              android:text = "A 图（默认）"
              android:textAlignment = "center"/>
    <TextView android:layout_width = "wrap_content"
              android:layout_height = "fill_parent"
              android:background = "#00aa00"
              android:gravity = "center"
              android:text = "B 图（拉伸）"/>
    <TextView android:layout_width = "wrap_content"
              android:layout_height = "fill_parent"
              android:background = "#00FFFF"
              android:gravity = "center"
              android:text = "C 图（默认）"/>
  </TableRow>
</TableLayout>
```

图　008.1

上面这段代码在 MyCode\MySample007\app\src\main\res\layout\activity_main.xml 文件中。在这段代码中，android：stretchColumns＝"1"表示拉伸第二列的控件（如果有剩余空间），android：stretchColumns＝"0"则表示拉伸第一列的控件（如果有剩余空间）。如果有多列控件需要拉伸，则应在属性值中使用逗号隔开列索引，如 android：stretchColumns＝"1,2"表示同时拉伸第二列和第三列的控件（如果有剩余空间）。stretchColumns 是布局管理器 TableLayout 的属性，用于设置允许被拉伸的列序号，TableLayout 继承自 LinearLayout，因此它完全可以支持 LinearLayout 的属性。此实例的完整项目在 MyCode\MySample007 文件夹中。

009　使用 TableLayout 缩小控件适应容器

此实例主要通过设置 TableLayout 布局管理器的 collapseColumns 属性，实现在屏幕不能显示全部控件时，折叠（隐藏）指定的控件。当实例运行之后，效果如图 009.1 的左图所示；该图表示由于屏幕限制，不能完全显示 A、B、C、D、E、F、G 图（控件），因此折叠（隐藏）B 图和 C 图，但是这样做之后就产生了一点剩余空间，因此又拉伸了 A 图以填满所有空间。默认情况下，如果不指定折叠控件，当屏幕不能完全显示所有控件时，则将按照顺序自动折叠（隐藏）最后的控件，如图 009.1 的右图所示。

主要代码如下：

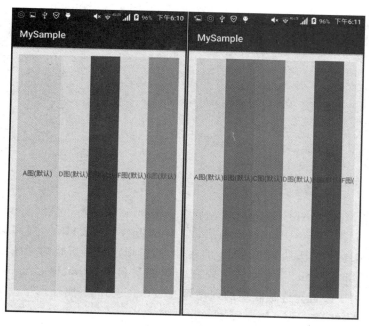

图 009.1

```
<!-- 定义 TableLayout 布局,折叠 1、2 列,拉伸第 0 列 -->
<TableLayout android:layout_width = "match_parent"
             android:layout_height = "match_parent"
             android:collapseColumns = "1,2"
             android:stretchColumns = "0">
  <TableRow>
    <TextView android:layout_width = "wrap_content"
              android:layout_height = "500dp"
              android:background = "#eee0a0"
              android:gravity = "center"
              android:text = "A 图(默认)"/>
    <TextView android:layout_width = "wrap_content"
              android:layout_height = "fill_parent"
              android:background = "#00aa00"
              android:gravity = "center"
              android:text = "B 图(默认)"/>
    <TextView android:layout_width = "wrap_content"
              android:layout_height = "fill_parent"
              android:background = "#00AAFF"
              android:gravity = "center"
              android:text = "C 图(默认)"/>
    <TextView android:layout_width = "wrap_content"
              android:layout_height = "fill_parent"
              android:background = "#CCFFFF"
              android:gravity = "center"
              android:text = "D 图(默认)"/>
    <TextView android:layout_width = "wrap_content"
              android:layout_height = "fill_parent"
              android:background = "#BB00FF"
              android:gravity = "center"
              android:text = "E 图(默认)"/>
```

```xml
<TextView android:layout_width = "wrap_content"
          android:layout_height = "fill_parent"
          android:background = "#FFFF00"
          android:gravity = "center"
          android:text = "F图(默认)"/>
<TextView android:layout_width = "wrap_content"
          android:layout_height = "fill_parent"
          android:background = "#00FF00"
          android:gravity = "center"
          android:text = "G图(默认)"/>
</TableRow></TableLayout>
```

上面这段代码在 MyCode\MySample008\app\src\main\res\layout\activity_main.xml 文件中。在这段代码中，android：collapseColumns = "1,2"表示折叠（隐藏）第二列和第三列中的子控件。collapseColumns 是布局管理器 TableLayout 的属性，用于指定需要折叠的列序号，TableLayout 继承自 LinearLayout。TableLayout 另外两个比较有特色的属性是 stretchColumns 和 shrinkColumns，shrinkColumns 属性用于设置允许被收缩的列序号，stretchColumns 属性用于设置允许被拉伸的列序号。此实例的完整项目在 MyCode\MySample008 文件夹中。

010　使用 LinearLayout 纵向居中对齐控件

此实例主要通过设置 LinearLayout 线性布局管理器的 gravity 属性，实现多个控件在屏幕中的不同位置对齐显示。当实例运行之后，单击"靠左对齐"按钮，则 3 幅图像靠左对齐的效果如图 010.1 的左图所示。单击"居中对齐"按钮，则 3 幅图像居中对齐的效果如图 010.1 的右图所示。单击"靠右对齐"按钮，则 3 幅图像将靠右对齐显示。

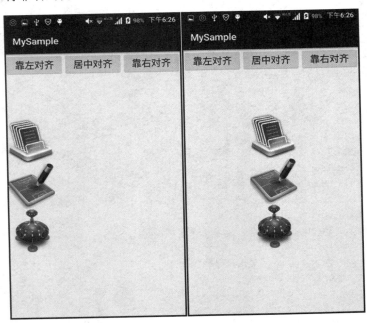

图　010.1

主要代码如下：

```xml
<LinearLayout xmlns:android = "http://schemas.android.com/apk/res/android"
    android:id = "@+id/myLinearLayout"
    android:layout_width = "match_parent"
    android:layout_height = "match_parent"
    android:gravity = "left|center_vertical"
    android:orientation = "vertical">
<ImageView android:layout_width = "96dp"
    android:layout_height = "96dp"
    android:src = "@mipmap/box"/>
<ImageView android:layout_width = "96dp"
    android:layout_height = "96dp"
    android:src = "@mipmap/pen"/>
<ImageView android:layout_width = "96dp"
    android:layout_height = "96dp"
    android:src = "@mipmap/airship"/>
</LinearLayout>
```

上面这段代码在 MyCode\MySample004\app\src\main\res\layout\activity_main.xml 文件中。在这段代码中，android：gravity＝"center|center_vertical"表示在水平方向和垂直方向均居中显示在 LinearLayout 中的控件，即 3 幅图像居中对齐。android：gravity＝"left|center_vertical"表示在水平方向上靠左对齐、在垂直方向均居中显示在 LinearLayout 中的控件，即 3 幅图像靠左对齐。android：orientation＝"vertical"表示 3 幅图像在一列中显示，如果设置 android：orientation＝"horizontal"，则 3 幅图像将在一行中显示。默认情况下，设置 LinearLayout 线性布局管理器的控件对齐方式均通过在 XML 文件中操作 gravity 属性实现；如果需要使用 Java 代码动态修改控件的对齐方式，则应该使用 setGravity()方法，主要代码如下：

```java
public class MainActivity extends Activity {
  LinearLayout myLinearLayout;
  @Override
  protected void onCreate(Bundle savedInstanceState) {
    super.onCreate(savedInstanceState);
    setContentView(R.layout.activity_main);
    myLinearLayout = (LinearLayout) findViewById(R.id.myLinearLayout);
  }
  public void myBtnLeftClick(View v) {                //响应单击"靠左对齐"按钮
    myLinearLayout.setGravity(Gravity.LEFT | Gravity.CENTER_VERTICAL);
  }
  public void myBtnCenterClick(View v) {              //响应单击"居中对齐"按钮
    myLinearLayout.setGravity(Gravity.CENTER | Gravity.CENTER_VERTICAL);
  }
  public void myBtnRightClick(View v) {               //响应单击"靠右对齐"按钮
    myLinearLayout.setGravity(Gravity.RIGHT | Gravity.CENTER_VERTICAL);
  } }
```

上面这段代码在 MyCode\MySample004\app\src\main\java\com\bin\luo\mysample\MainActivity.java 文件中。在这段代码中，setGravity()方法中的参数即是 gravity 属性值，该属性值可以有多个，使用"|"符号分隔；支持的属性值包括：TOP、AXIS_X_SHIFT、AXIS_Y_SHIFT、LEFT、RIGHT、NO_GRAVITY、RELATIVE_HORIZONTAL_GRAVITY_MASK、BOTTOM、HORIZONTAL_GRAVITY_MASK、VERTICAL_GRAVITY_MASK、AXIS_PULL_AFTER、CLIP_VERTICAL、CENTER、CENTER_HORIZONTAL、CENTER_VERTICAL、CLIP_HORIZONTAL、FILL_HORIZONTAL、DISPLAY_CLIP_

HORIZONTAL、DISPLAY_CLIP_VERTICAL、RELATIVE_LAYOUT_DIRECTION、START、FILL_VERTICAL。需要注意的是：当采用 LinearLayout 线性布局管理器管理控件时，如果其中的多个控件无法在单行中同时显示，则 LinearLayout 不会换行显示后面未显示的控件。此实例的完整项目在 MyCode\MySample004 文件夹中。

011　使用 LinearLayout 按权重分配控件空间

此实例主要通过为 LinearLayout 线性布局管理器的子控件设置不同的 layout_weight 属性（权重）值，从而实现按比例调节子控件的空间占比。当实例运行之后，"B 图"的 layout_weight 属性（权重）值最大，因此占用的空间最大，如图 011.1 所示。

主要代码如下：

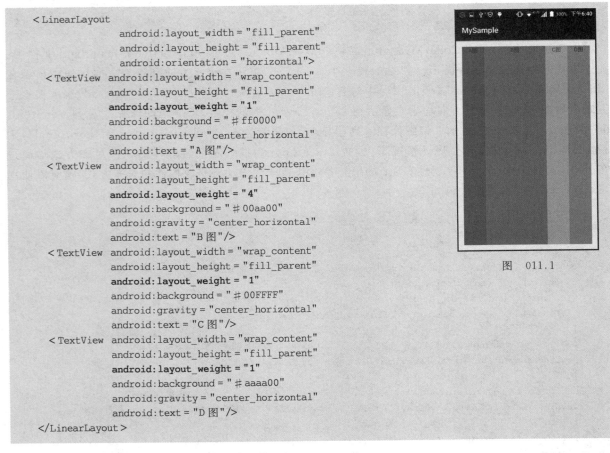

图　011.1

```
<LinearLayout
        android:layout_width="fill_parent"
        android:layout_height="fill_parent"
        android:orientation="horizontal">
<TextView android:layout_width="wrap_content"
        android:layout_height="fill_parent"
        android:layout_weight="1"
        android:background="#ff0000"
        android:gravity="center_horizontal"
        android:text="A 图"/>
<TextView android:layout_width="wrap_content"
        android:layout_height="fill_parent"
        android:layout_weight="4"
        android:background="#00aa00"
        android:gravity="center_horizontal"
        android:text="B 图"/>
<TextView android:layout_width="wrap_content"
        android:layout_height="fill_parent"
        android:layout_weight="1"
        android:background="#00FFFF"
        android:gravity="center_horizontal"
        android:text="C 图"/>
<TextView android:layout_width="wrap_content"
        android:layout_height="fill_parent"
        android:layout_weight="1"
        android:background="#aaaa00"
        android:gravity="center_horizontal"
        android:text="D 图"/>
</LinearLayout>
```

上面这段代码在 MyCode\MySample005\app\src\main\res\layout\activity_main.xml 文件中。在这段代码中，android：layout_weight＝"4"表示该控件的权重是 4，它是一个相对数，总是相对相关的其他控件而言，此处权重为 4 的控件实际表示对额外空间的分配权重是：4/(4＋1＋1＋1)＝57％。在 Android 中，layout_weight 属性表示子控件对 LinearLayout 额外空间的划分，需要注意的是额外二字，先有额外的空间，然后才可以按比例将其分配给设置了 layout_weight 的子 View，所以如果 LinearLayout 的 layout_width 或 layout_height 属性设置了 WRAP_CONTENT，则 LinearLayout 就没有额外的空间，因此子控件的 layout_weight 就没有用处。此实例的完整项目在 MyCode\MySample005 文件夹中。

012　使用 ConstraintLayout 平分剩余空间

此实例主要通过使用 android.support.constraint.ConstraintLayout 布局控件，实现使三个控件 ImageView 平分在水平方向上的剩余空间。当实例运行之后，奥巴马头像、克林顿头像、普京头像将平分水平方向的剩余空间，如图 012.1 所示。

图　012.1

主要代码如下：

```xml
<?xml version = "1.0" encoding = "utf-8"?>
<android.support.constraint.ConstraintLayout
    xmlns:android = "http://schemas.android.com/apk/res/android"
    xmlns:app = "http://schemas.android.com/apk/res-auto"
    android:id = "@+id/activity_main"
    android:layout_marginTop = "200dp"
    android:layout_width = "match_parent"
    android:layout_height = "wrap_content">
<ImageView   android:id = "@+id/myImage1"
        android:layout_width = "wrap_content"
        android:layout_height = "wrap_content"
        android:src = "@mipmap/myimage1"
        app:layout_constraintLeft_toLeftOf = "parent"
        app:layout_constraintRight_toLeftOf = "@+id/myImage2"/>
<ImageView   android:id = "@+id/myImage2"
        android:layout_width = "wrap_content"
        android:layout_height = "wrap_content"
        android:src = "@mipmap/myimage2"
        app:layout_constraintLeft_toRightOf = "@+id/myImage1"
        app:layout_constraintRight_toLeftOf = "@+id/myImage3"/>
```

```xml
<ImageView android:id="@+id/myImage3"
    android:layout_width="wrap_content"
    android:layout_height="wrap_content"
    android:src="@mipmap/myimage3"
    app:layout_constraintLeft_toRightOf="@+id/myImage2"
    app:layout_constraintRight_toRightOf="parent" />
</android.support.constraint.ConstraintLayout>
```

上面这段代码在 MyCode\MySample364\app\src\main\res\layout\activity_main.xml 文件中。在这段代码中，app：layout_constraintLeft_toRightOf="@+id/myImage2" 表示当前子控件 myImage3 在子控件 myImage2 的右边。app：layout_constraintRight_toRightOf="parent" 表示当前子控件 myImage3 与 ConstraintLayout 布局控件右对齐。默认情况下，由于 3 个子控件 ImageView 在水平方向形成一个链式布局，因此它们会平分水平方向的剩余空间。需要说明的是，在 Android Studio 中使用 ConstraintLayout 之前，需要在 app 下的 gradle 文件中添加 ConstraintLayout 依赖项，如下：

```
dependencies {
    compile 'com.android.support.constraint:constraint-layout:1.0.2'}
```

此实例的完整项目在 MyCode\MySample364 文件夹中。

013 使用 ConstraintLayout 无间隙布局控件

此实例主要通过使用 android.support.constraint.ConstraintLayout 布局控件，并设置子控件的 app：layout_constraintHorizontal_chainStyle 属性为 packed，实现清除子控件之间的间隙。当实例运行之后，奥巴马头像、克林顿头像、普京头像 3 个子控件之间没有间隙，水平剩余空间全部分配到奥巴马头像和普京头像与 ConstraintLayout 布局控件（屏幕左右两侧）的毗邻处，如图 013.1 所示。

图 013.1

主要代码如下：

```xml
<?xml version="1.0" encoding="utf-8"?>
<android.support.constraint.ConstraintLayout
        xmlns:android="http://schemas.android.com/apk/res/android"
        xmlns:app="http://schemas.android.com/apk/res-auto"
        android:id="@+id/activity_main"
        android:layout_width="match_parent"
        android:layout_height="wrap_content"
        android:layout_marginTop="200dp">
    <ImageView app:layout_constraintHorizontal_chainStyle="packed"
        android:id="@+id/myImage1"
        android:layout_width="wrap_content"
        android:layout_height="wrap_content"
        android:src="@mipmap/myimage1"
        app:layout_constraintLeft_toLeftOf="parent"
        app:layout_constraintRight_toLeftOf="@+id/myImage2"/>
    <ImageView android:id="@+id/myImage2"
        android:layout_width="wrap_content"
        android:layout_height="wrap_content"
        android:src="@mipmap/myimage2"
        app:layout_constraintLeft_toRightOf="@+id/myImage1"
        app:layout_constraintRight_toLeftOf="@+id/myImage3"/>
    <ImageView android:id="@+id/myImage3"
        android:layout_width="wrap_content"
        android:layout_height="wrap_content"
        android:src="@mipmap/myimage3"
        app:layout_constraintLeft_toRightOf="@+id/myImage2"
        app:layout_constraintRight_toRightOf="parent"/>
</android.support.constraint.ConstraintLayout>
```

上面这段代码在 MyCode\MySample365\app\src\main\res\layout\activity_main.xml 文件中。在这段代码中，app:layout_constraintHorizontal_chainStyle="packed" 表示 view 之间紧挨着显示，并且全体居中显示；如果删除此行代码，在图 013.1 中，奥巴马头像与克林顿头像之间、克林顿头像与普京头像之间就会出现缝隙。需要说明的是，在使用 ConstraintLayout 之前，需要在 app 下的 gradle 文件中添加 ConstraintLayout 依赖项，如下：

```
dependencies {
    compile 'com.android.support.constraint:constraint-layout:1.0.2'}
```

此实例的完整项目在 MyCode\MySample365 文件夹中。

014 使用 TabLayout 和适配器创建选项卡

此实例主要通过使用 TabLayout 和 PagerAdapter，从而创建滑动选项卡。当实例运行之后，单击"京东"选项卡，则在选项卡中加载的京东(手机网站)页面效果如图 014.1 的左图所示；单击"360"选项卡，则在选项卡中加载的 360(手机网站)页面效果如图 014.1 的右图所示。单击其他选项卡将实现类似的功能。

图 014.1

主要代码如下：

```xml
<?xml version = "1.0" encoding = "utf-8"?>
<LinearLayout xmlns:android = "http://schemas.android.com/apk/res/android"
    android:id = "@ + id/activity_main"
    android:layout_width = "match_parent"
    android:layout_height = "match_parent"
    android:orientation = "vertical">
<android.support.design.widget.TabLayout
    android:id = "@ + id/myTabLayout"
    android:layout_width = "match_parent"
    android:layout_height = "wrap_content"/>
<android.support.v4.view.ViewPager
    android:id = "@ + id/myViewPager"
    android:layout_width = "match_parent"
    android:layout_height = "wrap_content"/>
</LinearLayout>
```

上面这段代码在 MyCode\MySample677\app\src\main\res\layout\activity_main.xml 文件中。在这段代码中，TabLayout 控件主要用于管理选项卡的标签，ViewPager 控件则用于管理选项卡的视图（网站页面）。两者可以通过 TabLayout 的 setupWithViewPager（ViewPager）关联起来。

主要代码如下：

```java
public class MainActivity extends Activity {
    private TabLayout myTabLayout;
    private ViewPager myViewPager;
    private LayoutInflater myInflater;
    private List<String> myTitles = new ArrayList<String>();       //选项卡标题集合
    private List<View> myViews = new ArrayList<View>();            //选项卡页面集合
    @Override
```

```java
protected void onCreate(Bundle savedInstanceState) {
    super.onCreate(savedInstanceState);
    setContentView(R.layout.activity_main);
    myViewPager = (ViewPager) findViewById(R.id.myViewPager);
    myTabLayout = (TabLayout) findViewById(R.id.myTabLayout);
    InitView();
    //添加选项卡标题
    myTitles.add("京东");
    myTitles.add("天猫");
    myTitles.add("360");
    myTitles.add("搜狐");
    myTitles.add("百度");
    MyAdapter myAdapter = new MyAdapter(myViews,myTitles);
    myViewPager.setAdapter(myAdapter);                    //给 ViewPager 设置适配器
    //将 TabLayout 和 ViewPager 关联起来
    myTabLayout.setupWithViewPager(myViewPager);
}
    private class MyWebViewClient extends WebViewClient {
        @Override
        public boolean shouldOverrideUrlLoading(WebView view, String url) {
            view.loadUrl(url);
            return true;
        }
    }
    private void InitView() {
        myInflater = LayoutInflater.from(this);
        String[] myWebsites = {"https://www.jd.com","http://www.tmall.com",
                "http://www.360.cn","http://www.sohu.com","http://www.baidu.com"};
        for(int i = 1;i <= 5;i++){
            View myView = myInflater.inflate(R.layout.myitem,null);
            WebView myWebView = (WebView) myView.findViewById(R.id.myWebView);
            myWebView.loadUrl(myWebsites[i - 1]);
            myWebView.getSettings().setJavaScriptEnabled(true);
            myWebView.setWebViewClient(new MyWebViewClient());
            myViews.add(myView);
}}}
```

上面这段代码在 MyCode\MySample677\app\src\main\java\com\bin\luo\mysample\MainActivity.java 文件中。在这段代码中，myViewPager.setAdapter(myAdapter)表示使用自定义适配器 MyAdapter 管理选项卡的标题(标签)和内容(即 web 页面)。自定义适配器 MyAdapter 的主要代码如下：

```java
public class MyAdapter extends PagerAdapter {
    private List<View> myViews;
    private List<String> myTitles;
    public MyAdapter(List<View> mViewList, List<String> titles) {
        this.myViews = mViewList;
        myTitles = titles;
    }
    @Override
    public int getCount() { return myViews.size(); }
    @Override
```

```
public boolean isViewFromObject(View view, Object object){ return view == object;}
@Override
public Object instantiateItem(ViewGroup container, int position) {
    container.addView(myViews.get(position));                    //添加选项卡
    return myViews.get(position);
}
@Override
public void destroyItem(ViewGroup container, int position, Object object) {
    container.removeView(myViews.get(position));                 //删除选项卡
}
@Override
public CharSequence getPageTitle(int position) {return myTitles.get(position); }
}
```

上面这段代码在 MyCode\MySample677\app\src\main\java\com\bin\luo\mysample\MyAdapter.java 文件中。在 MainActivity.java 文件中，myView = myInflater.inflate（R.layout.myitem,null)用于在选项卡视图上加载 myitem 布局，myitem 布局的主要内容如下：

```xml
<?xml version = "1.0" encoding = "utf-8"?>
<LinearLayout
            xmlns:android = "http://schemas.android.com/apk/res/android"
            android:id = "@+id/activity_main"
            android:layout_width = "match_parent"
            android:layout_height = "match_parent">
<WebView  android:id = "@+id/myWebView"
            android:layout_width = "match_parent"
            android:layout_height = "match_parent"/>
</LinearLayout>
```

上面这段代码在 MyCode\MySample677\app\src\main\res\layout\myitem.xml 文件中。需要说明的是，使用此实例的相关类需要在 gradle 中引入 compile 'com.android.support：design：23.3.0' 依赖项。此外，访问网络需要相关的权限，因此需要在 AndroidManifest.xml 文件中添加< uses-permission android：name ="android.permission.INTERNET"/>权限。如果在编译和安装时出现错误，请注意检查主题是否有错误，即可能需要将 app\src\main\res\values\styles.xml 文件的主题 < style name = "AppTheme" parent = "android：Theme.Material.Light.DarkActionBar">修改为 < style name= "AppTheme" parent = "Theme.AppCompat.Light.DarkActionBar">。此实例的完整项目在 MyCode\MySample677 文件夹中。

015　使用 TabLayout 和 Fragment 创建选项卡

此实例主要通过使用 TabLayout 和 Fragment，实现创建选项卡风格的界面。当实例运行之后，单击"亲密的陌生人"标签，则将显示该标签对应的选项卡，效果如图 015.1 的左图所示。单击"善意的谎言"标签，则将显示该标签对应的选项卡，效果如图 015.1 的右图所示。单击"光荣岁月"标签，也将显示该标签对应的选项卡。

图 015.1

主要代码如下:

```xml
<?xml version = "1.0" encoding = "utf-8"?>
<LinearLayout xmlns:android = "http://schemas.android.com/apk/res/android"
    xmlns:app = "http://schemas.android.com/apk/res-auto"
    android:layout_width = "match_parent"
    android:layout_height = "match_parent"
    android:orientation = "vertical">
    <android.support.design.widget.TabLayout
        android:id = "@+id/myTabLayout"
        android:layout_width = "match_parent"
        android:layout_height = "55dp"
        app:tabGravity = "center"
        app:tabIndicatorColor = "@color/colorAccent"
        app:tabMode = "scrollable"
        app:tabSelectedTextColor = "@color/colorPrimaryDark"
        app:tabTextColor = "@color/colorPrimary"/>
    <android.support.v4.view.ViewPager
        android:id = "@+id/myViewPager"
        android:layout_width = "match_parent"
        android:layout_height = "match_parent"/>
</LinearLayout>
```

上面这段代码在 MyCode\MySample357\app\src\main\res\layout\activity_main.xml 文件中。在这段代码中,TabLayout 控件主要用于管理选项卡的标签,ViewPager 控件则用于管理选项卡的视图(电影海报图像)。当在 Activity 中加载 TabLayout 时,Activity 通常需要从 AppCompatActivity 继承,主要代码如下:

```java
public class MainActivity extends AppCompatActivity {
    String[] myTitle = {"亲密的陌生人","善意的谎言","光荣岁月"};
    String[] myData = {"亲密的陌生人","善意的谎言","光荣岁月"};
```

```java
TabLayout myTabLayout;
ViewPager myViewPager;
@Override
protected void onCreate(Bundle savedInstanceState) {
    super.onCreate(savedInstanceState);
    setContentView(R.layout.activity_main);
    initView();
}
private void initView() {
    myTabLayout = (TabLayout) findViewById(R.id.myTabLayout);
    myViewPager = (ViewPager) findViewById(R.id.myViewPager);
    myViewPager.setAdapter(new FragmentPagerAdapter(getSupportFragmentManager()){
        @Override
        public CharSequence getPageTitle(int position) {
            return myTitle[position % myTitle.length];                //此方法负责管理选项卡标签
        }
        @Override
        public Fragment getItem(int position) {                        //创建 Fragment 并返回
            TabFragment myTabFragment = new TabFragment();
            myTabFragment.setTitle(myData[position % myTitle.length]);
            return myTabFragment;
        }
        @Override
        public int getCount() {return myTitle.length; }
    });
    myTabLayout.setupWithViewPager(myViewPager);                       //将 ViewPager 关联 TabLayout
    myTabLayout.setOnTabSelectedListener(new TabLayout.OnTabSelectedListener() {
        @Override
        public void onTabSelected(TabLayout.Tab tab) {
            myViewPager.setCurrentItem(tab.getPosition());             //切换 ViewPager
        }
        @Override
        public void onTabUnselected(TabLayout.Tab tab) { }
        @Override
        public void onTabReselected(TabLayout.Tab tab) { }
    });} }
```

上面这段代码在 MyCode\MySample357\app\src\main\java\com\bin\luo\mysample\MainActivity.java 文件中。在这段代码中，TabFragment myTabFragment = new TabFragment()用于在 ViewPager 中显示控件（此处是电影海报图像）。TabFragment 类的主要代码如下：

```java
public class TabFragment extends Fragment {
    private String myTitle;
    String[] myMovies = {"亲密的陌生人","善意的谎言","光荣岁月"};
    public void setTitle(String title) { this.myTitle = title; }
    @Override
    public View onCreateView(LayoutInflater inflater,
                    ViewGroup container, Bundle savedInstanceState) {
        ImageView myImageView = new ImageView(getContext());
        myImageView.setScaleType(ImageView.ScaleType.FIT_XY);
        if(myTitle.contains(myMovies[0])){
            myImageView.setImageResource(R.mipmap.myimage1);
```

```
    }else if(myTitle.contains(myMovies[1])){
     myImageView.setImageResource(R.mipmap.myimage2);
    }else if(myTitle.contains(myMovies[2])){
     myImageView.setImageResource(R.mipmap.myimage3);
    }
    return myImageView;
} }
```

上面这段代码在 MyCode\MySample357\app\src\main\java\com\bin\luo\mysample\TabFragment.java 文件中。需要说明的是，TabLayout 是 design 库提供的控件，它既可以单独使用，也可以配合 ViewPager 控件来使用，在 Android Studio 中使用此控件时只需要在 gradle 中引入 compile 'com.android.support:design:23.3.0' 即可使用。具体操作步骤如下：在 Android Studio 环境中打开 app\build.gradle 文件，在 dependencies 中添加 compile 'com.android.support:design:23.3.0'，此时，在其上面将立即弹出一个黄色的提示框，单击右侧的"Sync Now"超链接，Android Studio 将会自动添加此依赖项。

另外，由于默认创建 MainActivity 继承自 Activity，如果像实例这样 MainActivity 继承自 AppCompatActivity，则需要将 app\src\main\res\values\styles.xml 文件的主题< style name = "AppTheme" parent = "android：Theme.Material.Light.DarkActionBar">修改为< style name = "AppTheme" parent="Theme.AppCompat.Light.DarkActionBar">，否则可能无法创建此应用。

此实例的完整项目在 MyCode\MySample357 文件夹中。

016　使用 FrameLayout 创建纵向选项卡

此实例主要通过动态创建 FrameLayout，从而创建纵向风格的选项卡。当实例运行之后，单击"百度"选项卡，则效果如图 016.1 的左图所示；单击"京东"选项卡，则效果如图 016.1 的右图所示。单击其他选项卡将实现类似的功能。

图　016.1

主要代码如下：

```java
public class MainActivity extends Activity {
  String[] myTitles = {"百度","京东","天猫","搜狗","微软","新浪","淘宝"};
  @Override
  protected void onCreate(Bundle savedInstanceState){
    super.onCreate(savedInstanceState);
    setContentView(R.layout.activity_main);
    getFragmentManager().beginTransaction().add(R.id.myFrame,
                                                new WebFragment()).commit();
    LinearLayout myLayout = (LinearLayout)findViewById(R.id.myLayout);
    for(int i = 0; i< myTitles.length; i++){
      FrameLayout myFrameLayout = new FrameLayout(this);
      myFrameLayout.setLayoutParams(new FrameLayout.LayoutParams(
        ViewGroup.LayoutParams.MATCH_PARENT, ViewGroup.LayoutParams.WRAP_CONTENT));
      //自动生成 myFrameLayout 帧布局 ID 值
      myFrameLayout.setId(View.generateViewId());
      //动态添加 myFrameLayout 布局(选项卡的标签)
      myLayout.addView(myFrameLayout);
      TitleFragment myTitleFragment = new TitleFragment();
      Bundle myBundle = new Bundle();
      myBundle.putString("myTitle", myTitles[i]);
      myTitleFragment.setArguments(myBundle);                  //传递选项卡的标签
      getFragmentManager().beginTransaction().
                add(myFrameLayout.getId(),myTitleFragment).commit();
} } }
```

上面这段代码在 MyCode\MySample813\app\src\main\java\com\bin\luo\mysample\MainActivity.java 文件中。在这段代码中，myTitleFragment = new TitleFragment()中的 TitleFragment 是自定义类，用于设置选项卡的标题。TitleFragment 类的主要内容如下：

```java
public class TitleFragment extends Fragment {
  String[] myTitles = {"百度","京东","天猫","搜狗","微软","新浪","淘宝"};
  String[] myUrls = {"http://www.baidu.com", "http://www.jd.com",
    "http://www.tmall.com", "http://www.sougou.com", "http://www.microsoft.com",
    "http://www.sina.com.cn", "http://www.taobao.com"};
  String myTitle;
  TextView myTextView;
  List<String> myTitleList = new ArrayList(), myUrlList = new ArrayList();
  @Override
  public void setArguments(Bundle args) { myTitle = args.getString("myTitle"); }
  @Override
  public View onCreateView(LayoutInflater inflater,
             ViewGroup container, Bundle savedInstanceState) {
    myTextView = new TextView(inflater.getContext());
    myTextView.setLayoutParams(new LinearLayout.LayoutParams(ViewGroup.
          LayoutParams.MATCH_PARENT, ViewGroup.LayoutParams.WRAP_CONTENT));
    myTextView.setText(myTitle);
    myTextView.setTextSize(18);
    myTextView.setPadding(0, 25, 0, 25);
    myTextView.setGravity(Gravity.CENTER_HORIZONTAL);
    return myTextView;
  }
```

```java
@Override
public void onActivityCreated(Bundle savedInstanceState) {
    super.onActivityCreated(savedInstanceState);
    for (int i = 0; i < myTitles.length; i++) {                //循环初始化数据
        myTitleList.add(myTitles[i]);
        myUrlList.add(myUrls[i]);
    }
    myTextView.setOnClickListener(new View.OnClickListener() {  //响应单击标签
        @Override
        public void onClick(View v) {
            //通过 FragmentManager 对象获取指定 Fragment
            WebFragment myWebFragment =
                    (WebFragment) getFragmentManager().findFragmentById(R.id.myFrame);
            Bundle myBundle = new Bundle();
            //将与标题对应的网址通过 Bundle 传入 WebFragment
            myBundle.putString("myUrl",
                    myUrlList.get(myTitleList.indexOf(myTextView.getText())));
            myWebFragment.setArguments(myBundle);
            myWebFragment.reloadView();                         //重新加载页面
        }});}}
```

上面这段代码在 MyCode \ MySample813 \ app \ src \ main \ java \ com \ bin \ luo \ mysample \ TitleFragment.java 文件中。在这段代码中，myWebFragment =（WebFragment）getFragmentManager(). findFragmentById(R. id. myFrame) 中的 WebFragment 是自定义类，用于加载 WebView 控件。WebFragment 类的主要内容如下：

```java
public class WebFragment extends Fragment {
    String myUrl = "http://www.baidu.com";
    WebView myWebView;
    @Override
    public void setArguments(Bundle args) { myUrl = args.getString("myUrl"); }
    @Override
    public View onCreateView(LayoutInflater inflater,
                ViewGroup container, Bundle savedInstanceState) {
        myWebView = new WebView(inflater.getContext());
        myWebView.getSettings().setJavaScriptEnabled(true);
        myWebView.loadUrl(myUrl);
        myWebView.setWebViewClient(new WebViewClient());
        return myWebView;
    }
    public void reloadView() { myWebView.loadUrl(myUrl); }
}
```

上面这段代码在 MyCode \ MySample813 \ app \ src \ main \ java \ com \ bin \ luo \ mysample \ WebFragment.java 文件中。需要说明的是，访问网络需要相关的权限，因此需要在 AndroidManifest.xml 文件中添加 < uses-permission android：name = " android. permission. INTERNET"/>权限。此实例的完整项目在 MyCode\MySample813 文件夹中。

017　使用 TabHost 创建横向选项卡

此实例主要通过使用 TabHost 控件，实现创建选项卡风格的界面。当实例运行之后，选择"黎明之战"选项卡的效果如图 017.1 的左图所示，选择"权力之眼"选项卡的效果如图 017.1 的右图所示。

图　017.1

主要代码如下：

```xml
<LinearLayout android:layout_width = "match_parent"
        android:layout_height = "match_parent">
  <TabHost android:layout_width = "match_parent"
        android:layout_height = "match_parent">
    <!-- 定义第一个选项卡的内容 -->
    <LinearLayout  android:id = "@ + id/tab01"
        android:layout_width = "fill_parent"
        android:layout_height = "fill_parent"
        android:background = "＃121212">
      < ImageView  android:layout_width = "wrap_content"
        android:layout_height = "wrap_content"
        android:layout_gravity = "center"
        android:layout_margin = "50dp"
        android:src = "@mipmap/myimage1"/>
    </LinearLayout>
    <!-- 定义第二个选项卡的内容 -->
    <LinearLayout  android:id = "@ + id/tab02"
        android:layout_width = "fill_parent"
        android:layout_height = "fill_parent"
        android:background = "＃121212">
      < ImageView  android:layout_width = "wrap_content"
        android:layout_height = "wrap_content"
```

```xml
            android:layout_gravity = "center"
            android:layout_margin = "50dp"
            android:src = "@mipmap/myimage2"/>
</LinearLayout>
<!-- 定义第三个选项卡的内容 -->
<LinearLayout android:id = "@ + id/tab03"
            android:layout_width = "fill_parent"
            android:layout_height = "fill_parent"
            android:background = "#121212">
    <ImageView android:layout_width = "wrap_content"
            android:layout_height = "wrap_content"
            android:layout_gravity = "center"
            android:layout_margin = "50dp"
            android:src = "@mipmap/myimage3"/>
</LinearLayout></TabHost></LinearLayout>
```

上面这段代码在 MyCode\MySample283\app\src\main\res\layout\activity_main.xml 文件中。在这段代码中，TabHost 控件主要实现了设置每个选项卡的内容，每个选项卡通常还需要使用 TabHost 的 addTab()方法进行添加才能进行有效的管理，主要代码如下：

```java
public class MainActivity extends TabActivity{
    @Override
    public void onCreate(Bundle savedInstanceState){
        super.onCreate(savedInstanceState);
        TabHost myTabhost = getTabHost();
        //加载 TabHost 布局
        LayoutInflater.from(this).inflate(R.layout.activity_main,
                myTabhost.getTabContentView(), true);
        //添加第一个标签页
        myTabhost.addTab(myTabhost.newTabSpec("tab1")
                .setIndicator("黎明之战").setContent(R.id.tab01));
        //添加第二个标签页
        myTabhost.addTab(myTabhost.newTabSpec("tab2")
                .setIndicator("权力之眼").setContent(R.id.tab02));
        //添加第三个标签页
        myTabhost.addTab(myTabhost.newTabSpec("tab3")
                .setIndicator("圆梦巨人").setContent(R.id.tab03));
    }}
```

上面这段代码在 MyCode\MySample283\app\src\main\java\com\bin\luo\mysample\MainActivity.java 文件中。在这段代码中，public class MainActivity extends TabActivity 表示当前承载 TabHost 的 Activity 继承自 TabActivity，而不是常用的 Activity，否则就无法使用 getTabHost()方法。myTabhost.addTab(myTabhost.newTabSpec("tab1").setIndicator("黎明之战").setContent(R.id.tab01))表示向 TabHost 添加 R.id.tab01 选项卡并设置其标签为"黎明之战"。此实例的完整项目在 MyCode\MySample283 文件夹中。

018 使用 AbsoluteLayout 实现平移控件

此实例主要通过设置在 AbsoluteLayout 布局管理器中的子控件坐标，并使用定时器不断改变坐标值，从而使子控件产生不断移动的动画效果。当实例运行之后，降落伞（TextView 控件）位于屏幕的左上角,如图018.1的左图所示;立即沿着对角线向屏幕右下角移动,如图018.1的右图所示,并且周而复始,永不停歇。

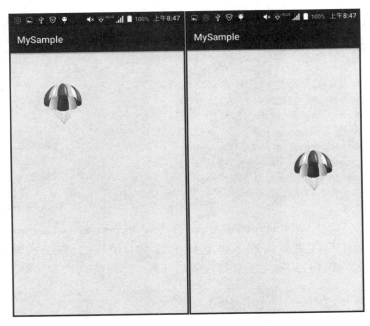

图 018.1

主要代码如下:

```xml
<AbsoluteLayout android:layout_width = "match_parent"
                android:layout_height = "match_parent">
<!-- 使用绝对定位定义 TextView 控件 -->
<TextView  android:id = "@ + id/myTextView"
           android:layout_width = "wrap_content"
           android:layout_height = "wrap_content"
           android:layout_x = "0dp"
           android:layout_y = "200dp"
           android:background = "@mipmap/myimage1"/>
</AbsoluteLayout>
```

上面这段代码在 MyCode\MySample012\app\src\main\res\layout\activity_main.xml 文件中。在这段代码中,android:layout_x = "0dp"和 android:layout_y = "200dp"用于设置 TextView 控件(降落伞)的左上角坐标,它表示控件位于 AbsoluteLayout 布局管理器的左上角。dp 是长度(距离)单位,Android 主要支持如下常用的单位。

(1) px(像素),每个 px 对应屏幕上的一个像素。

(2) dip 或 dp(device independent pixels,设备独立像素),一种基于屏幕密度的抽象单位。在每 in160 点的屏幕上,1dip=1px;但是随着屏幕密度的改变,dip 与 px 的换算会发生变化。

(3) sp(scaled pixels,比例像素),主要处理字体的大小,可以根据用户的字体大小首选项进行缩放。

(4) in(英寸),标准长度单位。

(5) mm(毫米),标准长度单位。

(6) pt(磅),标准长度单位,1/72in。

如果需要使用 Java 代码动态改变控件的坐标值,则应该使用 setLayoutParams()方法。

主要代码如下:

```java
public class MainActivity extends Activity {
    public TextView myTextView;
    public static int x = 0;
    public static int y = 0;
    Handler myHandler = new Handler() {
        public void handleMessage(Message msg) {
            if (msg.what == 0x123) {
                AbsoluteLayout.LayoutParams params =
                        new AbsoluteLayout.LayoutParams(256, 256, x++ % 756, y++ % 756);
                myTextView.setLayoutParams(params);
            }
            super.handleMessage(msg);
        }
    };
    @Override
    protected void onCreate(Bundle savedInstanceState) {
        super.onCreate(savedInstanceState);
        setContentView(R.layout.activity_main);
        myTextView = (TextView) findViewById(R.id.myTextView);
        new Timer().schedule(new TimerTask() {            //周期性地改变控件的坐标
            @Override
            public void run() { myHandler.sendEmptyMessage(0x123); }
        }, 0, 10);
    }
}
```

上面这段代码在 MyCode \ MySample012 \ app \ src \ main \ java \ com \ bin \ luo \ mysample \ MainActivity.java 文件中。在这段代码中,定时器每隔 10ms 调用 handleMessage()方法改变 TextView 控件(背景图像是降落伞)的坐标值。AbsoluteLayout.LayoutParams(256,256,x++ % 756,y++ % 756)中的两个 256 分别表示降落伞图像的宽度和高度,x++ % 756 和 y++ % 756 表示 TextView 控件(降落伞)的坐标值。myTextView.setLayoutParams(myParams)即是根据改变后的 x、y 值设置 TextView 控件(降落伞)的坐标值。大多数情况下,使用 AbsoluteLayout 布局管理器可能不是最优选择,因为运行 Android 应用的手机屏幕大小、分辨率千差万别,使用绝对布局不可能兼顾所有不同的手机,因此 AbsoluteLayout 布局管理器已经不在推荐使用之列,但是在需要绝对定位的情况下,AbsoluteLayout 也是一个不错的选择。此实例的完整项目在 MyCode\MySample012 文件夹中。

019 使用 FrameLayout 实现闪烁控件

此实例主要通过使用 FrameLayout 布局管理器叠加其中的子控件,并使用定时器不断改变其颜色,从而实现闪烁控件的效果。当实例运行之后,不同大小的实心矩形的背景颜色将不断改变,形成闪烁,效果如图 019.1 所示。

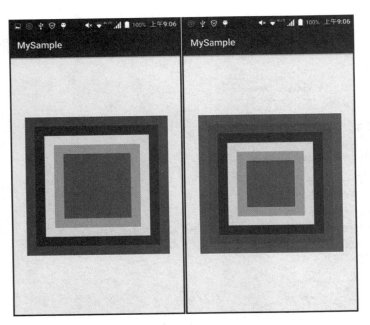

图 019.1

主要代码如下：

```xml
<!-- 使用 FrameLayout 布局管理器实现叠加 6 个 TextView 控件 -->
<FrameLayout  android:layout_width = "wrap_content"
              android:layout_height = "wrap_content">
<!-- 依次定义 6 个 TextView 控件,先定义控件位于底层,后定义控件位于上层 -->
    <TextView android:id = "@ + id/myView01"
              android:layout_width = "wrap_content"
              android:layout_height = "wrap_content"
              android:layout_gravity = "center"
              android:width = "300dp"
              android:height = "300dp"
              android:background = " # f00"/>
    <TextView android:id = "@ + id/myView02"
              android:layout_width = "wrap_content"
              android:layout_height = "wrap_content"
              android:layout_gravity = "center"
              android:width = "260dp"
              android:height = "260dp"
              android:background = " # 0f0"/>
    <TextView android:id = "@ + id/myView03"
              android:layout_width = "wrap_content"
              android:layout_height = "wrap_content"
              android:layout_gravity = "center"
              android:width = "220dp"
              android:height = "220dp"
              android:background = " # 00f"/>
    <TextView android:id = "@ + id/myView04"
              android:layout_width = "wrap_content"
              android:layout_height = "wrap_content"
              android:layout_gravity = "center"
```

```
            android:width = "180dp"
            android:height = "180dp"
            android:background = "#ff0"/>
    <TextView android:id = "@+id/myView05"
            android:layout_width = "wrap_content"
            android:layout_height = "wrap_content"
            android:layout_gravity = "center"
            android:width = "140dp"
            android:height = "140dp"
            android:background = "#f0f"/>
    <TextView android:id = "@+id/myView06"
            android:layout_width = "wrap_content"
            android:layout_height = "wrap_content"
            android:layout_gravity = "center"
            android:width = "100dp"
            android:height = "100dp"
            android:background = "#0ff"/>
</FrameLayout>
```

上面这段代码在 MyCode\MySample009\app\src\main\res\layout\activity_main.xml 文件中。在这段代码中，6 个 TextView 控件按照顺序，其宽度和高度依次变小，形成宝塔形的叠加；FrameLayout 布局管理器在这里只是容器，它为其中的子控件创建了一个空白的区域（称为一帧），每个子控件占据一帧，这些帧根据 gravity 属性执行自动对齐。实例在实现霓虹灯的动态闪烁效果时，主要是通过使用定时器按照顺序不断改变子控件的背景颜色，由于这些子控件的大小不同，因此各级子控件的背景颜色交替改变，形成闪烁效果。

主要代码如下：

```java
public class MainActivity extends Activity {
    private int myCurrentColor = 0;
    final int[] myColors = new int[]{                    //定义一个颜色数组
        Color.rgb(122, 122, 0),Color.rgb(0, 122, 122),Color.rgb(122, 0, 0),
        Color.rgb(255, 255, 0),Color.rgb(0, 255, 255),Color.rgb(255, 0, 0) };
    //此处的 6 个 ID 是在 activity_main.xml 文件中添加的 TextView 的 ID
    final int[] myIDs = new int[]{R.id.myView01, R.id.myView02, R.id.myView03,
                      R.id.myView04, R.id.myView05, R.id.myView06};
    TextView[] myViews = new TextView[myIDs.length];
    Handler myHandler = new Handler() {
    public void handleMessage(Message msg) {
     if (msg.what == 0x123) {
      for (int i = 0; i < myIDs.length; i++) {
       myViews[i].setBackgroundColor(myColors[(i + myCurrentColor) % myIDs.length]);
      }
      myCurrentColor++;
     }
     super.handleMessage(msg);
    } };
    @Override
    protected void onCreate(Bundle savedInstanceState) {
     super.onCreate(savedInstanceState);
     setContentView(R.layout.activity_main);
     for (int i = 0; i < myIDs.length; i++) {
      myViews[i] = (TextView) findViewById(myIDs[i]);
     }
```

```
new Timer().schedule(new TimerTask() {          //周期性改变控件的背景颜色
    @Override
    public void run() {
        //发送消息通知系统改变 6 个 TextView 控件的背景色
        myHandler.sendEmptyMessage(0x123);
    } }, 0, 300);
} }
```

上面这段代码在 MyCode \ MySample009 \ app \ src \ main \ java \ com \ bin \ luo \ mysample \ MainActivity.java 文件中。此实例的完整项目在 MyCode\MySample009 文件夹中。

020　自定义 FrameLayout 创建翻页卷边动画

此实例主要通过在自定义 FrameLayout 中使用 quadTo() 方法绘制贝塞尔曲线，从而实现图书翻页的卷边动画效果。当实例运行之后，在右上（或下）半部分使用手指从右向左滑动，则书页从右上（或下）角产生卷边效果，如图 020.1 的左图所示；松开手指，自动显示下一页，如图 020.1 的右图所示。在左上（或下）半部分使用手指从左向右滑动，则书页从左上（或下）角产生卷边效果；松开手指，自动显示上一页。

图　020.1

主要代码如下：

```
public class MainActivity extends Activity {
    @Override
    public void onCreate(Bundle savedInstanceState) {
        super.onCreate(savedInstanceState);
        setContentView(R.layout.activity_main);
        PageView myPageView = (PageView) findViewById(R.id.myPageView);
        myPageView.setAdapter(new PageAdapter(this));
    } }
```

上面这段代码在 MyCode \ MySample784 \ app \ src \ main \ java \ com \ bin \ luo \ mysample \ MainActivity.java 文件中。在这段代码中，myPageView ＝（PageView）findViewById（R. id. myPageView）中的 PageView 是以 FrameLayout 为基类创建的自定义类，关于该自定义类的主要代码请参考源代码中的 MyCode \ MySample784 \ app \ src \ main \ java \ com \ bin \ luo \ mysample \ PageView.java 文件。myPageView. setAdapter(new PageAdapter(this))表示使用 PageAdapter 为 PageView 创建数据适配器，PageAdapter 类的主要代码如下：

```
public class PageAdapter extends BaseAdapter {
 Context myContext;
  Integer[] myImages = {R.mipmap.myimage12, R.mipmap.myimage34,
                       R.mipmap.myimage56};          //总共 3 幅图像，代表 6 页
 public PageAdapter(Context context) { myContext = context; }
 @Override
 public int getCount() { return myImages.length; }
 @Override
 public Object getItem(int position) { return myImages[position]; }
 @Override
 public long getItemId(int position) { return position; }
 @Override
 public View getView(int position, View convertView, ViewGroup parent) {
  ViewGroup myLayout;
  if (convertView == null)
   myLayout = (ViewGroup) LayoutInflater.
                       from(myContext).inflate(R.layout.myitem, null);
  else myLayout = (ViewGroup)convertView;
  ImageView myImageView = (ImageView) myLayout.findViewById(R.id.myImageView);
  myImageView.setImageResource(myImages[position]);
  return myLayout;
 } }
```

上面这段代码在 MyCode \ MySample784 \ app \ src \ main \ java \ com \ bin \ luo \ mysample \ PageAdapter.java 文件中。在这段代码中，myLayout ＝（ViewGroup）LayoutInflater. from (myContext). inflate(R. layout. myitem，null)用于根据 myitem 布局创建 PageAdapter 的每个 Item 容器，这里只有一幅图像（书页）。myitem 布局的主要内容如下：

```xml
<?xml version = "1.0" encoding = "utf - 8"?>
<LinearLayout xmlns:android = "http://schemas.android.com/apk/res/android"
              android:layout_width = "match_parent"
              android:layout_height = "match_parent"
              android:baselineAligned = "false"
              android:orientation = "horizontal">
<ImageView android:id = "@ + id/myImageView"
           android:layout_width = "match_parent"
           android:layout_height = "wrap_content"
           android:scaleType = "fitXY"/>
</LinearLayout>
```

上面这段代码在 MyCode\MySample784\app\src\main\res\layout\myitem.xml 文件中。此实例的完整项目在 MyCode\MySample784 文件夹中。

第 2 章　常用控件

021　在 TextView 中创建空心文字

此实例主要通过设置 TextView 控件的 shadowColor、shadowRadius、shadowDx、shadowDy 等属性，从而创建指定颜色的空心文字。当实例运行之后，创建的粉色空心文字效果如图 021.1 所示。

图　021.1

主要代码如下：

```xml
<!-- 创建空心文字 -->
<TextView android:layout_width = "wrap_content"
          android:layout_height = "wrap_content"
          android:shadowColor = "@color/colorAccent"
          android:shadowDx = "0"
          android:shadowDy = "0"
```

```
android:shadowRadius = "15"
android:text = "炫酷应用实例"
android:textColor = "#fff"
android:textSize = "52sp"/>
```

上面这段代码在 MyCode\MySample330\app\src\main\res\layout\activity_main.xml 文件中。在这段代码中，android：shadowRadius 用于设置阴影的模糊程度，该值越大，阴影越模糊。android：shadowDx 用于设置阴影在水平方向的偏移。android：shadowDy 用于设置阴影在垂直方向的偏移。android：shadowColor 用于设置阴影的颜色。此实例的完整项目在 MyCode\MySample330 文件夹中。

022　在 TextView 中实现上文下图的布局

此实例主要实现了在 drawable 中添加图像资源，并设置 TextView 控件的 android：drawableBottom 属性为该图像资源，实现在 TextView 控件的下端显示图像、上端显示文字的效果。当实例运行之后，在 TextView 下端显示图像、上端显示文字的效果如图 022.1 所示。

图　022.1

主要代码如下：

```
<TextView android:layout_width = "match_parent"
    android:layout_height = "wrap_content"
    android:drawableBottom = "@drawable/myimage"
    android:text = "重庆市人民大礼堂位于渝中区人民路学田湾,于 1951 年 6 月破土兴建,1954 年 4 月竣工,是一座仿古民族建筑群,也是重庆独具特色的标志建筑物之一.建筑气势雄伟,金碧辉煌,是中国传统宫殿建筑风格与西方建筑的大跨度结构巧妙结合的杰作,以其非凡的建筑艺术蜚声中外."
    android:textSize = "20dp"/>
```

上面这段代码在 MyCode\MySample021\app\src\main\res\layout\activity_main.xml 文件中。

在这段代码中,android:drawableBottom="@drawable/myimage"表示将图像myimage放置在文本的Bottom(下端),myimage是图像资源ID,它是在向drawable添加图像资源时根据图像文件名自动生成的。如果需要将图像放置在TextView控件的上端、左边、右边,则应该设置该控件的drawableTop、drawableLeft、drawableRight等属性。drawableLeft属性表示在TextView控件的文本左边显示指定图像;drawableRight属性表示在TextView控件的文本右边显示指定图像;drawableTop属性表示在TextView控件的文本上端显示指定图像。drawableEnd属性表示在TextView控件的文本结尾显示指定图像。drawableStart属性表示在TextView控件的文本开始显示指定图像。TextView控件的drawablePadding属性用于设置文本与图像之间的间距。此实例的完整项目在MyCode\MySample021文件夹中。

023 在TextView中为文本添加超链接

此实例主要通过设置TextView控件的autoLink属性为all,实现为该控件的网址文本添加超链接功能。当实例运行之后,"百度官网:https://www.baidu.com/"是一个TextView控件,如图023.1的左图所示;单击其中的超链接"https://www.baidu.com/",则将跳转到百度官网,如图023.1的右图所示。

图 023.1

主要代码如下:

```xml
<TextView android:layout_width = "match_parent"
          android:layout_height = "wrap_content"
          android:textSize = "20dp"
          android:autoLink = "all"
          android:text = "百度官网:https://www.baidu.com/"/>
```

上面这段代码在MyCode\MySample366\app\src\main\res\layout\activity_main.xml文件中。在这段代码中,android:autoLink="all"表示设置TextView控件的所有文本为超链接,但是如果文

本有特殊内容,如"https：//www."，则将自动将超链接定位到特殊内容位置;如果文本的内容不是有效的网址,虽然 TextView 控件显示为超链接,但是单击后会出现空白页,如同在浏览器的地址栏中输入了错误的网址。此实例的完整项目在 MyCode\MySample366 文件夹中。

024　在自定义 View 中实现垂直滚动文本

此实例主要通过在自定义 View 中处理文本的行高和行数,实现在垂直方向上逐行滚动显示多行文本。当实例运行之后,文本将从上到下逐行滚动显示,效果分别如图 024.1 的左图和右图所示。

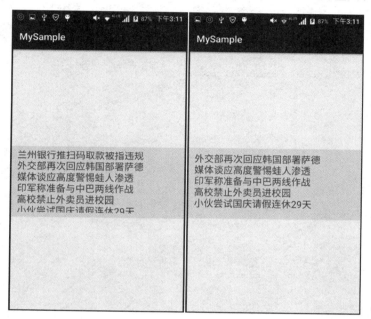

图　024.1

主要代码如下:

```xml
<com.bin.luo.mysample.ScrollTextView
    android:padding="10dp"
    android:background="#DEFFAC"
    android:layout_width="match_parent"
    android:layout_height="150dp"
    android:textSize="20dp"
    android:text="人民币八连涨逼近6.5\n
        湖南回应教材现"致命错误"\n
        北京市暂停共享自行车投放\n
        兰州银行推扫码取款被指违规\n
        外交部再次回应韩国部署萨德\n
        媒体谈应高度警惕蛙人渗透\n
        印军称准备与中巴两线作战\n
        高校禁止外卖员进校园\n
        小伙尝试国庆请假连休29天" />
```

上面这段代码在 MyCode\MySample226\app\src\main\res\layout\activity_main.xml 文件中。在这段代码中,com.bin.luo.mysample.ScrollTextView 即是用于实现文本在垂直方向上逐行滚动显示的自定义控件,自定义控件(类) ScrollTextView 的主要代码如下:

```java
public class ScrollTextView extends TextView {
    int curLines = 0, myLines = 0, perLines = 10;
    int lineHeight, myHeight, i = 0;
    boolean bScroll = true;
    Handler myHandler = new Handler() {
        @Override
        public void handleMessage(Message msg) {
            super.handleMessage(msg);
            scrollTo(0, lineHeight * curLines);
        }
    };
    public ScrollTextView(Context context) {
        super(context);
        init();
    }
    public ScrollTextView(Context context, AttributeSet attrs) {
        super(context, attrs);
        init();
    }
    public void init() {
        new Thread(new Runnable() {
            @Override
            public void run() {
                while (true) {
                    if (!bScroll) { continue; }
                    if (myHeight != 0) {
                        curLines = curLines + 1;
                        if (curLines == (myLines - 3) * perLines) { curLines = 0; }
                        Message message = myHandler.obtainMessage();
                        message.arg1 = curLines;
                        message.sendToTarget();
                        try {
                            if (curLines == 0) { Thread.sleep(1000); }
                            else Thread.sleep(150);
                        } catch (InterruptedException e) { e.printStackTrace(); }
                    } } }).start();
    }
    @Override
    protected void onSizeChanged(int w, int h, int oldw, int oldh) {
        super.onSizeChanged(w, h, oldw, oldh);
        curLines = 0;
        postInvalidate();
        lineHeight = getLineHeight() / perLines;              //获取行高
        myLines = getLineCount() - 1;                         //获取总行数
        myHeight = getLineCount() * lineHeight * perLines;    //获取总高度
        int height = getMeasuredHeight();                     //获取控件高度
        i = (int) (height / getTextSize());
        if (myLines <= i) { bScroll = false; }
        else { bScroll = true; }
    } }
```

上面这段代码在 MyCode \ MySample226 \ app \ src \ main \ java \ com \ bin \ luo \ mysample \ ScrollTextView.java 文件中。此实例的完整项目在 MyCode\MySample226 文件夹中。

025　在 EditText 中指定输入法的数字软键盘

此实例主要通过使用 EditText 的 setInputType()方法,实现在 EditText 控件获得焦点时,在输入法的软键盘上是显示数字软键盘,还是显示字母软键盘。当实例运行之后,单击"显示数字软键盘"按钮,则在输入法的软键盘上显示数字软键盘,如图 025.1 的左图所示;单击"显示字母软键盘"按钮,则在输入法的软键盘上显示字母软键盘,如图 025.1 的右图所示。

图　025.1

主要代码如下:

```
public void OnClickBtn1(View v) {                    //响应单击"显示数字软键盘"按钮
    myEditText.setInputType(EditorInfo.TYPE_CLASS_NUMBER);
}
public void OnClickBtn2(View v) {                    //响应单击"显示字母软键盘"按钮
    myEditText.setInputType(EditorInfo.TYPE_CLASS_TEXT);
}
```

上面这段代码在 MyCode \ MySample907 \ app \ src \ main \ java \ com \ bin \ luo \ mysample \ MainActivity.java 文件中。此实例的完整项目在 MyCode\MySample907 文件夹中。

026　禁止在 EditText 中插入非字符表情符号

此实例主要通过使用 EditText 的 setFilters()方法过滤输入的内容,实现禁止在输入框中插入非字符的表情符号。当实例运行之后,如果试图在输入框中输入表情符号,则在弹出的 Toast 中提示"此输入框禁止输入非纯文本字符!",如图 026.1 所示。

图 026.1

主要代码如下：

```java
public class MainActivity extends Activity {
 @Override
 protected void onCreate(Bundle savedInstanceState) {
  super.onCreate(savedInstanceState);
  setContentView(R.layout.activity_main);
  final EditText myEdit = (EditText) findViewById(R.id.myEdit);
  myEdit.setFilters(new InputFilter[]{new InputFilter(){             //自定义输入过滤器
   @Override
   public CharSequence filter(CharSequence source,
           int start, int end, Spanned dest, int dstart, int dend){
    Pattern myPattern = Pattern.compile(
     "[\ud83c\udc00-\ud83c\udfff]|[\ud83d\udc00-\ud83d\udfff]|[\u2600-\u27ff]",
     Pattern.UNICODE_CASE|Pattern.CASE_INSENSITIVE);
    //对输入框内容进行过滤
    Matcher myMatcher = myPattern.matcher(source);
    //删除输入框中非纯文本字符，并弹出 Toast 提示
    if(myMatcher.find()){
     Toast.makeText(MainActivity.this,
         "此输入框禁止输入非纯文本字符!", Toast.LENGTH_LONG).show();
     return "";
    }
    return null;
   }});
}}
```

上面这段代码在 MyCode \ MySample639 \ app \ src \ main \ java \ com \ bin \ luo \ mysample \ MainActivity.java 文件中。此实例的完整项目在 MyCode\MySample639 文件夹中。

027 使用 AutoCompleteTextView 实现自动提示

此实例主要通过使用 AutoCompleteTextView 控件,实现在输入框中输入内容时滑出列表框形式的提示。当实例运行之后,如果在输入框中输入"b",则在下面滑出一个以 b 开头的列表框,选择其中的列表项,则列表项内容将填充到输入框中,效果分别如图 027.1 的左图和右图所示。

图　027.1

主要代码如下:

```
public class MainActivity extends Activity {
  AutoCompleteTextView myAutoCompleteTextView;
  String[] myArray = {"BEIJING","SHANGHAI","TIANJIN","CHONGQING","AKESU",
      "ANNING","ANQING","ANSHAN","ANSHUN","ANYANG","BAICHENG","BAISHAN",
      "BAIYIN","BENGBU","BAODING","BAOJI","BAOSHAN"};
  @Override
  protected void onCreate(Bundle savedInstanceState) {
    super.onCreate(savedInstanceState);
    setContentView(R.layout.activity_main);
    myAutoCompleteTextView =
        (AutoCompleteTextView)findViewById(R.id.myAutoCompleteTextView);
    ArrayAdapter<String> adapter = new ArrayAdapter<String>(this,
        android.R.layout.simple_dropdown_item_1line, myArray);   //配置适配器
    myAutoCompleteTextView.setAdapter(adapter);
    myAutoCompleteTextView.setThreshold(1);     //表示从第1个字符开始显示提示
} }
```

上面这段代码在 MyCode\MySample258\app\src\main\java\com\bin\luo\mysample\MainActivity.java 文件中。在这段代码中,myComplete.setThreshold(1)表示从第1个字符开始滑出列表框内容,如果不设置此值,则 AutoCompleteTextView 控件不会滑出下拉列表框。此实例的完整项目在 MyCode\MySample258 文件夹中。

028　使用 SearchView 和 ListView 实现过滤输入

此实例主要通过综合使用 SearchView、ListView 等控件，实现在输入文字时过滤内容。当实例运行之后，如果在 SearchView 中输入文字（如"luo"），则在 ListView 中显示手机联系人中包含输入文字（如"luo"）的联系人名称，如图 028.1 的左图所示。单击 ListView 中的任一联系人名称，如"luobinbin"，则该联系人名称将显示在 SearchView 中，单击右侧的右箭头，则在弹出的 Toast 中显示 SearchView 中的内容，如图 028.1 的右图所示。

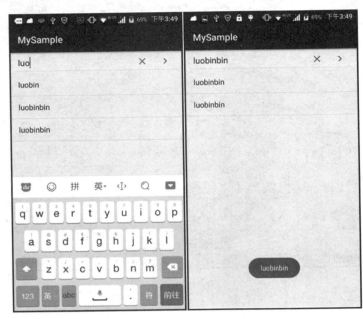

图　028.1

主要代码如下：

```java
public class MainActivity extends Activity {
    private SearchView mySearchView;
    private ListView myListView;
    private SimpleCursorAdapter myAdapter;
    private Cursor myCursor;
    static final String[] MYID = new String[]{ContactsContract.RawContacts._ID,
            ContactsContract.RawContacts.DISPLAY_NAME_PRIMARY };
    @Override
    protected void onCreate(Bundle savedInstanceState) {
        super.onCreate(savedInstanceState);
        setContentView(R.layout.activity_main);
        myCursor = getContentResolver().query(
                ContactsContract.RawContacts.CONTENT_URI, MYID, null, null, null);
        myAdapter = new SimpleCursorAdapter(this,
                android.R.layout.simple_list_item_1, myCursor,
                new String[] { ContactsContract.RawContacts.DISPLAY_NAME_PRIMARY },
                new int[] { android.R.id.text1 },0);
        myListView = (ListView) findViewById(android.R.id.list);
        myListView.setAdapter(myAdapter);
```

```
myListView.setOnItemClickListener(new AdapterView.OnItemClickListener() {
    @Override
    public void onItemClick(AdapterView<?> parent, View view, int position, long id){
            mySearchView.setQuery(((Cursor)
            myAdapter.getItem(position)).getString(1),false);
    } });
mySearchView = (SearchView) findViewById(R.id.mySearchView);
mySearchView.setIconifiedByDefault(true);
mySearchView.onActionViewExpanded();
mySearchView.setSubmitButtonEnabled(true);
//监听搜索框文本的改变情况
mySearchView.setOnQueryTextListener(new SearchView.OnQueryTextListener() {
    @Override
    public boolean onQueryTextChange(String queryText) {
     String mySQL = ContactsContract.RawContacts.DISPLAY_NAME_PRIMARY + " LIKE '%"
        + queryText + "%'" + " OR "
        + ContactsContract.RawContacts.SORT_KEY_PRIMARY
        + " LIKE '%" + queryText + "%'";
     myCursor = getContentResolver().query(
        ContactsContract.RawContacts.CONTENT_URI, MYID, mySQL, null, null);
     myAdapter.swapCursor(myCursor);
     return true;
    }
    @Override
    public boolean onQueryTextSubmit(String str) {//在Toast中显示搜索框文本内容
     Toast.makeText(MainActivity.this, str, Toast.LENGTH_SHORT).show();
     return false;
    } });
} }
```

上面这段代码在 MyCode\MySample382\app\src\main\java\com\bin\luo\mysample\MainActivity.java 文件中。在这段代码中，SearchView 是 Android 原生的搜索框控件，它提供了一个用户界面，用于用户搜索查询。SearchView 默认显示一个搜索图标，点击搜索图标则展开搜索框，如果需要搜索框默认展开，可以通过 setIconifiedByDefault（false）实现。onQueryTextChange（String queryText）重载方法在搜索框内容发生变化时响应，此实例在此实现根据搜索框内容过滤手机联系人。onQueryTextSubmit（String str）重载方法在单击了右侧的右箭头时响应，此实例在此实现在 Toast 中显示搜索框内容。此外，读取手机联系人信息需要相关的权限。因此需要在 AndroidManifest.xml 文件中添加< uses-permission android：name = "android.permission.READ_CONTACTS"/>权限。此实例的完整项目在 MyCode\MySample382 文件夹中。

029 在 EditText 右端设置输入提示内容和图标

此实例主要通过使用 EditText 的 setError（）方法，实现在 EditText 的右端设置提示内容和图标。当实例运行之后，在 EditText 的右端将显示灯泡图标和提示文字"必须填写"，如图 029.1 的左图所示。如果不设置自己的图标，Android 将显示一个默认的感叹号图标，如图 029.1 的右图所示。

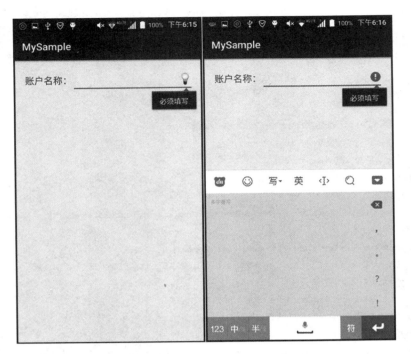

图 029.1

主要代码如下：

```
public class MainActivity extends Activity {
 EditText myEditText;
 @Override
 protected void onCreate(Bundle savedInstanceState) {
  super.onCreate(savedInstanceState);
  setContentView(R.layout.activity_main);
  myEditText = (EditText) findViewById(R.id.myEditText);
  Drawable myDrawable = getResources().getDrawable(R.mipmap.myimage1);
  myDrawable.setBounds(0, 0,72,72);
  myEditText.setError("必须填写", myDrawable);
  //myEditText.setError(" * * * ");
 } }
```

上面这段代码在 MyCode\MySample629\app\src\main\java\com\bin\luo\mysample\MainActivity.java 文件中。在这段代码中，myEditText.setError("必须填写"，myDrawable)用于在 EditText 上设置错误提示内容及图标，其实 TextView 也可以如此设置错误提示内容，不过它要获取焦点(TextView.requestFocus())才能显示错误信息，而 EditText 不用获取焦点是因为自己能够抢到焦点。此实例的完整项目在 MyCode\MySample629 文件夹中。

030　通过自定义 Shape 创建不同的圆角按钮

此实例主要通过在选择器(selector)文件中配置按钮在按下和抬起时的 Shape,从而创建不同的圆角按钮。当实例运行之后，所有按钮的右上角和右下角都是圆角，而左上角和左下角都是直角；单击第一个按钮，效果如图 030.1 的左图所示；单击第三个按钮，效果如图 030.1 的右图所示；单击其他按钮会实现类似的功能。

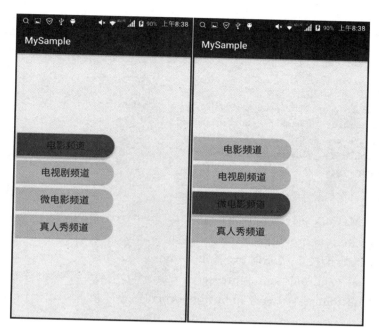

图 030.1

主要代码如下：

```xml
<LinearLayout android:orientation = "vertical"
            android:layout_width = "wrap_content"
            android:layout_height = "wrap_content">
    <Button android:layout_margin = "5dp"
            android:background = "@drawable/buttonselector"
            android:layout_width = "200dp"
            android:layout_height = "wrap_content"
            android:textSize = "20dp"
            android:text = "电影频道"/>
    <Button android:layout_margin = "5dp"
            android:background = "@drawable/buttonselector"
            android:layout_width = "200dp"
            android:layout_height = "wrap_content"
            android:textSize = "20dp"
            android:text = "电视剧频道"/>
    <Button android:layout_margin = "5dp"
            android:background = "@drawable/buttonselector"
            android:layout_width = "200dp"
            android:layout_height = "wrap_content"
            android:textSize = "20dp"
            android:text = "微电影频道"/>
    <Button android:layout_margin = "5dp"
            android:background = "@drawable/buttonselector"
            android:layout_width = "200dp"
            android:layout_height = "wrap_content"
            android:textSize = "20dp"
            android:text = "真人秀频道"/>
</LinearLayout>
```

上面这段代码在 MyCode\MySample352\app\src\main\res\layout\activity_main.xml 文件中。在这段代码中，android:background="@drawable/buttonselector"表示按钮采用buttonselector资源进行背景设置。buttonselector是一个XML文件，该文件的主要内容如下：

```xml
<?xml version="1.0" encoding="utf-8"?>
<selector xmlns:android="http://schemas.android.com/apk/res/android">
    <item android:drawable="@drawable/buttonpressed"
        android:state_pressed="true"/>
    <item android:drawable="@drawable/buttonnormal"
        android:state_pressed="false"/>
</selector>
```

上面这段代码在 MyCode\MySample352\app\src\main\res\drawable\buttonselector.xml 文件中。在这段代码中，< item android:drawable="@drawable/buttonpressed" android:state_pressed="true"/>表示当按钮处于按下状态时，设置buttonpressed资源。< item android:drawable="@drawable/buttonnormal" android:state_pressed="false"/>表示当按钮处于正常状态时，设置buttonnormal资源。buttonpressed资源和buttonnormal资源都是XML文件，buttonpressed资源的主要内容如下：

```xml
<?xml version="1.0" encoding="utf-8"?>
<!-- 按钮按下时候的背景 -->
<shape xmlns:android="http://schemas.android.com/apk/res/android">
    <!-- 按钮右上角和右下角的圆角半径值 -->
    <corners android:topRightRadius="30dp"
        android:bottomRightRadius="30dp"/>
    <!-- 按钮的填充色 -->
    <solid android:color="#CD00CD"/>
</shape>
```

上面这段代码在 MyCode\MySample352\app\src\main\res\drawable\buttonpressed.xml 文件中。在这段代码中，android:topRightRadius 表示设置按钮右上角的圆角半径，android:bottomRightRadius 表示设置按钮右下角的圆角半径。topLeftRadius 表示设置按钮左上角的圆角半径，bottomLeftRadius 表示设置按钮左下角的圆角半径，radius 表示设置按钮的4个角的圆角半径。如果未设置这些属性，则按钮的4个角将显示为直角。buttonnormal资源的主要内容如下：

```xml
<?xml version="1.0" encoding="utf-8"?>
<!-- 按钮正常时候的背景 -->
<shape xmlns:android="http://schemas.android.com/apk/res/android">
    <!-- 按钮的右上角和右下角圆角半径 -->
    <corners android:topRightRadius="30dp"
        android:bottomRightRadius="30dp"/>
    <!-- 按钮的填充色 -->
    <solid android:color="#98F5FF"/>
    <!--按钮边框的宽度,每段虚线的长度和两段虚线之间的颜色-->
    <!--<stroke-->
    <!--android:width="1dp"-->
    <!--android:dashWidth="8dp"-->
    <!--android:dashGap="4dp"-->
    <!--android:color="#4eb621"/>-->
</shape>
```

上面这段代码在 MyCode\MySample352\app\src\main\res\drawable\buttonnormal.xml 文件中。此实例的完整项目在 MyCode\MySample352 文件夹中。

031　通过设置背景图像创建立体的质感按钮

此实例主要通过设置 Button 控件的 background 属性为指定的按钮图像，创建具有立体效果的质感按钮。当实例运行之后，"确认"和"取消"按钮的背景是使用图像设置的背景，单击"确认"按钮，则在弹出的 Toast 中显示"同意此协议"，如图 031.1 的左图所示；单击"取消"按钮，则在弹出的 Toast 中显示"此协议待修正"，如图 031.1 的右图所示。

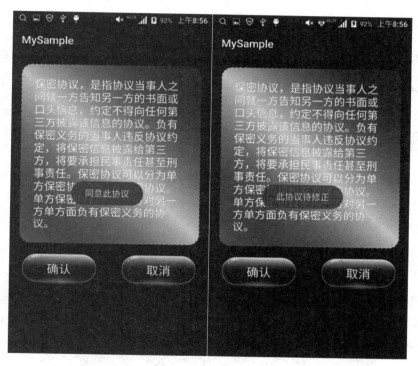

图　031.1

主要代码如下：

```xml
<LinearLayout  android:layout_width = "match_parent"
               android:layout_height = "wrap_content"
               android:gravity = "center"
               android:orientation = "horizontal">
    <Button android:layout_width = "180dp"
            android:layout_height = "100dp"
            android:background = "@drawable/mypng"
            android:onClick = "onClickmyBtn1"
            android:paddingBottom = "5dp"
            android:text = "确认"
            android:textColor = "#fff"
            android:textSize = "22dp"/>
    <Button android:id = "@+id/myBtn2"
            android:layout_width = "180dp"
            android:layout_height = "100dp"
```

```
            android:background = "@drawable/mypng"
            android:onClick = "onClickmyBtn2"
            android:paddingBottom = "5dp"
            android:text = "取消"
            android:textColor = "#fff"
            android:textSize = "22dp"/>
</LinearLayout>
```

上面这段代码在 MyCode\MySample048\app\src\main\res\layout\activity_main.xml 文件中。在这段代码中，android：background＝"@drawable/mypng"表示采用 drawable 中的 mypng 图像资源设置按钮的背景。android：onClick＝"onClickmyBtn2"用于将按钮单击事件与 onClickmyBtn2 方法关联起来。onClickmyBtn2 方法的主要代码如下：

```
//响应单击"取消"按钮
public void onClickmyBtn2(View v) {
    Toast myToast = Toast.makeText(getApplicationContext(),
                                    "此协议待修正", Toast.LENGTH_LONG);
    myToast.setGravity(Gravity.RIGHT | Gravity.TOP, 360, 870);
    myToast.show();
}
```

上面这段代码在 MyCode\MySample048\app\src\main\java\com\bin\luo\mysample\MainActivity.java 文件中。在这段代码中，onClickmyBtn2 方法的主要作用就是在 Toast 中显示提示信息。此实例的完整项目在 MyCode\MySample048 文件夹中。

032　使用 FloatingActionButton 创建悬浮按钮

此实例主要通过使用 FloatingActionButton 控件，实现悬浮按钮的效果。当实例运行之后，将在屏幕中心显示一个悬浮按钮，如图 032.1 的左图所示；单击该悬浮按钮，则在下面弹出一个 Toast，如图 032.1 的右图所示。

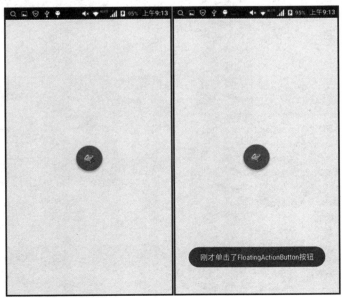

图　032.1

主要代码如下:

```
<android.support.design.widget.FloatingActionButton
    android:layout_width = "wrap_content"
    android:layout_height = "wrap_content"
    android:onClick = "onClickmyBtn1"
    android:src = "@mipmap/myimage"
    android:layout_centerInParent = "true"
    app:fabSize = "normal"/>
```

上面这段代码在 MyCode\MySample374\app\src\main\res\layout\activity_main.xml 文件中。在这段代码中,android:layout_centerInParent="true"表示悬浮按钮位于屏幕正中,如果需要悬浮按钮位于右下角,则应该设置 android:layout_alignParentRight = "true" 和 android:layout_alignParentBottom= "true"。此外,FloatingActionButton 可以通过 fabSize 属性设置大小,正常 size 是 normal,还有一个更小的模式为 mini。需要说明的是,在 Android Studio 中使用 FloatingActionButton 控件需要在 gradle 中引入 compile 'com.android.support:design:24.2.0'依赖项,24 是当前开发环境的 SDK 版本号。此实例的完整项目在 MyCode\MySample374 文件夹中。

033　以全屏效果显示在 ImageView 中的图像

此实例主要通过设置 ImageView 控件的 android:scaleType 属性为 fitXY,从而实现以全屏效果显示图像。当实例运行之后,图像全屏显示的效果如图 033.1 所示。

图　033.1

主要代码如下：

```xml
< ImageView  android:layout_width = "match_parent"
             android:layout_height = "match_parent"
             android:scaleType = "fitXY"
             android:src = "@mipmap/myimage"/>
```

上面这段代码在 MyCode\MySample059\app\src\main\res\layout\activity_main.xml 文件中。在这段代码中，android：src＝"@mipmap/myimage"用于设置 ImageView 控件显示的图像，myimage 是图像资源，它是以资源的形式将同名文件添加到此实例项目的"app\src\main\res\mipmap"节点（目录）下。android：scaleType＝"fitXY"表示将图像无条件填满控件的窗口。默认情况下，Android 应用会显示标题栏和通知栏等装饰栏，全屏显示通常应该隐藏这些装饰栏，这可以通过 Java 代码来实现，主要代码如下：

```java
public class MainActivity extends Activity {
 @Override
 protected void onCreate(Bundle savedInstanceState) {
  super.onCreate(savedInstanceState);
  //全屏设置，隐藏窗口所有装饰
  getWindow().setFlags(WindowManager.LayoutParams.FLAG_FULLSCREEN,
                       WindowManager.LayoutParams.FLAG_FULLSCREEN);
  //标题是属于 View 的，所以窗口所有的装饰部分被隐藏后标题依然有效，须隐藏
  requestWindowFeature(Window.FEATURE_NO_TITLE);
  setContentView(R.layout.activity_main);
 } }
```

上面这段代码在 MyCode\MySample059\app\src\main\java\com\bin\luo\mysample\MainActivity.java 文件中。在这段代码中，getWindow().setFlags(WindowManager.LayoutParams.FLAG_FULLSCREEN，WindowManager.LayoutParams.FLAG_FULLSCREEN)和 requestWindowFeature(Window.FEATURE_NO_TITLE)代码必须在 setContentView(R.layout.activity_main)之前调用，否则会报错。此实例的完整项目在 MyCode\MySample059 文件夹中。

034　在自定义 ImageView 中显示圆形图像

此实例主要通过以 ImageView 类为基类创建自定义控件 CircleImageView，实现在 ImageView 控件中以圆形的风格显示图像。当实例运行之后，圆形图像的显示效果如图 034.1 所示。

主要代码如下：

```xml
< com.bin.luo.mysample.CircleImageView
     android:layout_gravity = "center"
     android:layout_margin = "35dp"
     android:id = "@ + id/myView"
     android:layout_width = "wrap_content"
     android:layout_height = "wrap_content"
     android:src = "@mipmap/myimage"/>
```

上面这段代码在 MyCode\MySample282\app\src\main\res\layout\activity_main.xml 文件中。在这段代码中，com.bin.luo.mysample.CircleImageView 即是用于实现图像以圆形的风格显示的自

图 034.1

定义控件，CircleImageView 类的主要代码如下：

```java
public class CircleImageView extends ImageView {
  public CircleImageView(Context context) { super(context); }
  public CircleImageView(Context context, AttributeSet attrs) {
      super(context, attrs);
  }
  public CircleImageView(Context context, AttributeSet attrs, int defStyle) {
    super(context, attrs, defStyle);
  }
  @Override
  protected void onDraw(Canvas myCanvas) {
    Drawable myDrawable = getDrawable();
    if (myDrawable == null) { return; }
    if (getWidth() == 0 || getHeight() == 0) { return; }
    Bitmap myBitmap = ((BitmapDrawable) myDrawable).getBitmap();
    Bitmap oldBmp = myBitmap.copy(Bitmap.Config.ARGB_8888, true);
    int myRadius = getWidth();
    Bitmap CircleBmp = getCroppedBitmap(oldBmp, myRadius);
    myCanvas.drawBitmap(CircleBmp, 0, 0, null);
  }
  public static Bitmap getCroppedBitmap(Bitmap oldBmp, int myRadius) {
    Bitmap myBmp;
    if (oldBmp.getWidth() != myRadius || oldBmp.getHeight() != myRadius){
     myBmp = Bitmap.createScaledBitmap(oldBmp, myRadius, myRadius, false);
    } else{ myBmp = oldBmp; }
    Bitmap newBmp = Bitmap.createBitmap(myBmp.getWidth(),
               myBmp.getHeight(), Bitmap.Config.ARGB_8888);
```

```
    Canvas myCanvas = new Canvas(newBmp);
    final Paint myPaint = new Paint();
    final Rect myRect = new Rect(0, 0, myBmp.getWidth(), myBmp.getHeight());
    myPaint.setAntiAlias(true);
    myCanvas.drawCircle(myBmp.getWidth() / 2,
            myBmp.getHeight() / 2, myBmp.getWidth() / 2, myPaint);
    myPaint.setXfermode(new PorterDuffXfermode(PorterDuff.Mode.SRC_IN));
    myCanvas.drawBitmap(myBmp, myRect, myRect, myPaint);
    return newBmp;
} }
```

上面这段代码在 MyCode\MySample282\app\src\main\java\com\bin\luo\mysample\CircleImageView.java 文件中。在这段代码中，myBitmap=((BitmapDrawable) myDrawable).getBitmap()用于获取 CircleImageView 已加载图像。newBmp = Bitmap.createBitmap(myBmp.getWidth(),Bmp.getHeight(),Bitmap.Config.ARGB_8888)用于根据图像的大小创建空白的新位图。myCanvas=new Canvas(newBmp)用于根据空白的新位图创建新画布。myCanvas.drawCircle(myBmp.getWidth()/2,myBmp.getHeight()/2，myBmp.getWidth() / 2，myPaint)用于在画布上绘制与图像大小相同的圆形。myPaint.setXfermode(new PorterDuffXfermode(PorterDuff.Mode.SRC_IN))用于控制圆形如何与将要绘制的图像进行交互(裁剪)。myCanvas.drawBitmap(myBmp,myRect，myRect，myPaint)用于根据指定的 PorterDuff.Mode.SRC_IN 模式在圆形中绘制图像。此实例的完整项目在 MyCode\MySample282 文件夹中。

035　使用单指滑动拖曳 ImageView 的图像

此实例主要通过在 ImageView 的 setOnTouchListener()方法中监听手指滑动的位置变化，实现拖曳 ImageView 控件的图像。当实例运行之后，使用手指按住图像，然后在屏幕上滑动，则图像将跟随滑动，效果分别如图 035.1 的左图和右图所示。

图　035.1

主要代码如下：

```java
public class MainActivity extends Activity {
    private PointF myStartPoint = new PointF();
    private Matrix myMatrix = new Matrix();
    private Matrix myCurrentMatrix = new Matrix();
    private int MODE = 0;                            //用于判断当前手势是否为拖曳操作
    private static final int DRAG = 1;               //拖曳操作标志
    @Override
    protected void onCreate(Bundle savedInstanceState) {
        super.onCreate(savedInstanceState);
        setContentView(R.layout.activity_main);
        final ImageView myImageView = (ImageView) findViewById(R.id.myImageView);
        myImageView.setScaleType(ImageView.ScaleType.MATRIX);
        myImageView.setOnTouchListener(new View.OnTouchListener() {
            @Override
            public boolean onTouch(View v, MotionEvent event) {
                switch (event.getAction() & MotionEvent.ACTION_MASK) {
                    case MotionEvent.ACTION_DOWN:
                        MODE = DRAG;
                        myCurrentMatrix.set(myImageView.getImageMatrix());
                        myStartPoint.set(event.getX(), event.getY());     //记录起始位置
                        break;
                    case MotionEvent.ACTION_MOVE:
                        if (MODE == DRAG) {                               //拖曳发生
                            float dx = event.getX() - myStartPoint.x;     //x轴移动距离
                            float dy = event.getY() - myStartPoint.y;     //y轴移动距离
                            myMatrix.set(myCurrentMatrix);
                            myMatrix.postTranslate(dx, dy);               //矩阵位移操作
                        }
                        break;
                    case MotionEvent.ACTION_UP:
                        MODE = 0;
                        break;
                    case MotionEvent.ACTION_POINTER_UP:
                        MODE = 0;
                        break;
                }
                myImageView.setImageMatrix(myMatrix);
                return true;
            }
        });
    }
}
```

上面这段代码在 MyCode\MySample676\app\src\main\java\com\bin\luo\mysample\MainActivity.java 文件中。在这段代码中，myImageView.setImageMatrix(myMatrix)用于根据 myMatrix 参数平移（滑动）图像。myMatrix.postTranslate(dx, dy)则用于根据参数定制 myMatrix，dx 参数表示在 x 轴上的平移值，dy 参数表示在 y 轴上的平移值。注意，在 ImageView 控件中使用 Matrix 平移图像时，通常应用调用 ImageView 的 setScaleType（ImageView.ScaleType.MATRIX）方法。此实例的完整项目在 MyCode\MySample676 文件夹中。

036　使用 Gallery 实现滑动浏览多幅图像

此实例主要通过使用 Gallery 控件和 ImageView 控件,实现以手指左右滑动的方式浏览画廊的多幅图像的功能。当实例运行之后,如果手指按住图像向左滑动,则滑出右侧隐藏的图像;如果手指按住图像向右滑动,则滑出左侧隐藏的图像,效果分别如图 036.1 的左图和右图所示。

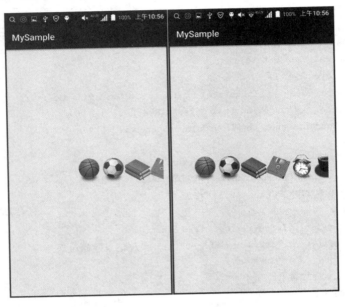

图　036.1

主要代码如下:

```xml
<Gallery android:layout_width = "match_parent"
        android:layout_height = "wrap_content"
        android:id = "@ + id/myGallery" />
```

上面这段代码在 MyCode\MySample058\app\src\main\res\layout\activity_main.xml 文件中。在这段代码中,Gallery 控件用于显示画廊图像,它主要通过自定义的 ImageAdapter 管理多幅图像。自定义 ImageAdapter 的主要代码如下:

```java
public class MainActivity extends Activity {
    @Override
    protected void onCreate(Bundle savedInstanceState) {
        super.onCreate(savedInstanceState);
        setContentView(R.layout.activity_main);
        Gallery myGallery = (Gallery) findViewById(R.id.myGallery);
        //添加一个 ImageAdapter 并配置给 Gallery
        myGallery.setAdapter(new ImageAdapter(MainActivity.this));
    }
}
public class ImageAdapter extends BaseAdapter {
    private Context myContext;
    private int[] myImages = {R.drawable.mypng1, R.drawable.mypng2,
            R.drawable.mypng3, R.drawable.mypng4, R.drawable.mypng5,
            R.drawable.mypng6, R.drawable.mypng7, R.drawable.mypng8};
    public ImageAdapter(Context c) { this.myContext = c; }
```

```
//获取已定义的图像总数量
public int getCount() { return myImages.length; }
//获取指定位置的图像
public Object getItem(int position) { return position; }
//获取指定位置的图像 ID
public long getItemId(int position) { return position; }
public View getView(int position, View convertView, ViewGroup parent) {
    ImageView myView = new ImageView(myContext);              //创建 ImageView 对象
    myView.setImageResource(myImages[position]);              //设置 ImageView 的图像
    myView.setScaleType(ImageView.ScaleType.FIT_XY);          //重新设置图像的宽高
    myView.setPadding(25, 25, 25, 25);                        //重新设置图像的内边距
    myView.setLayoutParams(
            new Gallery.LayoutParams(220, 220));              //重新设置 Layout 的宽高
    return myView;
}
//根据距离中央的位移量,利用 getScale()返回大小
public float getScale(boolean focused, int offset) {
    return Math.max(0, 1.0f / (float) Math.pow(2, Math.abs(offset)));
} } }
```

上面这段代码在 MyCode\MySample058\app\src\main\java\com\bin\luo\mysample\MainActivity.java 文件中。在这段代码中,R.drawable.mypng1,R.drawable.mypng2,R.drawable.mypng3,R.drawable.mypng4,R.drawable.mypng5,R.drawable.mypng6,R.drawable.mypng7,R.drawable.mypng8 是画廊的图像资源。它是以资源的形式添加到此实例项目的"app\src\main\res\drawable"节点下。此实例的完整项目在 MyCode\MySample058 文件夹中。

037　使用 SwipeRefreshLayout 切换图像

此实例主要通过使用 SwipeRefreshLayout,实现切换图像。当实例运行之后,按住屏幕下滑,则将出现一个转圈的动画表示正在刷新,1 秒之后转圈动画停止,然后显示一幅不同的图像(电影海报),如图 037.1 的左图所示。再次按住屏幕下滑,也将出现一个转圈的动画表示正在刷新,1 秒之后转圈动画停止,然后又显示一幅不同的图像(电影海报),如图 037.1 的右图所示。

图　037.1

主要代码如下:

```java
public class MainActivity extends AppCompatActivity {
 ImageView myImageView;
 Boolean bChecked = true;
 @Override
 protected void onCreate(@Nullable Bundle savedInstanceState) {
  super.onCreate(savedInstanceState);
  setContentView(R.layout.activity_main);
  final SwipeRefreshLayout mySwipeRefreshLayout = (SwipeRefreshLayout)
          findViewById(R.id.mySwipeRefreshLayout);
  myImageView = (ImageView) findViewById(R.id.myImageView);
  mySwipeRefreshLayout.setOnRefreshListener(
          new SwipeRefreshLayout.OnRefreshListener() {           //下拉时执行更新操作
   @Override
   public void onRefresh() {
    mySwipeRefreshLayout.setRefreshing(true);
    (new Handler()).postDelayed(new Runnable() {
     @Override
     public void run() {
      mySwipeRefreshLayout.setRefreshing(false);
      if (bChecked) {                                            //显示第一幅图像
       myImageView.setImageDrawable(getDrawable(R.mipmap.myimage1));
      } else {                                                   //显示第二幅图像
       myImageView.setImageDrawable(getDrawable(R.mipmap.myimage2));
      }
      bChecked = !bChecked;
     } }, 1000);                                                 //更新间隔时间 1 秒
} }); } }
```

上面这段代码在 MyCode\MySample645\app\src\main\java\com\bin\luo\mysample\MainActivity.java 文件中。在 Android 中，SwipeRefrshLayout 控件主要用于实现下拉刷新的动态效果，基本方法说明如下。

（1）setOnRefreshListener(OnRefreshListener)，该方法用于添加下拉刷新监听器。

（2）setRefreshing(boolean)，该方法用于显示或者隐藏刷新进度条。

（3）isRefreshing()，该方法检查是否处于刷新状态。

（4）setColorSchemeResources()，该方法设置进度条的颜色主题，最多设置 4 种。

大多数情况下，使用 SwipeRefrshLayout 控件均需要处理 setOnRefreshListener()方法，在其中实现刷新的目的（如此实例的切换图像）。需要说明的是，使用此实例的相关类需要在 gradle 中引入 compile 'com.android.support：design：25.0.1' 依赖项。此实例的完整项目在 MyCode\MySample645 文件夹中。

038 使用 AdapterViewFlipper 自动播放图像

此实例主要通过使用 AdapterViewFlipper 控件，实现自动播放在图库中的多幅图像。当实例运行之后，单击"上一幅图像"按钮，则将显示当前图像的上一幅图像，如图 038.1 的左图所示。单击"下一幅图像"按钮，则将显示当前图像的下一幅图像，如图 038.1 的右图所示。单击"自动播放"按钮，则将依次循环显示在图库中的多幅图像；如果已经是图库的最后一幅图像，则下一幅图像即是图库的第一幅图像。

第 2 章 常用控件

图 038.1

主要代码如下：

```java
public class MainActivity extends Activity {
    int[] myImages = new int[]{R.mipmap.myimage1, R.mipmap.myimage2,
            R.mipmap.myimage3,R.mipmap.myimage4,R.mipmap.myimage5,
            R.mipmap.myimage6,R.mipmap.myimage7 };
    AdapterViewFlipper myAdapterViewFlipper;
    @Override
    protected void onCreate(Bundle savedInstanceState) {
        super.onCreate(savedInstanceState);
        setContentView(R.layout.activity_main);
        myAdapterViewFlipper =
                (AdapterViewFlipper)findViewById(R.id.myAdapterViewFlipper);
        //创建 BaseAdapter 对象,该对象负责提供图库的列表项
        BaseAdapter myBaseAdapter = new BaseAdapter(){
            @Override
            public int getCount() { return myImages.length; }
            @Override
            public Object getItem(int position) { return position; }
            @Override
            public long getItemId(int position) { return position; }
            //该方法返回的 View 代表了每个列表项
            @Override
            public View getView(int position, View convertView, ViewGroup parent) {
                ImageView myImageView = new ImageView(MainActivity.this);
                myImageView.setImageResource(myImages[position]);
                //设置缩放类型
```

```
        myImageView.setScaleType(ImageView.ScaleType.FIT_XY);
        //为 ImageView 设置布局参数
        myImageView.setLayoutParams(new ViewGroup.LayoutParams(
                ViewGroup.LayoutParams.MATCH_PARENT,
                ViewGroup.LayoutParams.MATCH_PARENT));
        return myImageView;
      }
    };
    myAdapterViewFlipper.setAdapter(myBaseAdapter);
}
//响应单击"上一幅图像"按钮
public void onClickmyBtn1(View v){
    myAdapterViewFlipper.showPrevious();
    myAdapterViewFlipper.stopFlipping();
}
//响应单击"下一幅图像"按钮
public void onClickmyBtn2(View v){
    myAdapterViewFlipper.showNext();
    myAdapterViewFlipper.stopFlipping();
}
//响应单击"自动播放"按钮
public void onClickmyBtn3(View v){
    myAdapterViewFlipper.startFlipping();
} }
```

上面这段代码在 MyCode\MySample384\app\src\main\java\com\bin\luo\mysample\MainActivity.java 文件中。在这段代码中，AdapterViewFlipper 控件可以显示 Adapter 提供的 View 控件，但是每次只能显示一个，可以通过该控件的 showPrevious() 和 showNext() 方法实现显示上一个或下一个 View 控件；除此之外，它还可以通过 startFlipping() 方法实现自动播放控件，也可以通过 stopFlipping() 方法停止自动播放。AdapterViewFlipper 控件常用的属性如下。

（1）android：animateFirstView 属性，该属性用于设置在显示该控件的第一个 View 时是否使用动画。

（2）android：inAnimation 属性，该属性用于设置控件在显示时使用的动画。

（3）android：loopViews 属性，该属性用于设置循环到最后一个控件后是否自动"转头"到第一个控件。

（4）android：outAnimation 属性，该属性用于设置控件在隐藏时使用的动画。

（5）android：autoStart 属性，该属性用于设置该控件是否是自动播放动画。

（6）android：flipInterval 属性，该属性用于设置自动播放的时间间隔。

此实例的完整项目在 MyCode\MySample384 文件夹中。

039　使用两幅图像定制 ToggleButton 开关状态

此实例主要通过在选择器（selector）文件中配置 ToggleButton 的 state_checked 属性的不同（true/false）值对应的图像，从而实现定制 ToggleButton 的开关状态。当实例运行之后，ToggleButton 处于关闭状态，没有显示背景图像，如图 039.1 的左图所示；单击 ToggleButton，则 ToggleButton 处于打开状态，立即显示背景图像，如图 039.1 右图所示。

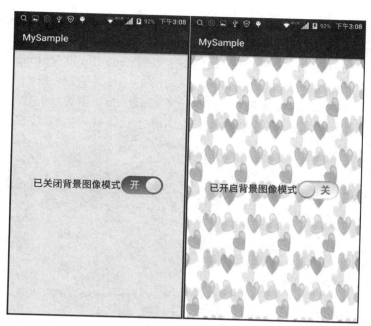

图 039.1

主要代码如下：

```
<LinearLayout android:layout_width = "match_parent"
        android:layout_height = "match_parent"
        android:gravity = "center"
        android:id = "@ + id/myLinearLayout"
        android:orientation = "horizontal">
  <TextView android:id = "@ + id/myTextView"
        android:layout_width = "wrap_content"
        android:layout_height = "wrap_content"
        android:text = "已关闭背景图像模式"
        android:textColor = "@android:color/black"
        android:textSize = "20.0dp"/>
  <ToggleButton android:id = "@ + id/myToggleButton"
        android:layout_width = "wrap_content"
        android:layout_height = "wrap_content"
        android:background = "@drawable/myselector"
        android:checked = "false"
        android:text = ""
        android:textOff = ""
        android:textOn = "" />
</LinearLayout>
```

上面这段代码在 MyCode\MySample053\app\src\main\res\layout\activity_main.xml 文件中。在这段代码中，android：background＝"@drawable/myselector"表示 ToggleButton 采用 myselector 资源进行背景配置。myselector 是一个 XML 文件，该文件的主要内容如下：

```
<?xml version = "1.0" encoding = "utf-8"?>
<selector xmlns:android = "http://schemas.android.com/apk/res/android">
  <item android:drawable = "@drawable/toggleopen" android:state_checked = "true"/>
  <item android:drawable = "@drawable/toggleclose" android:state_checked = "false"/>
</selector>
```

上面这段代码在 MyCode\MySample053\app\src\main\res\drawable\myselector.xml 文件中。在这段代码中,< item android：drawable = "@drawable/toggleopen" android：state_checked = "true"/>表示当 ToggleButton 的 state_checked 属性值为 true 时,设置 toggleopen 资源。< item android：drawable = "@drawable/toggleclose" android：state_checked = "false"/>表示当 ToggleButton 的 state_checked 属性值为 false 时,设置 toggleclose 资源。

ToggleButton 的 CheckedChange 事件响应方法则通过 Java 代码实现,主要代码如下:

```java
public class MainActivity extends Activity {
    private ToggleButton myToggleButton;
    private TextView myTextView;
    private LinearLayout myLinearLayout;
    @Override
    protected void onCreate(Bundle savedInstanceState) {
        super.onCreate(savedInstanceState);
        setContentView(R.layout.activity_main);
        myLinearLayout = (LinearLayout) findViewById(R.id.myLinearLayout);
        myTextView = (TextView) findViewById(R.id.myTextView);
        myToggleButton = (ToggleButton) findViewById(R.id.myToggleButton);
        //响应单击 ToggleButton
        myToggleButton.setOnCheckedChangeListener(
                    new CompoundButton.OnCheckedChangeListener() {
            public void onCheckedChanged(CompoundButton buttonView,
                                            boolean isChecked) {
                if (isChecked) {
                    myTextView.setText("已开启背景图像模式");
                    myLinearLayout.setBackground(
                            getResources().getDrawable(R.mipmap.mybackground));
                } else {
                    myTextView.setText("已关闭背景图像模式");
                    myLinearLayout.setBackground(null);
                } } });
    }
}
```

上面这段代码在 MyCode\MySample053\app\src\main\java\com\bin\luo\mysample\MainActivity.java 文件中。在这段代码中,myLinearLayout.setBackground(null)表示取消屏幕(LinearLayout)背景图像。myLinearLayout.setBackground(getResources().getDrawable(R.mipmap.mybackground))表示使用 mybackground 资源设置屏幕的背景图像。此实例的完整项目在 MyCode\MySample053 文件夹中。

040 使用 GridView 创建网格显示多幅图像

此实例主要通过使用 GridView 控件,实现在网格中显示多幅图像。当实例运行之后,在 GridView 控件中将显示多幅图像,单击其中任意一幅图像,则在弹出的 Toast 中显示哪幅图像被单击了,效果分别如图 040.1 的左图和右图所示。

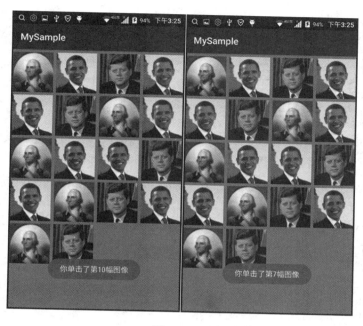

图 040.1

主要代码如下：

```java
public class MainActivity extends Activity {
  private GridView myGridView = null;
  @Override
  protected void onCreate(Bundle savedInstanceState) {
    super.onCreate(savedInstanceState);
    setContentView(R.layout.activity_main);
    myGridView = (GridView) findViewById(R.id.myGridView);
    myGridView.setAdapter(new ImageAdapter(MainActivity.this));
    //响应单击网格,即指明单击了哪一个网格
    myGridView.setOnItemClickListener(new AdapterView.OnItemClickListener() {
    public void onItemClick(AdapterView<?> arg0, View arg1, int arg2, long arg3){
      arg2 = arg2 + 1;
      Toast.makeText(MainActivity.this,
              "你单击了第" + arg2 + "幅图像",Toast.LENGTH_SHORT).show();
    } }); }
public class ImageAdapter extends BaseAdapter {
  private Context myContext;
  private int[] ImageBox = {R.drawable.sport1, R.drawable.movie1,
      R.drawable.book1, R.drawable.movie1,R.drawable.movie1,
      R.drawable.book1,R.drawable.sport1, R.drawable.movie1,
      R.drawable.sport1, R.drawable.movie1, R.drawable.movie1,
      R.drawable.book1,R.drawable.movie1, R.drawable.sport1,
      R.drawable.book1, R.drawable.movie1,R.drawable.sport1, R.drawable.book1};
  public ImageAdapter(Context context) { this.myContext = context; }
  public int getCount() { return ImageBox.length; }
  public Object getItem(int arg0) { return null; }
  public long getItemId(int position) { return 0; }
  //设置网格内容
  public View getView(int position, View convertView, ViewGroup parent) {
```

```
    ImageView myImageView;
    if (convertView == null) {
     myImageView = new ImageView(myContext);
     myImageView.setLayoutParams(new GridView.LayoutParams(260, 260));
     myImageView.setScaleType(ImageView.ScaleType.CENTER_CROP);
     myImageView.setPadding(8, 0, 0, 0);
    } else { myImageView = (ImageView) convertView; }
    myImageView.setImageResource(ImageBox[position]);
    return myImageView;
} } }
```

上面这段代码在 MyCode\MySample257\app\src\main\java\com\bin\luo\mysample\MainActivity.java 文件中。在这段代码中，myGridView.setAdapter（new ImageAdapter(MainActivity.this)）用于设置 GridView 控件的数据适配器，也可以说是数据源，即 GridView 控件将要加载的内容。除此之外，GridView 控件通常还需要设置列数和行数等基本属性。

主要代码如下：

```
<GridView android:id="@+id/myGridView"
          android:layout_width="match_parent"
          android:layout_height="match_parent"
          android:columnWidth="80dp"
          android:gravity="center"
          android:horizontalSpacing="6dp"
          android:numColumns="auto_fit"
          android:paddingTop="6dp"
          android:stretchMode="columnWidth"
          android:verticalSpacing="6dp"/>
```

上面这段代码在 MyCode\MySample257\app\src\main\res\layout\activity_main.xml 文件中。在这段代码中，android：columnWidth＝"80dp"用于设置列宽，也可以通过 android：numColumns 属性设置具体的列数。当某个属性有替代方案时，GridView 控件能够进行自动计算。此实例的完整项目在 MyCode\MySample257 文件夹中。

041 使用 ViewPager 实现缩放轮播多幅图像

此实例主要通过使用 onPageScrolled（int position，float positionOffset，int positionOffsetPixels）方法的 positionOffset 参数调整图像的内边距，实现 ViewPager 在轮播多幅图像时，进场图像根据当前位置放大，退场图像根据当前位置缩小的动画效果。当实例运行之后，7 个红色的圆点表示总共有 7 幅图像，白色的圆点表示当前显示的图像；使用手指向左滑动，则左边的退场图像缩小至消失，右边的进场图像放大至满屏，效果分别如图 041.1 的左图和右图所示。

主要代码如下：

```
public class MainActivity extends Activity {
 private ViewPager myViewPager;
 @Override
 protected void onCreate(Bundle savedInstanceState) {
  super.onCreate(savedInstanceState);
  setContentView(R.layout.activity_main);
  myViewPager = (ViewPager) findViewById(R.id.myViewPager);
```

```
//加载7个圆点指示符
LinearLayout myIndicators = (LinearLayout) findViewById(R.id.myIndicators);
for (int i = 0; i < 7; i++) {
  ImageView myPoint = new ImageView(this);
  LinearLayout.LayoutParams myParams = new LinearLayout.LayoutParams(
                    LinearLayout.LayoutParams.WRAP_CONTENT,
                    LinearLayout.LayoutParams.WRAP_CONTENT);
  myParams.setMargins(10, 10, 10, 10);
  myPoint.setImageResource(R.drawable.mypointer_inactive);
  myPoint.setLayoutParams(myParams);
  myIndicators.addView(myPoint);
}
//第一个圆点为白点表示当前显示的是第一幅图像
myIndicators.getChildAt(0).setBackgroundResource(
                    R.drawable.mypointer_active);
MyAdapter myAdapter = new MyAdapter(this, myIndicators);
myViewPager.setAdapter(myAdapter);
myViewPager.addOnPageChangeListener(myAdapter);
} }
```

图　041.1

上面这段代码在 MyCode\MySample696\app\src\main\java\com\bin\luo\mysample\MainActivity.java 文件中。在这段代码中，myViewPager.setAdapter(myAdapter)表示使用自定义适配器 MyAdapter 管理 myViewPager 的所有图像，自定义适配器 MyAdapter 的主要代码如下：

```
public class MyAdapter extends PagerAdapter implements OnPageChangeListener{
  public int myPosition = 0;
  public int myWidthPadding;
  public int myHeightPadding;
  public List<View> myViews = new ArrayList<View>();
  public Context myContext;
```

```java
    public LinearLayout myIndicator;
    public MyAdapter(Context context, LinearLayout indicator) {
        myContext = context;
        InitViews();
        myIndicator = indicator;
        //进场和退场缩放图像的(内边距)宽度和高度
        myWidthPadding = 100;
        myHeightPadding = 150;
    }
    public Integer[] myImages = {R.mipmap.myimage1, R.mipmap.myimage2,
            R.mipmap.myimage3, R.mipmap.myimage4, R.mipmap.myimage5,
            R.mipmap.myimage6, R.mipmap.myimage7};
private void InitViews() {
    for (int i = 0; i < 7; i++) {
        ImageView myImage = new ImageView(myContext);
        Bitmap myBitmap =
                BitmapFactory.decodeResource(myContext.getResources(), myImages[i]);
        myImage.setImageBitmap(myBitmap);
        myImage.setScaleType(ImageView.ScaleType.FIT_XY);
        myViews.add(myImage);
    } }
    @Override
    public int getCount() {return myViews.size();}
    @Override
    public boolean isViewFromObject(View view, Object object) { return view == object;}
    @Override
    public void onPageScrolled(int position, float positionOffset,
        int positionOffsetPixels) {
      for (int i = 0; i < myViews.size(); i++) {
          myIndicator.getChildAt(i).setBackgroundResource(
                                        R.drawable.mypointer_inactive);
      }
      if (position < myViews.size()) {
       myPosition = position;
       myIndicator.getChildAt(position).setBackgroundResource(
                                        R.drawable.mypointer_active);
       //退场缩小图像
       int outHeightPadding = (int) (positionOffset * myHeightPadding);
       int outWidthPadding = (int) (positionOffset * myWidthPadding);
       myViews.get(position).setPadding(outWidthPadding,
                        outHeightPadding, outWidthPadding, outHeightPadding);
       //进场放大图像
       if (position < myViews.size() - 1) {
        int inWidthPadding = (int) ((1 - positionOffset) * myWidthPadding);
        int inHeightPadding = (int) ((1 - positionOffset) * myHeightPadding);
        myViews.get(position + 1).setPadding(inWidthPadding, inHeightPadding,
                                    inWidthPadding, inHeightPadding);
    } } }
    @Override
    public void onPageSelected(int position) { }
    @Override
    public void destroyItem(ViewGroup container, int position, Object object) {
     container.removeView(myViews.get(position));
```

```
    }
    @Override
    public void onPageScrollStateChanged(int state) { }
    @Override
    public Object instantiateItem(ViewGroup container, int position) {
      View myView = myViews.get(position);
      container.addView(myView);
      return myView;
    }
}
```

上面这段代码在 MyCode\MySample696\app\src\main\java\com\bin\luo\mysample\MyAdapter.java 文件中。在这段代码中，myIndicator.getChildAt(i).setBackgroundResource(R.drawable.mypointer_inactive)用于设置红色的圆点（非当前图像）指示器，mypointer_inactive 是 XML 文件，它是采用标签的形式直接绘制圆形实现的。mypointer_inactive 文件的主要内容如下：

```xml
<?xml version = "1.0" encoding = "utf-8"?>
<shape xmlns:android = "http://schemas.android.com/apk/res/android"
       android:shape = "oval">
    <size android:width = "15dp" android:height = "15dp" />
    <solid android:color = "#44FF0000" />
</shape>
```

上面这段代码在 MyCode\MySample696\app\src\main\res\drawable\mypointer_inactive.xml 文件中。在 MyAdapter.java 文件中，myIndicator.getChildAt(position).setBackgroundResource(R.drawable.mypointer_active)用于设置白色的圆点（当前图像）指示器，mypointer_active 是一个 XML 文件，它也是采用标签的形式直接绘制圆形实现的。mypointer_active 文件的主要内容如下：

```xml
<?xml version = "1.0" encoding = "utf-8"?>
<shape xmlns:android = "http://schemas.android.com/apk/res/android"
       android:shape = "oval">
    <size android:width = "15dp" android:height = "15dp"/>
    <solid android:color = "#FFF"/>
</shape>
```

上面这段代码在 MyCode\MySample696\app\src\main\res\drawable\mypointer_active.xml 文件中。需要说明的是，使用此实例的相关类需要在 gradle 中引入 compile 'com.android.support：design：23.3.0'依赖项。此实例的完整项目在 MyCode\MySample696 文件夹中。

042 使用 Handler 实现自动轮播 ViewPager

此实例主要通过使用 Handler 的 postDelayed()方法延迟显示图像，实现自动轮播在 ViewPager 中的多幅图像。当实例运行之后，在 ViewPager 中的多幅图像将周而复始地自动轮播，效果分别如图 042.1 的左图和右图所示。

图 042.1

主要代码如下：

```java
public class MainActivity extends Activity {
    ViewPager myViewPager;
    ViewPagerAdapter myViewPagerAdapter;
    Handler myHandler = new Handler();
    int[] myImages;
    @Override
    protected void onCreate(Bundle savedInstanceState) {
        super.onCreate(savedInstanceState);
        setContentView(R.layout.activity_main);
        myViewPager = (ViewPager) findViewById(R.id.myViewPager);
        myImages = new int[]{R.mipmap.myimage1,R.mipmap.myimage2,R.mipmap.myimage3,
                R.mipmap.myimage4,R.mipmap.myimage5,R.mipmap.myimage6};
        myViewPagerAdapter = new ViewPagerAdapter(MainActivity.this, myImages);
        myViewPager.setAdapter(myViewPagerAdapter);
        myViewPager.setOffscreenPageLimit(myImages.length);         //设置缓存数量
        myHandler.postDelayed(new Runnable(){
            @Override
            public void run(){
                myViewPager.setCurrentItem((myViewPager.getCurrentItem() + 1) % myViewPager.
                        getChildCount());                            //自动显示下一幅图像,并做越界处理
                myHandler.postDelayed(this,1500);
            } },1500);                                               //每间隔 1500ms 执行一次
    }
public class ViewPagerAdapter extends PagerAdapter {
    private Context myContext;
    private int[] myImages;
    public ViewPagerAdapter(Context context, int[] datas){
        myImages = datas;
        myContext = context;
```

```
    }
    @Override
    public int getCount() { return myImages.length; }
    @Override
    public boolean isViewFromObject(View view, Object object){return view == object;}
    @Override
    public Object instantiateItem(ViewGroup container, int position) {
      ImageView myImage = createImageView(myContext, position);
      container.addView(myImage);
      return myImage;
    }
    @Override
    public void destroyItem(ViewGroup container, int position, Object object) {
      container.removeView((View)object);
    }
    private ImageView createImageView(Context mContext, int position) {
      ImageView myImage = new ImageView(mContext);
      ViewPager.LayoutParams layoutParams = new ViewPager.LayoutParams();
      myImage.setLayoutParams(layoutParams);
      myImage.setImageResource(myImages[position]);
      myImage.setScaleType(ImageView.ScaleType.CENTER_CROP);
      return myImage;
    } }
}
```

上面这段代码在 MyCode\MySample786\app\src\main\java\com\bin\luo\mysample\MainActivity.java 文件中。在这段代码中，myHandler.postDelayed(this,1500)表示间隔1500ms显示下一幅图像，postDelayed()方法的语法声明如下：

```
public final boolean postDelayed(Runnable r, long delayMillis)
```

其中，参数 Runnable r 表示要延迟执行的具体任务。参数 long delayMillis 表示要延迟的具体时长。

需要说明的是，在 Android Studio 中，使用此实例的相关控件需要在 gradle 中引入 compile 'com.android.support：design：23.3.0'依赖项。此实例的完整项目在 MyCode\MySample786 文件夹中。

043 使用 ViewPager 实现苹果风格的 cover flow

此实例主要通过在 ViewPager 控件的自定义适配器中为图像添加倒影，并在该控件的 setPageTransformer()方法中将图像围绕 y 轴旋转指定的角度，使图像在左右滑动时实现苹果风格的 cover flow。cover flow 是苹果首创的将多首歌曲的封面以 3D 界面的形式显示出来的方式。当实例运行之后，左右滑动图像，则可出现如图 043.1 的左图和右图所示的 cover flow 效果。

主要代码如下：

```
public class MainActivity extends Activity{
  @Override
  protected void onCreate(Bundle savedInstanceState){
    super.onCreate(savedInstanceState);
    setContentView(R.layout.activity_main);
```

```
ViewPager myViewPager = (ViewPager)findViewById(R.id.myViewPager);
MyAdapter myAdapter = new MyAdapter(this);
myViewPager.setAdapter(myAdapter);
ViewGroup.LayoutParams myParams = myViewPager.getLayoutParams();
myParams.width = getWindowManager().getDefaultDisplay().getWidth()/2;
myViewPager.setLayoutParams(myParams);
myViewPager.setOnPageChangeListener(myAdapter);            //设置滑动监听
myViewPager.setOffscreenPageLimit(3);                      //设置缓存图像数量
myViewPager.setPageTransformer(false,new ViewPager.PageTransformer(){
  @Override
  public void transformPage(View page,float position){
    //添加 3D 偏移特效(使图像以指定的角度围绕 y 轴放置)
    page.setRotationY(position * - 30f);
  } }); }
}
```

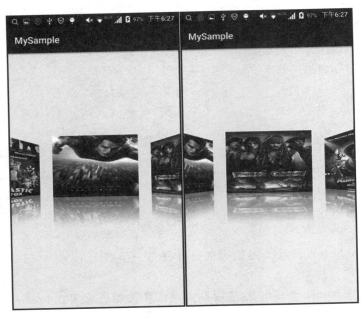

图　043.1

上面这段代码在 MyCode\MySample766\app\src\main\java\com\bin\luo\mysample\MainActivity.java 文件中。在这段代码中，myViewPager.setAdapter(myAdapter)表示使用自定义适配器 MyAdapter 管理 myViewPager 的所有图像，自定义适配器 MyAdapter 的主要代码如下：

```
public class MyAdapter extends
                PagerAdapter implements ViewPager.OnPageChangeListener {
public List< ImageView > myImageViewList;
Integer[] myImageResources = {R.mipmap.myimage1, R.mipmap.myimage2,
            R.mipmap.myimage3,R.mipmap.myimage4, R.mipmap.myimage5,
            R.mipmap.myimage6, R.mipmap.myimage7};
public MyAdapter(Context context) { InitImageViewList(context); }
  @Override
  public int getCount() { return myImageViewList.size(); }
  @Override
```

```java
        public Object instantiateItem(ViewGroup container, int position) {
            int myPosition = position % myImageViewList.size();
            ImageView myImageView = myImageViewList.get(myPosition);
            container.addView(myImageView);
            return myImageView;
        }
        @Override
        public void destroyItem(ViewGroup container, int position, Object object) {
            container.removeView((View) object);
        }
        @Override
        public boolean isViewFromObject(View view, Object object){ return view == object; }
        public void InitImageViewList(Context context) {
            myImageViewList = new ArrayList<ImageView>();
            for (int i = 0; i < myImageResources.length; i++) {
                ImageView myImageView = new ImageView(context);
                myImageView.setScaleType(ImageView.ScaleType.FIT_XY);
                Bitmap myBitmap = BitmapFactory.decodeResource(
                                    context.getResources(), myImageResources[i]);
                Bitmap myReflectedBitmap = GenerateReflectedBitmap(myBitmap);
                myImageView.setImageBitmap(myReflectedBitmap);
                myImageViewList.add(myImageView);                   //添加图像及其倒影
            } }
        @Override
        public void onPageScrolled(int position,
                        float positionOffset, int positionOffsetPixels) {
        }
        @Override
        public void onPageSelected(int position) {
        }
        @Override
        public void onPageScrollStateChanged(int state) {
        }
        public Bitmap GenerateReflectedBitmap(Bitmap bitmap) {          //生成倒影图像
            int myWidth = bitmap.getWidth();
            int myHeight = bitmap.getHeight();
            Matrix myMatrix = new Matrix();
            myMatrix.preScale(1, -1);                                   //创建缩放矩阵
            Bitmap myReflectImagePart = Bitmap.createBitmap(bitmap, 0,
                            myHeight / 2, myWidth, myHeight / 2, myMatrix, false);
            Bitmap myReflectedImage = Bitmap.createBitmap(myWidth,
                    (myHeight + myHeight / 2), Bitmap.Config.ARGB_8888);
            Canvas myCanvas = new Canvas(myReflectedImage);             //初始化空白画布
            myCanvas.drawBitmap(bitmap, 0, 0, null);                    //绘制原图
            myCanvas.drawBitmap(myReflectImagePart, 0, myHeight, null); //绘制倒影部分
            Paint myShaderPaint = new Paint();
            LinearGradient myShader = new LinearGradient(0, bitmap.getHeight(), 0,
            myReflectedImage.getHeight(), 0x70ffffff, 0x00ffffff, Shader.TileMode.CLAMP);
            myShaderPaint.setShader(myShader);                          //添加渐变透明效果
            myShaderPaint.setXfermode(new PorterDuffXfermode(PorterDuff.Mode.DST_IN));
                myCanvas.drawRect(0, myHeight, myWidth,
                    myReflectedImage.getHeight(), myShaderPaint);
            return myReflectedImage;
    } }
```

上面这段代码在 MyCode \ MySample766 \ app \ src \ main \ java \ com \ bin \ luo \ mysample \

MyAdapter.java 文件中。需要说明的是，使用此实例的相关类需要在 gradle 中引入 compile 'com.android.support:design:23.3.0' 依赖项。此实例的完整项目在 MyCode\MySample766 文件夹中。

044 使用 RecyclerView 创建水平瀑布流图像

此实例主要通过使用随机数设置在列表项中的 ImageView 控件的宽度，从而实现使用 RecyclerView 显示水平瀑布流样式的网格。当实例运行之后，单击"显示水平瀑布流网格对话框"按钮，则将从屏幕底部滑出一个瀑布流样式的网格对话框，如图 044.1 的左图所示。在水平方向上拖动图像，则网格中的图像将像瀑布流一样在水平方向上滚动，如图 044.1 的右图所示。

图　044.1

主要代码如下：

```java
public class MainActivity extends AppCompatActivity {
    int myImages[] = {R.mipmap.myimage1,R.mipmap.myimage2,R.mipmap.myimage3,
                R.mipmap.myimage4,R.mipmap.myimage5,R.mipmap.myimage6,
                R.mipmap.myimage7,R.mipmap.myimage8,R.mipmap.myimage9};
    @Override
    protected void onCreate(Bundle savedInstanceState) {
        super.onCreate(savedInstanceState);
        setContentView(R.layout.activity_main); }
    public void onClickmyBtn1(View v) {                     //响应单击"显示水平瀑布流网格对话框"按钮
        RecyclerView myRecyclerView = (RecyclerView)LayoutInflater.
                from(this).inflate(R.layout.dialogrecyclerview, null);
        List<Item> myList = new ArrayList<>();
        for (int i = 0; i < 100; i++) {
            Item myItem = new Item();
            int min = 0;
            int max = 8;
            Random random = new Random();
```

```java
        int num = random.nextInt(max) % (max - min + 1) + min;
        myItem.setId(myImages[num]);
        myList.add(myItem);
    }
    ItemAdapter myAdapter = new ItemAdapter(myList);                //加载数据
    myRecyclerView.setAdapter(myAdapter);
    myRecyclerView.setItemAnimator(new DefaultItemAnimator());
    myRecyclerView.setLayoutManager(new StaggeredGridLayoutManager(
            5,StaggeredGridLayoutManager.HORIZONTAL));               //创建5行网格布局
    final BottomSheetDialog myBottomSheetDialog = new BottomSheetDialog(this);
    myBottomSheetDialog.setContentView(myRecyclerView);              //加载 RecyclerView
    myBottomSheetDialog.show();                                      //显示底部滑出的对话框
}
private class ItemHolder extends RecyclerView.ViewHolder {
    private ImageView myImageView;
    public ItemHolder(View itemView) {
        super(itemView);
        int min = 50;
        int max = 350;
        Random random = new Random();
        int num = random.nextInt(max) % (max - min) + min;
        //使用随机数设置myitem布局文件的ImageView控件的宽度
        myImageView = (ImageView) itemView.findViewById(R.id.myImageView);
        ViewGroup.LayoutParams myParams = myImageView.getLayoutParams();
        myParams.width = num;
        myImageView.setLayoutParams(myParams);
} }
public class Item {
    private int myID;
    public int getId() { return myID; }
    public void setId(int id) { this.myID = id; }
}
private class ItemAdapter extends RecyclerView.Adapter<ItemHolder> {
    private List<Item> myItems;
    public ItemAdapter(List<Item> items) { myItems = items; }
    @Override
    public ItemHolder onCreateViewHolder(ViewGroup viewGroup, int i) {
        LayoutInflater myInflater = LayoutInflater.from(MainActivity.this);
        View myView = myInflater.inflate(R.layout.myitem,null);
        return new ItemHolder(myView);
    }
    @Override
    public void onBindViewHolder(final ItemHolder myHolder, int i) {
        final Item item = myItems.get(i);
        myHolder.myImageView.setImageResource(item.getId());         //设置图像资源
    }
    @Override
    public int getItemCount() { return myItems.size();}
} }
```

上面这段代码在 MyCode\MySample492\app\src\main\java\com\bin\luo\mysample\ MainActivity. java 文件中。在这段代码中，myRecyclerView.setLayoutManager(new StaggeredGridLayoutManager(5,StaggeredGridLayoutManager.HORIZONTAL))表示在水平方向上创建5行网格。myRecyclerView =

（RecyclerView）LayoutInflater.from(this).inflate(R.layout.dialogrecyclerview,null)表示根据 dialogrecyclerview.xml 布局创建 myRecyclerView。关于 dialogrecyclerview 布局的详细内容请参考源代码中的 MyCode\MySample492\app\src\main\res\layout\dialogrecyclerview.xml 文件。

在 MainActivity.java 文件中，myView = myInflater.inflate(R.layout.myitem,null)表示根据 myitem 布局设置列表项内容。关于 myitem 布局的详细内容请参考源代码中的 MyCode\MySample492\app\src\main\res\layout\myitem.xml 文件。需要说明的是，使用此实例的相关类需要在 gradle 中引入 compile 'com.android.support：design：25.0.1'依赖项。此实例的完整项目在 MyCode\MySample492 文件夹中。

045　以网格或列表显示 RecyclerView 列表项

此实例主要通过在 RecyclerView 的 setLayoutManager()方法中使用不同的参数，从而实现 RecyclerView 的列表项以网格样式显示，或者以列表样式显示。当实例运行之后，单击"网格显示"按钮，则 RecyclerView 的列表项将以网格显示，如图 045.1 的左图所示；单击"列表显示"按钮，则 RecyclerView 的列表项将以列表显示，如图 045.1 的右图所示。

图　045.1

主要代码如下：

```java
public class MainActivity extends Activity {
 public RecyclerView myRecyclerView;
 @Override
 protected void onCreate(Bundle savedInstanceState) {
  super.onCreate(savedInstanceState);
  setContentView(R.layout.activity_main);
  myRecyclerView = (RecyclerView) findViewById(R.id.myRecyclerView);
  myRecyclerView.setLayoutManager(new GridLayoutManager(this,2));
  myRecyclerView.setAdapter(new MyAdapter(this,true));
 }
```

```java
public void onClickmyBtn1(View v) {                    //响应单击"网格显示"按钮
    myRecyclerView.setLayoutManager(new GridLayoutManager(this,2));
    myRecyclerView.setAdapter(new MyAdapter(this,true));
}
public void onClickmyBtn2(View v) {                    //响应单击"列表显示"按钮
    myRecyclerView.setLayoutManager(new LinearLayoutManager(this));
    myRecyclerView.setAdapter(new MyAdapter(this,false));
} }
```

上面这段代码在 MyCode\MySample653\app\src\main\java\com\bin\luo\mysample\MainActivity.java 文件中。在这段代码中,myRecyclerView.setAdapter(new MyAdapter(this, true))表示使用自定义数据适配器 MyAdapter 设置 RecyclerView 的数据源,自定义数据适配器 MyAdapter 的主要代码如下:

```java
public class MyAdapter extends RecyclerView.Adapter<MyAdapter.MyViewHolder> {
    public boolean bGrid = false;
    public Context myContext;
    public String[] myFolders = {"手机备份","我的下载","我的文档","我的照片"};
    public Integer[] myIcons = {R.mipmap.myicon_phone, R.mipmap.myicon_download,
                                R.mipmap.myicon_folder, R.mipmap.myicon_camera};
    public MyAdapter(Context context, boolean isGrid) {
        myContext = context;
        bGrid = isGrid;
    }
    @Override
    public MyViewHolder onCreateViewHolder(ViewGroup parent, int viewType) {
        MyViewHolder holder = null;
        if (bGrid) {                //加载网格时的 Item 布局
            holder = new MyViewHolder(LayoutInflater.from(myContext).
                inflate(R.layout.myrecyclerview_griditem, parent,false));
        } else {                    //加载列表时的 Item 布局
            holder = new MyViewHolder(LayoutInflater.from(myContext).
                inflate(R.layout.myrecyclerview_linearitem, parent,false));
        }
        return holder;
    }
    @Override
    //加载每个 Item
    public void onBindViewHolder(MyViewHolder holder, int position) {
        holder.myTextView.setText(myFolders[position]);
        holder.myImageView.setImageResource(myIcons[position]);
    }
    @Override
    public int getItemCount() { return myFolders.length; }
    class MyViewHolder extends RecyclerView.ViewHolder {
        TextView myTextView;
        ImageView myImageView;
        public MyViewHolder(View view) {
            super(view);
            myImageView = (ImageView) view.findViewById(R.id.myImageView);
            myTextView = (TextView) view.findViewById(R.id.myTextView);
} } }
```

上面这段代码在 MyCode\MySample653\app\src\main\java\com\bin\luo\mysample\MyAdapter.java 文件中。在这段代码中，holder = new MyViewHolder（LayoutInflater.from(myContext).inflate(R.layout.myrecyclerview_griditem, parent, false))用于加载 myrecyclerview_griditem 布局，实现 Item 以网格样式显示。myrecyclerview_griditem 布局的主要内容如下：

```xml
<?xml version="1.0" encoding="utf-8"?>
<LinearLayout xmlns:android="http://schemas.android.com/apk/res/android"
    android:layout_width="match_parent"
    android:layout_height="wrap_content"
    android:layout_margin="10dp"
    android:gravity="center"
    android:orientation="vertical">
    <ImageView android:id="@+id/myImageView"
        android:layout_width="128dp"
        android:layout_height="128dp"
        android:scaleType="fitXY"
        android:src="@mipmap/myicon_folder"/>
    <TextView android:id="@+id/myTextView"
        android:layout_width="128dp"
        android:layout_height="wrap_content"
        android:gravity="center"/>
</LinearLayout>
```

上面这段代码在 MyCode\MySample653\app\src\main\res\layout\myrecyclerview_griditem.xml 文件中。在 MyAdapter.java 文件中，holder = new MyViewHolder（LayoutInflater.from(myContext).inflate(R.layout.myrecyclerview_linearitem, parent, false))用于加载 myrecyclerview_linearitem 布局，从而实现 Item 以列表样式显示。myrecyclerview_linearitem 布局的主要内容如下：

```xml
<?xml version="1.0" encoding="utf-8"?>
<LinearLayout xmlns:android="http://schemas.android.com/apk/res/android"
    android:layout_width="match_parent"
    android:layout_height="wrap_content"
    android:layout_margin="10dp"
    android:orientation="horizontal">
    <ImageView android:id="@+id/myImageView"
        android:layout_width="wrap_content"
        android:layout_height="32dp"
        android:scaleType="fitCenter"
        android:src="@mipmap/myicon_folder"/>
    <TextView android:id="@+id/myTextView"
        android:layout_width="match_parent"
        android:layout_height="32dp"
        android:textSize="18dp"
        android:gravity="center_vertical"/>
</LinearLayout>
```

上面这段代码在 MyCode\MySample653\app\src\main\res\layout\myrecyclerview_linearitem.xml 文件中。需要说明的是，使用此实例的相关类需要在 gradle 中引入 compile 'com.android.support:recyclerview-v7:25.2.0'依赖项。此实例的完整项目在 MyCode\MySample653 文件夹中。

046 使用 RecyclerView 仿表情包插入输入框

此实例主要通过使用自定义 RecyclerView.Adapter,并实现 View.OnClickListener 接口,从而实现把在 RecyclerView 中选择的表情图像插入到输入框的效果。当实例运行之后,在输入框中输入文字"好极了",再在下面的 RecyclerView 中任意选择一个表情,则该表情图像将会显示在输入框中,效果分别如图 046.1 的左图和右图所示。

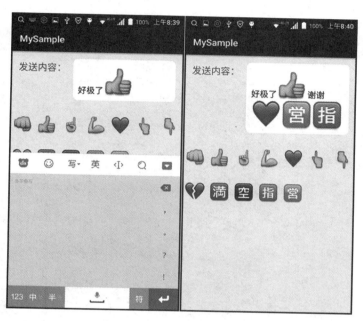

图 046.1

主要代码如下:

```java
public class MainActivity extends Activity {
    @Override
    protected void onCreate(Bundle savedInstanceState) {
        super.onCreate(savedInstanceState);
        setContentView(R.layout.activity_main);
        final EditText myEdit = (EditText) findViewById(R.id.myEdit);
        RecyclerView myRecyclerView = (RecyclerView) findViewById(R.id.myRecyclerView);
        myRecyclerView.setLayoutManager(new StaggeredGridLayoutManager(7,
                StaggeredGridLayoutManager.VERTICAL));        //限制每行显示 7 个表情图像
        MyAdapter myAdapter = new MyAdapter(this);
        myAdapter.setOnItemClickListener(
                        new MyAdapter.OnRecyclerViewItemClickListener() {
            @Override
            public void onItemClick(View view,Integer data){
                Bitmap myBitmap = BitmapFactory.decodeResource(getResources(),
                                        data);              //将表情图像转换为 Bitmap 对象
                ImageSpan myImageSpan = new ImageSpan(MainActivity.this,myBitmap);
                SpannableString myString = new SpannableString("face");
                myString.setSpan(myImageSpan,0,4,
                        Spannable.SPAN_EXCLUSIVE_EXCLUSIVE);  //向输入框插入表情图像
```

```
        myEdit.append(myString);
    } });
    myRecyclerView.setAdapter(myAdapter);
} }
```

上面这段代码在 MyCode\MySample638\app\src\main\java\com\bin\luo\mysample\MainActivity.java 文件中。在这段代码中，MyAdapter 是自定义的 RecyclerView.Adapter，它主要用于为 RecyclerView 管理表情图像，MyAdapter 的主要内容如下所示：

```
public class MyAdapter extends RecyclerView.Adapter<MyAdapter.MyViewHolder>
implements View.OnClickListener {
  public Context myContext;
  private OnRecyclerViewItemClickListener myItemClickListener = null;
  public Integer[] myEmoji = {R.mipmap.myemoji1, R.mipmap.myemoji2,
    R.mipmap.myemoji3, R.mipmap.myemoji4, R.mipmap.myemoji5, R.mipmap.myemoji6,
    R.mipmap.myemoji7, R.mipmap.myemoji8, R.mipmap.myemoji9,
    R.mipmap.myemoji10, R.mipmap.myemoji11, R.mipmap.myemoji12 };       //表情图像
  public MyAdapter(Context context) { myContext = context; }
  @Override
  public MyAdapter.MyViewHolder onCreateViewHolder(ViewGroup parent, int viewType){
    View myView =
        LayoutInflater.from(myContext).inflate(R.layout.myitem, parent, false);
    MyViewHolder myHolder = new MyViewHolder(myView);
    myView.setOnClickListener(this);
    return myHolder;
  }
  @Override
  public void onClick(View v) {                                         //获取选择的表情图像
    if (myItemClickListener != null) {
      myItemClickListener.onItemClick(v, (Integer) v.getTag());
    } }
  @Override
  public void onBindViewHolder(MyAdapter.MyViewHolder holder, int position) {
    holder.myImageView.setImageResource(myEmoji[position]);
    holder.itemView.setTag(myEmoji[position]);
  }
  @Override
  public int getItemCount() { return myEmoji.length; }
  public void setOnItemClickListener(OnRecyclerViewItemClickListener listener) {
    this.myItemClickListener = listener;
  }
  public class MyViewHolder extends RecyclerView.ViewHolder {
    ImageView myImageView;
    public MyViewHolder(View view) {                                    //加载所有表情图像
      super(view);
      myImageView = (ImageView) view.findViewById(R.id.myImageView);
    } }
  public interface OnRecyclerViewItemClickListener {
    void onItemClick(View view, Integer data);
  }
} }
```

上面这段代码在 MyCode\MySample638\app\src\main\java\com\bin\luo\mysample\

MyAdapter.java 文件中。在这段代码中,myView = LayoutInflater.from(myContext).inflate(R.layout.myitem,parent,false)用于加载 RecyclerView 的每个 Item 的布局,此处的 myitem 非常简单,它只有一个 ImageView 控件,即显示一个表情图像;实际上,如果将 Item 布局设计得较为复杂,RecyclerView 就会呈现更加丰富的效果。需要说明的是,使用此实例的相关类需要在 gradle 中引入 compile 'com.android.support:recyclerview-v7:25.3.1'依赖项。此实例的完整项目在 MyCode\MySample638 文件夹中。

047 使用 CardView 显示 RecyclerView 列表项

此实例主要实现了在单击 RecyclerView 的任一列表项时,以动画的形式动态在 CardView 中显示单击的列表项。当实例运行之后,RecyclerView 的所有列表项如图 047.1 的左图所示;单击 RecyclerView 的任一列表项,则将以动画的形式弹出 CardView,并在其中显示此列表项,同时背景模糊,如图 047.1 的右图所示。

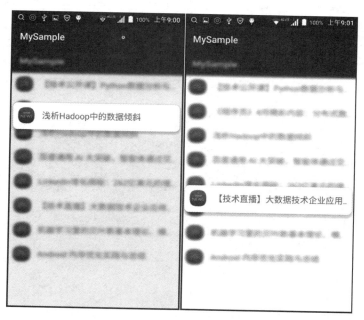

图 047.1

主要代码如下:

```java
public class MainActivity extends Activity {
    private FrameLayout myRootView;
    private CardView myCardView;
    private MyAdapter myAdapter;
    private RecyclerView myRecyclerView;
    @Override
    protected void onCreate(Bundle savedInstanceState) {
        super.onCreate(savedInstanceState);
        setContentView(R.layout.activity_main);
        myCardView = (CardView) findViewById(R.id.MyCardView);
        myRootView = (FrameLayout) findViewById(R.id.activity_main);
        myRecyclerView = (RecyclerView) findViewById(R.id.MyRecyclerView);
```

```java
myAdapter = new MyAdapter(this);
myRecyclerView.setAdapter(myAdapter);
myRecyclerView.setLayoutManager(new LinearLayoutManager(this));
myAdapter.setOnItemClickListener(new MyAdapter.OnItemClickListener() {
    @Override
    public void onItemClick(View view, int position){               //单击列表项
        final View myMainView = getWindow().getDecorView();
        myMainView.setDrawingCacheEnabled(true);
        myMainView.buildDrawingCache();
        Bitmap myOriginBitmap = myMainView.getDrawingCache();
        Rect myFrame = new Rect();
        getWindow().getDecorView().getWindowVisibleDisplayFrame(myFrame);
        int myStatusBarHeight = myFrame.top;
        int myScreenWidth = getWindowManager().getDefaultDisplay().getWidth();
        int myScreenHeight = getWindowManager().getDefaultDisplay().getHeight();
        Bitmap myClipImage = Bitmap.createBitmap(myOriginBitmap,0,
                myStatusBarHeight,myScreenWidth,myScreenHeight - myStatusBarHeight);
        myRootView.setBackground(new BitmapDrawable(getResources(),
                MyUtils.doBlur(myClipImage,20)));                   //使用模糊的截图设置背景
        myMainView.setDrawingCacheEnabled(false);
        myRecyclerView.setVisibility(GONE);                         //隐藏 myRecyclerView
        showView(view,position);                                    //使用 CardView 显示选择的 Item
    } }); }
@Override
public boolean onKeyDown(int keyCode, KeyEvent event) {
    if (keyCode == KeyEvent.KEYCODE_BACK) {
        myRecyclerView.setVisibility(View.VISIBLE);                 //显示 myRecyclerView
        myRootView.setBackgroundColor(Color.WHITE);                 //主窗口设置白色背景
        myCardView.setVisibility(GONE);                             //隐藏 myCardView
        myCardView.removeAllViews();
        return true;
    }
    return super.onKeyDown(keyCode,event);
}
    //使用 CardView 显示对应的 Item
private void showView(View view,int position){
    View myItemView = LayoutInflater.from(this).inflate(R.layout.myitem,null);
    TextView myTextView = (TextView)myItemView.findViewById(R.id.myTextView);
    myTextView.setText(myAdapter.getMyData().get(position));
    //设置 myCardView 样式,位置通过 margintop 来计算
    FrameLayout.LayoutParams myParams =
            new FrameLayout.LayoutParams(view.getWidth() - 30, view.getHeight());
    myParams.topMargin = (int)view.getY();
    myParams.leftMargin = 15;
    myParams.rightMargin = 15;
    myCardView.setVisibility(View.VISIBLE);
    myCardView.setLayoutParams(myParams);
    //在 CardView 上加载 myItemView,并设置样式为 item 样式
    myCardView.addView(myItemView, view.getLayoutParams());
    startAnimate(myCardView);                                       //显示 myCardView 动画
}
    private void startAnimate(CardView cardView) {
        PropertyValuesHolder myOutScaleXHolder =
```

```
                PropertyValuesHolder.ofFloat("scaleX", 0.1f, 1.05f);
    PropertyValuesHolder myOutScaleYHolder =
                PropertyValuesHolder.ofFloat("scaleY", 0.1f, 1.05f);
    ObjectAnimator myOutAnim = ObjectAnimator.ofPropertyValuesHolder(
                myCardView, myOutScaleXHolder,myOutScaleYHolder);
    myOutAnim.setInterpolator(new AccelerateDecelerateInterpolator());
    myOutAnim.setDuration(350);
    PropertyValuesHolder myInScaleXHolder =
                PropertyValuesHolder.ofFloat("scaleX", 1.05f, 1f);
    PropertyValuesHolder myInScaleYHolder =
                PropertyValuesHolder.ofFloat("scaleY", 1.05f, 1f);
    ObjectAnimator myInAnim = ObjectAnimator.ofPropertyValuesHolder(
                myCardView, myInScaleXHolder,myInScaleYHolder);
    myInAnim.setInterpolator(new AccelerateDecelerateInterpolator());
    myInAnim.setDuration(100);
    AnimatorSet myAnimSet = new AnimatorSet();
    myAnimSet.playSequentially(myOutAnim, myInAnim);           // 按顺序执行两个动画
    myAnimSet.start();
} }
```

上面这段代码在 MyCode \ MySample731 \ app \ src \ main \ java \ com \ bin \ luo \ mysample \ MainActivity.java 文件中。在这段代码中，myRecyclerView.setAdapter(myAdapter)表示使用自定义适配器 MyAdapter 的实例 myAdapter 管理 myRecyclerView 的列表项，自定义适配器 MyAdapter 的主要代码如下：

```
public class MyAdapter extends RecyclerView.Adapter<MyAdapter.MyViewHolder> {
  public Context myContext;
  public interface OnItemClickListener {void onItemClick(View view, int position); }
  public String[] myItems = {"【技术公开课】Python 数据分析与机器学习实战!",
"《程序员》4月精彩内容:分布式数据库应用实践", "浅析 Hadoop 中的数据倾斜", "百度通用 AI 大突破,智能体通过交互式学习实现举一反三", "LinkedIn 增长揭秘:262 亿美元的增长引擎是如何练成的?", "【技术直播】大数据技术企业应用实战!!", "机器学习里的贝叶斯基本理论、模型和算法", "Android 内存优化实践与总结"};
  public OnItemClickListener myListener;
  public List<String> myData = new ArrayList<String>();
  public MyAdapter(Context context) {
    myContext = context;
    for (int i = 0; i < myItems.length; i++) { myData.add(myItems[i]); }
  }
  public List<String> getMyData() { return myData; }
  public void setOnItemClickListener(OnItemClickListener listener) {
    this.myListener = listener;
  }
  @Override
  public MyViewHolder onCreateViewHolder(ViewGroup parent, int viewType) {
    MyViewHolder myHolder = new MyViewHolder(LayoutInflater.from(
            myContext).inflate(R.layout.myitem, parent,false));
    return myHolder;
  }
  @Override
  public void onBindViewHolder(MyViewHolder holder, final int position) {
    holder.myTextView.setText(myData.get(position));
  }
  @Override
  public int getItemCount() { return myData.size(); }
```

```
class MyViewHolder extends RecyclerView.ViewHolder {
  TextView myTextView;
  public MyViewHolder(View view) {
    super(view);
    myTextView = (TextView) view.findViewById(R.id.myTextView);
    view.setOnClickListener(new View.OnClickListener() {
      @Override
      public void onClick(View view) {
        myListener.onItemClick(view, getAdapterPosition());
      }}); }}
```

上面这段代码在 MyCode\MySample731\app\src\main\java\com\bin\luo\mysample\MyAdapter.java 文件中。在这段代码中，myHolder = new MyViewHolder(LayoutInflater.from(myContext).inflate(R.layout.myitem, parent, false))用于加载 myitem 布局以创建 MyAdapter 的每个 Item 容器，即每个列表项的布局。关于 myitem 布局的详细内容请参考源代码中的 MyCode\MySample731\app\src\main\res\layout\myitem.xml 文件。在 MainActivity.java 文件中，MyUtils.doBlur(myClipImage,20)主要用于模糊主窗口截图，MyUtils 是自定义类，代码较多，详细内容请查看源代码中的 MyCode\MySample731\app\src\main\java\com\bin\luo\mysample\MyUtils.java 文件。需要说明的是，使用此实例的相关类需要在 gradle 中引入 compile 'com.android.support：recyclerview-v7：25.3.1'和 compile 'com.android.support：cardview-v7：25.3.1'依赖项。此实例的完整项目在 MyCode\MySample731 文件夹中。

048　在 ListView 中创建图文结合列表项

此实例主要通过在布局文件中为每个列表项添加图像和文本等控件，实现在 ListView 中显示图文混合的列表项。当实例运行之后，ListView 的每个列表项将混合显示图像和文本，每个列表项之间用细线分隔，单击任意列表项，则在弹出的 Toast 中显示该列表项的内容，效果分别如图 048.1 的左图和右图所示。

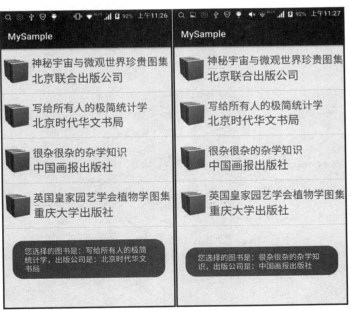

图　048.1

主要代码如下:

```xml
<RelativeLayout android:layout_width = "fill_parent"
                android:layout_height = "?android:attr/listPreferredItemHeight">
  <ImageView android:id = "@+id/myImage"
             android:layout_width = "wrap_content"
             android:layout_height = "fill_parent"
             android:layout_alignParentBottom = "true"
             android:layout_alignParentTop = "true"
             android:adjustViewBounds = "true"
             android:padding = "5dip"/>
  <TextView android:id = "@+id/myName"
            android:layout_width = "wrap_content"
            android:layout_height = "wrap_content"
            android:layout_above = "@+id/myPress"
            android:layout_alignParentRight = "true"
            android:layout_alignParentTop = "true"
            android:layout_alignWithParentIfMissing = "true"
            android:layout_toRightOf = "@+id/myImage"
            android:gravity = "center_vertical"
            android:textSize = "22dip"/>
  <TextView android:id = "@+id/myPress"
            android:layout_width = "fill_parent"
            android:layout_height = "wrap_content"
            android:layout_alignParentBottom = "true"
            android:layout_alignParentRight = "true"
            android:layout_toRightOf = "@+id/myImage"
            android:ellipsize = "marquee"
            android:singleLine = "true"
            android:textSize = "24dip"/>
</RelativeLayout>
```

上面这段代码在 MyCode\MySample068\app\src\main\res\layout\activity_main.xml 文件中。在这段代码中,所有控件组合在一起仅表示 ListView 的一个列表项,ListView 的所有列表项则通过 Adapter 来映射实现,主要代码如下:

```java
public class MainActivity extends ListActivity {
 private String[] myNameList = {"神秘宇宙与微观世界珍贵图集","写给所有人的极简统计学","很杂很杂的杂学知识","英国皇家园艺学会植物学图集"};
 private String[] myPressList = {"北京联合出版公司","北京时代华文书局","中国画报出版社","重庆大学出版社"};
 ArrayList<Map<String, Object>> myData = new ArrayList<Map<String, Object>>();
 @Override
 protected void onCreate(Bundle savedInstanceState) {
  int myLength = myNameList.length;
  for (int i = 0; i < myLength; i++) {                    //向每个列表项添加内容
   Map<String, Object> myItem = new HashMap<String, Object>();
   myItem.put("image", R.mipmap.mybook);
   myItem.put("name", myNameList[i]);
   myItem.put("press", myPressList[i]);
   myData.add(myItem);
  }
  SimpleAdapter myAdapter = new SimpleAdapter(this, myData,
```

```
            R.layout.activity_main,new String[]{"image", "name", "press"},
            new int[]{R.id.myImage, R.id.myName, R.id.myPress});
    setListAdapter(myAdapter);                          //实现列表项数据映射
    ListView myList = getListView();
    myList.setChoiceMode(ListView.CHOICE_MODE_SINGLE);  //设置列表项单选模式
    //为列表项添加单击事件响应方法
    myList.setOnItemClickListener(new OnItemClickListener() {
      @Override
      public void onItemClick(AdapterView<?> adapterView,
                              View view, int position, long id) {
        Toast.makeText(MainActivity.this, "您选择的图书是:" +
                       myNameList[position] + ",出版公司是:" +
                       myPressList[position], Toast.LENGTH_LONG).show();
      } });
      super.onCreate(savedInstanceState);
} }
```

上面这段代码在 MyCode\MySample068\app\src\main\java\com\bin\luo\mysample\MainActivity.java 文件中。在这段代码中，setListAdapter(myAdapter)用于把数据映射到界面里边，ListView 显示的数据必须借助 Adapter 来映射。getListView()方法用于获取当前环境的 ListView 对象。setChoiceMode(ListView.CHOICE_MODE_SINGLE)表示 ListView 支持单行选择。setOnItemClickListener()主要是为单行选择添加监听事件响应方法。此实例的完整项目在 MyCode\MySample068 文件夹中。

049　使用 ListPopupWindow 实现下拉选择

此实例主要通过在 ListPopupWindow 中添加列表选项，实现为 EditText 控件添加类似于自动完成的选择功能。当实例运行之后，单击"报考专业:"输入框，则将弹出一个列表选项窗口，单击其中的任意一个选项，如"计算机科学与技术"，则该选项将自动填充到"报考专业:"输入框中，如图 049.1 的左图和右图所示。

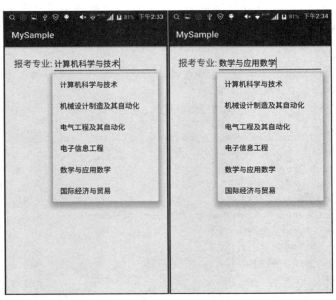

图　049.1

主要代码如下：

```java
public class MainActivity extends Activity {
    private EditText myEditText;
    private ListPopupWindow myPopupWindow;
    private List<String> myArray = new ArrayList();
    @Override
    protected void onCreate(Bundle savedInstanceState) {
        super.onCreate(savedInstanceState);
        setContentView(R.layout.activity_main);
        myArray.add("计算机科学与技术");
        myArray.add("机械设计制造及其自动化");
        myArray.add("电气工程及其自动化");
        myArray.add("电子信息工程");
        myArray.add("数学与应用数学");
        myArray.add("国际经济与贸易");
        myEditText = (EditText) findViewById(R.id.myEditText);
        myPopupWindow = new ListPopupWindow(this);
        myPopupWindow.setAdapter(new ArrayAdapter(this,
                    android.R.layout.simple_list_item_1, myArray));
        myPopupWindow.setWidth(ViewGroup.LayoutParams.WRAP_CONTENT);
        myPopupWindow.setHeight(ViewGroup.LayoutParams.WRAP_CONTENT);
        //设置 ListPopupWindow 的锚点，即关联 PopupWindow 的显示位置和这个锚点
        myPopupWindow.setAnchorView(myEditText);
        myPopupWindow.setModal(true);                            //设置模式
        myPopupWindow.setOnItemClickListener(new AdapterView.OnItemClickListener() {
            @Override
            public void onItemClick(AdapterView parent,
                            View view, int position, long id) {
                myEditText.setText(myArray.get(position).toString());
                myPopupWindow.dismiss();
            } });
        myEditText.setOnClickListener(new View.OnClickListener() {
            @Override
            public void onClick(View v) { myPopupWindow.show(); }
            });
    } }
```

上面这段代码在 MyCode\MySample431\app\src\main\java\com\bin\luo\mysample\MainActivity.java 文件中。在这段代码中，myPopupWindow.setAnchorView(myEditText) 用于将 ListPopupWindow 锚定在 EditText 控件上，即将两者关联起来。此实例的完整项目在 MyCode\MySample431 文件夹中。

050 使用 Elevation 创建阴影扩散的控件

此实例主要通过设置 elevation 属性或使用 setElevation() 方法，实现给控件设置扩散阴影。当实例运行之后，单击"设置扩散阴影"按钮，则图像的四周将有一层灰色的扩散阴影，如图 050.1 的左图

所示。单击"移除扩散阴影"按钮,则将移除图像四周的扩散阴影,如图 050.1 的右图所示。

图 050.1

主要代码如下:

```
//响应单击"设置扩散阴影"按钮
public void onClickmyBtn1(View v) {
    myImageView.setElevation(35);
}
//响应单击"移除扩散阴影"按钮
public void onClickmyBtn2(View v) {
    myImageView.setElevation(0);
}
```

上面这段代码在 MyCode \ MySample293 \ app \ src \ main \ java \ com \ bin \ luo \ mysample \ MainActivity.java 文件中。在这段代码中,myImageView.setElevation(35)用于设置图像四周的阴影扩散尺寸为 35 像素,在 XML 文件中,直接设置控件的 android:elevation 属性即可。需要注意的是,在设置 android:elevation 属性时,应该同时设置 android:background 属性。此实例的完整项目在 MyCode\MySample293 文件夹中。

051　在单击 CheckBox 时显示波纹扩散效果

此实例主要通过设置 CheckBox 控件的 android:background 属性为"?android:attr/selectableItemBackground",实现在单击 CheckBox 控件时产生有界矩形的波纹扩散动画效果。当实例运行之后,单击前面 3 个 CheckBox 控件,将会产生默认的圆形无界的中心向四周扩散的波纹动画效果,如图 051.1 的左图所示;单击第 4 个 CheckBox 控件,则产生定制的矩形有界的中心向四周扩散的波纹动画效果,如图 051.1 的右图所示。

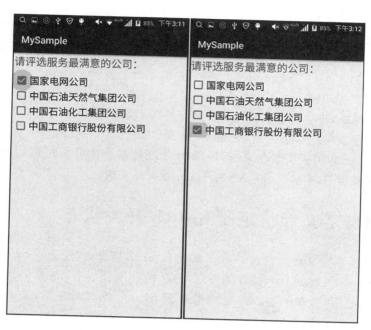

图 051.1

主要代码如下：

```xml
<LinearLayout android:layout_width = "match_parent"
        android:layout_height = "match_parent"
        android:orientation = "vertical"
        android:padding = "10px">
    <TextView android:layout_width = "wrap_content"
        android:layout_height = "wrap_content"
        android:layout_marginBottom = "10dp"
        android:text = "请评选服务最满意的公司："
        android:textSize = "22dp"/>
    <CheckBox android:id = "@ + id/checkbox1"
        android:layout_width = "wrap_content"
        android:layout_height = "wrap_content"
        android:textSize = "20dp"
        android:text = "国家电网公司"/>
    <CheckBox android:id = "@ + id/checkbox2"
        android:layout_width = "wrap_content"
        android:layout_height = "wrap_content"
        android:textSize = "20dp"
        android:text = "中国石油天然气集团公司"/>
    <CheckBox android:id = "@ + id/checkbox3"
        android:layout_width = "wrap_content"
        android:layout_height = "wrap_content"
        android:textSize = "20dp"
        android:background =
            "?android:attr/selectableItemBackgroundBorderless"
        android:text = "中国石油化工集团公司"/>
    <CheckBox android:id = "@ + id/checkbox4"
        android:layout_width = "wrap_content"
        android:layout_height = "wrap_content"
        android:textSize = "20dp"
        android:background = "?android:attr/selectableItemBackground"
```

```
        android:text="中国工商银行股份有限公司"/>
</LinearLayout>
```

上面这段代码在 MyCode\MySample570\app\src\main\res\layout\activity_main.xml 文件中。在这段代码中，android：background＝"？android：attr/selectableItem BackgroundBorderless"表示在单击 checkbox3 控件时，将会产生从圆形无界的中心向四周扩散的波纹动画效果，这也是默认的单击控件效果，不设置此属性也会产生这种效果。android：background＝"？android：attr/selectableItemBackground"用于实现在单击 checkbox4 控件时产生有界矩形边框限制的波纹扩散动画效果。此实例的完整项目在 MyCode\MySample570 文件夹中。

052　使用自定义形状定制 Switch 开关状态

此实例主要通过以自定义形状设置 thumb 属性和 track 属性，实现定制 Switch 控件的开关状态。当实例运行之后，空心圆位于 Switch 控件的右端，整个控件顶部居中，如图 052.1 的左图所示；单击 Switch 控件右端的空心圆，右端的空心圆立即滑动到 Switch 控件的左端，整个控件底部居中，如图 052.1 的右图所示。

图　052.1

主要代码如下：

```
<Switch android:id="@+id/mySwitch"
        android:layout_width="wrap_content"
        android:layout_height="wrap_content"
        android:checked="true"
        android:text="顶部居中"
        android:textSize="20dp"
        android:thumb="@drawable/mythumb"
        android:track="@drawable/mytrack"/>
```

上面这段代码在 MyCode\MySample054\app\src\main\res\layout\activity_main.xml 文件中。在这段代码中,android:thumb="@drawable/mythumb"表示 Switch 控件的 thumb(滑动的空心圆)采用 mythumb 资源进行配置。android:track="@drawable/mytrack"表示 Switch 控件的 track(紫色/灰色的轨道)采用 mytrack 资源进行配置。mythumb 是一个 XML 文件,该文件的主要内容如下:

```xml
<?xml version = "1.0" encoding = "utf-8"?>
<shape xmlns:android = "http://schemas.android.com/apk/res/android"
       android:shape = "oval">
<size android:width = "30dp" android:height = "30dp"/>
<solid android:color = "@android:color/white" />
</shape>
```

上面这段代码在 MyCode\MySample054\app\src\main\res\drawable\mythumb.xml 文件中。在这段代码中,android:shape="oval"指定 thumb 的形状为椭圆;当 android:width="30dp",android:height="30dp"时,此椭圆即是一个圆。android:color="@android:color/white"表示圆的填充颜色是白色,因此看起来像是一个空心圆。

mytrack 也是一个 XML 文件,而且是一个 selector 文件。该文件的主要内容如下:

```xml
<?xml version = "1.0" encoding = "utf-8"?>
<selector xmlns:android = "http://schemas.android.com/apk/res/android">
 <item android:state_checked = "true" android:drawable = "@drawable/mytrackon" />
 <item android:state_checked = "false" android:drawable = "@drawable/mytrackoff" />
</selector>
```

上面这段代码在 MyCode\MySample054\app\src\main\res\drawable\mytrack.xml 文件中。在这段代码中,<item android:state_checked="true" android:drawable="@drawable/mytrackon"/>表示当 Switch 的 state_checked 属性值为 true 时,设置 mytrackon 资源。<item android:state_checked="false" android:drawable="@drawable/mytrackoff"/>表示当 Switch 的 state_checked 属性值为 false 时,设置 mytrackoff 资源。mytrackoff 资源和 mytrackon 资源都是 XML 文件,mytrackon 文件的主要内容如下:

```xml
<?xml version = "1.0" encoding = "utf-8"?>
<shape xmlns:android = "http://schemas.android.com/apk/res/android"
    android:shape = "rectangle">
 <solid android:color = "@android:color/holo_purple"/>
 <corners android:radius = "30dp"/>
</shape>
```

上面这段代码在 MyCode\MySample054\app\src\main\res\drawable\mytrackon.xml 文件中。在这段代码中,android:shape="rectangle"表示 Switch 在打开状态时的 track 是一个 rectangle。<solid android:color="@android:color/holo_purple"/>表示该 rectangle 是一个紫色的 rectangle。<corners android:radius="30dp"/>表示该 rectangle 的 4 个直角的圆角半径是 30dp。mytrackoff 文件的主要内容如下:

```xml
<?xml version = "1.0" encoding = "utf-8"?>
<shape xmlns:android = "http://schemas.android.com/apk/res/android"
    android:shape = "rectangle">
```

```xml
<solid android:color = "@android:color/darker_gray"/>
<corners android:radius = "30dp"/>
</shape>
```

上面这段代码在 MyCode\MySample054\app\src\main\res\drawable\mytrackoff.xml 文件中。在这段代码中,android:shape="rectangle"表示 Switch 在关闭状态时的 track 是一个 rectangle。<solid android:color="@android:color/darker_gray"/>表示该 rectangle 是一个灰色的 rectangle。<corners android:radius="30dp"/>表示该 rectangle 的 4 个直角的圆角半径是 30dp。

Switch 的 CheckedChange 事件的响应方法则通过 Java 代码实现,主要代码如下:

```java
public class MainActivity extends Activity {
    private Switch mySwitch;
    private LinearLayout myLayout;
    @Override
    protected void onCreate(Bundle savedInstanceState) {
        super.onCreate(savedInstanceState);
        setContentView(R.layout.activity_main);
        mySwitch = (Switch) findViewById(R.id.mySwitch);
        myLayout = (LinearLayout) findViewById(R.id.myLayout);
        mySwitch.setOnCheckedChangeListener(
                new CompoundButton.OnCheckedChangeListener() {
            @Override
            public void onCheckedChanged(CompoundButton buttonView, boolean isChecked) {
                if (isChecked) {
                    myLayout.setGravity(Gravity.CENTER_HORIZONTAL);
                    mySwitch.setText("顶部居中");
                } else {
                    myLayout.setGravity(Gravity.CENTER_HORIZONTAL | Gravity.BOTTOM);
                    mySwitch.setText("底部居中");
                }}});}}
```

上面这段代码在 MyCode\MySample054\app\src\main\java\com\bin\luo\mysample\MainActivity.java 文件中。在这段代码中,myLayout.setGravity(Gravity.CENTER_HORIZONTAL)用于设置 myLayout 布局管理器顶部居中。myLayout.setGravity(Gravity.CENTER_HORIZONTAL | Gravity.BOTTOM)用于设置 myLayout 布局管理器底部居中。此实例的完整项目在 MyCode\MySample054 文件夹中。

053 自定义 selector 以渐变前景切换控件

此实例主要通过设置 CardView 控件的 android:foreground 属性为自定义 selector,实现在按下 CardView 控件时新增线性渐变前景层的效果。当实例运行之后,单击 CardView 控件,则在按下该控件时将在该控件上新增由蓝到绿的线性渐变(半)透明层,如图 053.1 的左图所示;在离开该控件时将移除该层(恢复)至正常状态,如图 053.1 的右图所示。

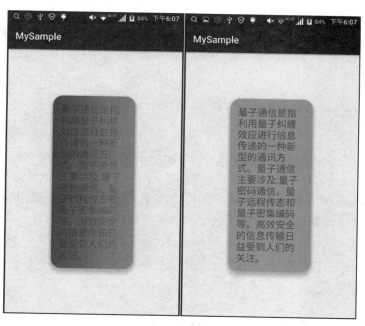

图 053.1

主要代码如下：

```xml
<android.support.v7.widget.CardView
    xmlns:card_view="http://schemas.android.com/apk/res-auto"
    android:layout_width="200dp"
    android:layout_height="wrap_content"
    android:layout_centerInParent="true"
    android:clickable="true"
    android:foreground="@drawable/myselector"
    card_view:cardBackgroundColor="#00EEEE"
    card_view:cardCornerRadius="20dp"
    card_view:cardElevation="10dp"
    card_view:cardPreventCornerOverlap="true"
    card_view:cardUseCompatPadding="true"
    card_view:contentPadding="15dp">
    <TextView android:layout_width="wrap_content"
        android:layout_height="wrap_content"
        android:text="量子通信是指利用量子纠缠效应进行信息传递的一种新型的通讯方式.量子通信主要涉及:量子密码通信、量子远程传态和量子密集编码等.高效安全的信息传输日益受到人们的关注."
        android:textSize="20dp"/>
</android.support.v7.widget.CardView>
```

上面这段代码在MyCode\MySample574\app\src\main\res\layout\activity_main.xml文件中。在这段代码中，android：foreground="@drawable/myselector"用于设置具有半透明线性渐变前景层的自定义selector。自定义selector的主要代码如下：

```xml
<?xml version="1.0" encoding="utf-8"?>
<selector xmlns:android="http://schemas.android.com/apk/res/android">
```

```xml
<item android:state_pressed="true">
  <shape><gradient android:startColor="#AA0000FF"
         android:endColor="#AA00FF00" android:type="linear"/>
  </shape></item>
<item>
  <shape><solid android:color="@android:color/transparent"/></shape>
</item></selector>
```

上面这段代码在 MyCode\MySample574\app\src\main\res\drawable\myselector.xml 文件中。在这段代码中，<item android:state_pressed="true">表示在控件被按下时添加此标签中的线性渐变前景层。android:startColor="#AA0000FF"表示渐变开始颜色是半透明蓝色。android:endColor="#AA00FF00"表示渐变结束颜色是半透明绿色。android:type="linear"表示渐变类型是线性渐变。此外，在 Android Studio 中，使用此实例的相关控件需要在 gradle 中引入 compile 'com.android.support:cardview-v7:25.3.1'依赖项。此实例的完整项目在 MyCode\MySample574 文件夹中。

054　使用 ViewSwitcher 平滑切换两个 View

此实例主要通过使用 ViewSwitcher 控件，实现以动画的形式在两个 View 之间进行平滑切换。当实例运行之后，单击"下一个视图"按钮，则将显示第二个 View，如图 054.1 的右图所示；再次单击"下一个视图"按钮，则将显示第一个 View，如图 054.1 的左图所示。如果在实例运行之后，单击"上一个视图"按钮，也将显示第二个 View，如图 054.1 的右图所示；再次单击"上一个视图"按钮，则将显示第一个 View，如图 054.1 的左图所示。由于 ViewSwitcher 控件只支持两个 View 的切换，并且周而复始地循环，因此两个按钮的功能完全相同。

图　054.1

主要代码如下:

```java
public class MainActivity extends Activity {
    ViewSwitcher myViewSwitcher;
    Animation mySlideInLeft, mySlideOutRight;
    @Override
    protected void onCreate(Bundle savedInstanceState) {
        super.onCreate(savedInstanceState);
        setContentView(R.layout.activity_main);
        myViewSwitcher = (ViewSwitcher) findViewById(R.id.viewswitcher);
        mySlideInLeft = AnimationUtils.loadAnimation(this,
                        android.R.anim.slide_in_left);          //加载从左边滑入的动画
        mySlideOutRight = AnimationUtils.loadAnimation(this,
                        android.R.anim.slide_out_right);         //加载从右边滑出的动画
        myViewSwitcher.setInAnimation(mySlideInLeft);
        myViewSwitcher.setOutAnimation(mySlideOutRight);
    }
    //响应单击"上一个视图"按钮
    public void onClickmyBtn1(View v) { myViewSwitcher.showPrevious();}
    //响应单击"下一个视图"按钮
    public void onClickmyBtn2(View v) { myViewSwitcher.showNext(); }
}
```

上面这段代码在 MyCode \ MySample386 \ app \ src \ main \ java \ com \ bin \ luo \ mysample \ MainActivity.java 文件中。在这段代码中,ViewSwitcher 是一个视图切换控件,它可以将两个 View 放在一起,每次只显示一个 View。当从一个 View 切换到另一个 View 时,ViewSwitcher 支持指定的动画效果。ViewSwitcher 的 showPrevious() 方法用于显示上一个 View。ViewSwitcher 的 showNext() 方法用于显示下一个 View。在 XML 文件中,ViewSwitcher 添加 View 的主要代码如下:

```xml
<ViewSwitcher android:id="@+id/viewswitcher"
              android:layout_width="match_parent"
              android:layout_height="match_parent">
    <ImageView android:layout_width="match_parent"
               android:layout_height="match_parent"
               android:scaleType="fitXY"
               android:src="@mipmap/myimage1"/>
    <ImageView android:scaleType="fitXY"
               android:layout_width="match_parent"
               android:layout_height="match_parent"
               android:src="@mipmap/myimage2"/>
</ViewSwitcher>
```

上面这段代码在 MyCode\MySample386\app\src\main\res\layout\activity_main.xml 文件中。在这段代码中,<ImageView>代表一个 View,也可以使用 LinearLayout 等布局控件创建更复杂的 View。此实例的完整项目在 MyCode\MySample386 文件夹中。

055 使用 SlidingDrawer 实现抽屉式滑动

此实例主要通过使用 SlidingDrawer 控件,在垂直方向实现抽屉式的拖曳滑动效果。当实例运行之后,按住底部的向上箭头向顶部移动,则下面的大桥图像也跟随向上移动,如图 055.1 的左图所示,如果松开手指,则大桥图像将自动滑向顶部,向上箭头变为向下箭头。当大桥图像位于顶部时,向上箭头自动变为向下箭头,此时拖动向下箭头向底部移动,则下面的大桥图像也跟随向底部移动,如图055.1的右图所示。

图 055.1

主要代码如下:

```xml
<SlidingDrawer  android:id = "@ + id/mySlidingDrawer"
                android:background = "@mipmap/myimage1"
                android:layout_width = "fill_parent"
                android:layout_height = "fill_parent"
                android:content = "@ + id/myLinearLayout"
                android:handle = "@ + id/myImageButton"
                android:orientation = "vertical">
    <ImageButton android:id = "@id/myImageButton"
                 android:layout_width = "match_parent"
                 android:layout_height = "wrap_content"
                 android:background = "#00FFFF"
                 android:src = "@mipmap/buttonup"/>
    <LinearLayout android:id = "@id/myLinearLayout"
                  android:background = "#ffffff"
                  android:orientation = "vertical"
                  android:layout_width = "match_parent"
                  android:layout_height = "match_parent">
        <TextView android:id = "@ + id/myTextView"
                  android:layout_width = "match_parent"
                  android:layout_height = "60dp"
                  android:gravity = "center_vertical|center_horizontal"
                  android:text = "滑动抽屉"
                  android:background = "#EE82EE"
                  android:textColor = "#000000"
                  android:textSize = "22dp"
```

```
                android:textStyle = "bold"/>
        < ImageView android:id = "@ + id/myImageView"
                android:layout_width = "match_parent"
                android:layout_height = "wrap_content"
                android:scaleType = "fitXY"
                android:src = "@mipmap/myimage2"/>
    </LinearLayout>
</SlidingDrawer>
```

上面这段代码在 MyCode\MySample359\app\src\main\res\layout\activity_main.xml 文件中。在这段代码中，SlidingDrawer 控件由两个 View 组成，一个是可以拖动的 handle，一个是隐藏内容的 View，它里面的控件必须设置布局，在布局文件中必须指定 handle 和 content。在 Java 代码中，可以通过监听器获取和设置 SlidingDrawer 在滑动过程中的信息，主要代码如下：

```java
public class MainActivity extends Activity {
    private SlidingDrawer mySlidingDrawer;
    private ImageButton myImageButton;
    private Boolean myFlag = false;
    private TextView TextView;
    @Override
    protected void onCreate(Bundle savedInstanceState) {
        super.onCreate(savedInstanceState);
        setContentView(R.layout.activity_main);
        myImageButton = (ImageButton)findViewById(R.id.myImageButton);
        mySlidingDrawer = (SlidingDrawer)findViewById(R.id.mySlidingDrawer);
        TextView = (TextView)findViewById(R.id.myTextView);
        mySlidingDrawer.setOnDrawerOpenListener(
                                new SlidingDrawer.OnDrawerOpenListener(){
            @Override
            public void onDrawerOpened() {
                myFlag = true;
                myImageButton.setImageResource(R.mipmap.buttondown);
            } });
        mySlidingDrawer.setOnDrawerCloseListener(
                                new SlidingDrawer.OnDrawerCloseListener(){
            @Override
            public void onDrawerClosed() {
                myFlag = false;
                myImageButton.setImageResource(R.mipmap.buttonup);
            } });
        mySlidingDrawer.setOnDrawerScrollListener(
                                new SlidingDrawer.OnDrawerScrollListener(){
            @Override
            public void onScrollEnded() { TextView.setText("结束拖动"); }
            @Override
            public void onScrollStarted() { TextView.setText("开始拖动"); }
}); } }
```

上面这段代码在 MyCode\MySample359\app\src\main\java\com\bin\luo\mysample\MainActivity.java 文件中。此实例的完整项目在 MyCode\MySample359 文件夹中。

056 自定义 ScrollView 实现下拉回弹动画

此实例主要通过在自定义 ScrollView 的 onLayout 事件响应方法中记录其子视图(控件)的初始位置,然后重写 dispatchTouchEvent 事件响应方法来监听手势滑动操作,并在重置子视图(控件)位置时为其添加平移动画,从而实现下拉回弹效果。当实例运行之后,手指按住图像向下滑动,则图像随之向下滑动,如图 056.1 的左图所示;然后松开手指,则图像回弹到初始位置(即顶端),如图 056.1 的右图所示。

图 056.1

主要代码如下:

```xml
<com.bin.luo.mysample.MyScrollView
    android:layout_width = "match_parent"
    android:layout_height = "match_parent">
  <ImageView
    android:layout_width = "match_parent"
    android:layout_height = "match_parent"
    android:scaleType = "fitXY"
    android:src = "@mipmap/myimage1"/>
</com.bin.luo.mysample.MyScrollView>
```

上面这段代码在 MyCode\MySample709\app\src\main\res\layout\activity_main.xml 文件中。在这段代码中,com.bin.luo.mysample.MyScrollView 表示自定义控件 MyScrollView,实现下拉回弹的 ImageView 控件应该放置在其里面。MyScrollView 的自定义代码如下:

```java
public class MyScrollView extends ScrollView {
    //拉出屏幕时的拖曳系数,如果是1,类似拖动效果
    private static final float MOVE_DELAY = 0.3f;
```

```java
private static final int ANIM_TIME = 300;                    //回弹动画持续时间
  private View myChildView;
  private boolean bMoved;
  private Rect myOriginalView = new Rect();
  private float startY;
  public MyScrollView(Context context, AttributeSet attrs, int defStyle) {
    super(context, attrs, defStyle);
}
  public MyScrollView(Context context, AttributeSet attrs){super(context, attrs); }
  public MyScrollView(Context context) {super(context);}
  @Override
  protected void onFinishInflate() {
    super.onFinishInflate();
    if (getChildCount() > 0) { myChildView = getChildAt(0); }
}
  @Override
  protected void onLayout(boolean changed, int l, int t, int r, int b) {
    super.onLayout(changed, l, t, r, b);
    if (myChildView == null) return;
    myOriginalView.set(myChildView.getLeft(), myChildView.getTop(),
                      myChildView.getRight(), myChildView.getBottom());
}
  @Override
  public boolean dispatchTouchEvent(MotionEvent ev) {
    if (myChildView == null) { return super.dispatchTouchEvent(ev);}
    int myAction = ev.getAction();
    switch (myAction) {
     case MotionEvent.ACTION_DOWN:
       startY = ev.getY();
       break;
     case MotionEvent.ACTION_UP:
     case MotionEvent.ACTION_CANCEL:
       if (!bMoved) break;
       TranslateAnimation myAnim = new TranslateAnimation(0, 0,
         myChildView.getTop(), myOriginalView.top);              //执行回弹动画
       myAnim.setDuration(ANIM_TIME);
       myChildView.startAnimation(myAnim);
       bMoved = false;
       resetViewLayout();
       break;
     case MotionEvent.ACTION_MOVE:
       float nowY = ev.getY();
       int deltaY = (int) (nowY - startY);
       int offset = (int) (deltaY * MOVE_DELAY);
       myChildView.layout(myOriginalView.left, myOriginalView.top + offset,
              myOriginalView.right, myOriginalView.bottom + offset);
       bMoved = true;
       break;
     default:
       break;
   }
   return super.dispatchTouchEvent(ev);
}
```

```
public void resetViewLayout() {
    myChildView.layout( myOriginalView.left,
                       myOriginalView.top,
                       myOriginalView.right,
                       myOriginalView.bottom);
} }
```

上面这段代码在 MyCode \ MySample709 \ app \ src \ main \ java \ com \ bin \ luo \ mysample \ MyScrollView.java 文件中。此实例的完整项目在 MyCode\MySample709 文件夹中。

057　使用 CollapsingToolbarLayout 实现滚动折叠

此实例主要通过使用 AppBarLayout.LayoutParams 的 setScrollFlags()方法设置滚动标志,实现在 CollapsingToolbarLayout 中的控件出现滚动折叠效果。当实例运行之后,向下拖动工具栏,则工具栏图像展开,如图 057.1 的左图所示;向上拖动工具栏,则工具栏图像折叠,直到图像消失,如图 057.1 的右图所示。

图　057.1

主要代码如下:

```
public class MainActivity extends Activity {
    @Override
    protected void onCreate(Bundle savedInstanceState) {
        super.onCreate(savedInstanceState);
        setContentView(R.layout.activity_main);
        CollapsingToolbarLayout myCollapsingToolbarLayout = (CollapsingToolbarLayout)
                findViewById(R.id.myCollapsingToolbarLayout);
        AppBarLayout.LayoutParams myLayoutParams = (AppBarLayout.LayoutParams)
                myCollapsingToolbarLayout.getLayoutParams();
        //启用滚动折叠功能
        myLayoutParams.setScrollFlags(
```

```
            AppBarLayout.LayoutParams.SCROLL_FLAG_SCROLL|
            AppBarLayout.LayoutParams.SCROLL_FLAG_EXIT_UNTIL_COLLAPSED);
    WebView myWebView = (WebView)findViewById(R.id.myWebView);
    myWebView.getSettings().setJavaScriptEnabled(true);
    myWebView.setWebViewClient(new WebViewClient());
    myWebView.loadUrl("http://www.baidu.com");
    }
}
```

上面这段代码在 MyCode\MySample910\app\src\main\java\com\bin\luo\mysample\MainActivity.java 文件中。在这段代码中，myLayoutParams.setScrollFlags（AppBarLayout.LayoutParams.SCROLL_FLAG_SCROLL|AppBarLayout.LayoutParams.SCROLL_FLAG_EXIT_UNTIL_COLLAPSED）用于实现 CollapsingToolbarLayout 的滚动折叠功能，它实际上等效于布局文件（MyCode\MySample910\app\src\main\res\layout\activity_main.xml）的属性设置：app：layout_scrollFlags＝"scroll｜exitUntilCollapsed"。需要说明的是，在 Android Studio 中，使用 CollapsingToolbarLayout 控件需要在 gradle 中引入 compile 'com.android.support：design：25.0.1'依赖项。另外，如果应用的主题是 android：Theme.Material.Light.DarkActionBar，可能在运行时会崩溃，因此需要将 app\src\main\res\values\styles.xml 文件的主题<style name＝"AppTheme" parent＝"android：Theme.Material.Light.DarkActionBar">修改为<style name＝"AppTheme" parent＝"Theme.AppCompat.Light.DarkActionBar">。此实例的完整项目在 MyCode\MySample910 文件夹中。

058 使用 BottomNavigationView 实现底部导航

此实例主要演示了使用 BottomNavigationView 控件，在屏幕底部实现菜单风格的导航效果。当实例运行之后，在屏幕底部将显示 5 个导航菜单项，并且每个菜单项同时显示图标和文本；单击第 1 个菜单项，将显示该菜单项对应的图像，如图 058.1 的左图所示；单击第 5 个菜单项，也将显示该菜单项对应的图像，如图 058.1 的右图所示；单击其他菜单项将实现类似的功能。

图 058.1

主要代码如下：

```java
public class MainActivity extends Activity {
 BottomNavigationView myBottomNavigationView;
 ImageView myImageView;
 @Override
 protected void onCreate(Bundle savedInstanceState) {
  super.onCreate(savedInstanceState);
  setContentView(R.layout.activity_main);
  myImageView = (ImageView) findViewById(R.id.myImageView);
  myBottomNavigationView =
          (BottomNavigationView) findViewById(R.id.myBottomNavigationView);
  myBottomNavigationView.setOnNavigationItemSelectedListener(
          new BottomNavigationView.OnNavigationItemSelectedListener() {
           @Override
           public boolean onNavigationItemSelected(MenuItem item) {
            switch (item.getItemId()) {
             case R.id.myItem1:
              myImageView.setImageResource(R.mipmap.myimage1);
               break;
             case R.id.myItem2:
              myImageView.setImageResource(R.mipmap.myimage2);
              break;
             case R.id.myItem3:
              myImageView.setImageResource(R.mipmap.myimage3);
              break;
             case R.id.myItem4:
              myImageView.setImageResource(R.mipmap.myimage4);
              break;
             case R.id.myItem5:
              myImageView.setImageResource(R.mipmap.myimage5);
              break;
            }
            return true;
}});}}
```

上面这段代码在 MyCode \ MySample433 \ app \ src \ main \ java \ com \ bin \ luo \ mysample \ MainActivity.java 文件中。在这段代码中，onNavigationItemSelected()方法主要用于响应用户单击在导航栏中的菜单项，即确定用户单击了导航栏的哪一个菜单项。导航栏的菜单项设置则是在布局文件中实现的，主要代码如下：

```xml
<LinearLayout  android:layout_width = "match_parent"
              android:layout_height = "match_parent"
              android:orientation = "vertical">
  <ImageView  android:id = "@ + id/myImageView"
              android:layout_width = "match_parent"
              android:layout_height = "match_parent"
              android:layout_weight = "10"
              android:scaleType = "fitXY"
              android:src = "@mipmap/myimage1"/>
  <android.support.design.widget.BottomNavigationView
              android:id = "@ + id/myBottomNavigationView"
```

```
            android:layout_width = "match_parent"
            android:layout_height = "56dp"
            android:layout_alignParentBottom = "true"
            android:layout_weight = "1"
            android:background = "#00FFFF"
            app:itemIconTint = "#FF0000"
            app:itemTextColor = "@android:color/black"
            app:menu = "@menu/mynavigation"/>
</LinearLayout>
```

上面这段代码在 MyCode\MySample433\app\src\main\res\layout\activity_main.xml 文件中。在这段代码中,app:menu="@menu/mynavigation"的属性值 mynavigation 即是导航菜单布局文件;如果在 res 目录中没有 menu 目录,请首先创建 menu 目录,然后再添加 mynavigation.xml 文件。mynavigation.xml 文件的主要内容如下:

```
<?xml version = "1.0" encoding = "utf-8"?>
<menu xmlns:android = "http://schemas.android.com/apk/res/android">
    <item
        android:id = "@+id/myItem1"
        android:icon = "@mipmap/mymovie"
        android:title = "我是传奇"/>
    <item
        android:id = "@+id/myItem2"
        android:icon = "@mipmap/mymovie"
        android:title = "银翼杀手"/>
    <item
        android:id = "@+id/myItem3"
        android:icon = "@mipmap/mymovie"
        android:title = "三百斯巴达勇士"/>
    <item
        android:id = "@+id/myItem4"
        android:icon = "@mipmap/mymovie"
        android:title = "功夫熊猫"/>
    <item
        android:id = "@+id/myItem5"
        android:icon = "@mipmap/mymovie"
        android:title = "怒火救援"/>
</menu>
```

上面这段代码在 MyCode\MySample433\app\src\main\res\menu\mynavigation.xml 文件中。需要说明的是,在 Android Studio 中使用 BottomNavigationView 控件需要在 gradle 中引入 compile 'com.android.support:design:25.0.0'依赖项,25 是开发环境的 SDK 版本号。另外,如果应用的主题是 android:Theme.Material.Light.DarkActionBar,应用可能在运行时会崩溃,因此需要将 app\src\main\res\values\styles.xml 文件的主题<style name="AppTheme" parent="android:Theme.Material.Light.DarkActionBar">修改为<style name="AppTheme" parent="Theme.AppCompat.Light.DarkActionBar">。此实例的完整项目在 MyCode\MySample433 文件夹中。

059 在 ProgressBar 上同时显示两种进度

此实例主要通过使用 setSecondaryProgress()方法设置第二进度值,实现在 ProgressBar 控件上同时显示两种进度。当实例运行之后,单击"增加进度"按钮,则在 ProgressBar 控件上第一进度(青色)和第二进度(红色)将同时增加10%;单击"减少进度"按钮,则在ProgressBar控件上第一进度和第二进度将同时减少10%,效果分别如图059.1的左图和右图所示。需要注意的是:如果第二进度达到100%,第一进度未达到100%,则单击"增加进度"按钮仅增加第一进度。

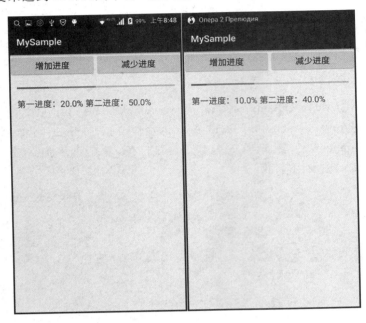

图 059.1

主要代码如下:

```
public class MainActivity extends Activity {
 TextView myTextView;
 ProgressBar myProgressBar;
 @Override
 protected void onCreate(Bundle savedInstanceState) {
  super.onCreate(savedInstanceState);
  setContentView(R.layout.activity_main);
  myTextView = (TextView) findViewById(R.id.myTextView);
  myProgressBar = (ProgressBar) findViewById(R.id.myProgressBar);
  //设置第二进度的显示颜色
  myProgressBar.setSecondaryProgressTintList(
          getResources().getColorStateList (android.R.color.holo_red_dark));
  myTextView.setText("第一进度:" + myProgressBar.getProgress() * 100.0/ myProgressBar.getMax() +
"%" + " 第二进度:" + myProgressBar.getSecondaryProgress() * 100.0/ myProgressBar.getMax() + "%");
 }
//响应单击按钮"增加进度"
public void onClickBtn1(View v) {
  myProgressBar.incrementProgressBy(10);
  myProgressBar.setSecondaryProgress(myProgressBar.getProgress() + 30);
  myTextView.setText("第一进度:" + myProgressBar.getProgress() * 100.0 / myProgressBar.getMax() +
"%" +" 第二进度:" + myProgressBar.getSecondaryProgress() * 100.0 / myProgressBar.getMax()
+ "%");
 }
//响应单击按钮"减少进度"
public void onClickBtn2(View v) {
```

```
    myProgressBar.incrementProgressBy(-10);
    myProgressBar.setSecondaryProgress(myProgressBar.getProgress() + 30);
    myTextView.setText("第一进度:" + myProgressBar.getProgress() * 100.0 / myProgressBar.getMax() +
"%" + " 第二进度:" + myProgressBar.getSecondaryProgress() * 100.0 / myProgressBar.getMax()
+ "%");
    }}
```

上面这段代码在 MyCode\MySample797\app\src\main\java\com\bin\luo\mysample\ MainActivity. java 文件中。在这段代码中，myProgressBar.incrementProgressBy(10)用于设置第一进度增量，myProgressBar.setSecondaryProgress（myProgressBar.getProgress() + 30）用于设置第二进度，myProgressBar.getProgress()用于获取第一进度的当前值，myProgressBar.getSecondaryProgress()用于获取第二进度的当前值。此实例的完整项目在 MyCode\MySample797 文件夹中。

060 使用 ViewOutlineProvider 创建圆角控件

此实例主要通过创建圆角矩形风格的 ViewOutlineProvider 自定义样式，并使用 setOutlineProvider()方法和 setClipToOutline()方法，从而实现为 ImageView 控件添加圆角效果。当实例运行之后，单击"设置圆角效果"按钮，则 ImageView 控件（电影海报图像）的圆角效果如图 060.1 的左图所示；单击"取消圆角效果"按钮，则 ImageView 控件（电影海报图像）的正常显示效果如图 060.1 的右图所示。

图 060.1

主要代码如下：

```
public class MainActivity extends Activity {
  ImageView myImageView;
  @Override
  protected void onCreate(Bundle savedInstanceState) {
    super.onCreate(savedInstanceState);
```

```
setContentView(R.layout.activity_main);
myImageView = (ImageView)findViewById(R.id.myImageView);
ViewOutlineProvider myProvider = new ViewOutlineProvider() {
  public void getOutline(View view, Outline outline) {
    outline.setRoundRect(0, 0, view.getWidth(),
      view.getHeight(), 60);                         //60 表示圆角半径
  } };
myImageView.setOutlineProvider(myProvider);
}
public void onClickmyBtn1(View v) {                  //响应单击"设置圆角效果"按钮
    myImageView.setClipToOutline(true);
}
public void onClickmyBtn2(View v) {                  //响应单击"取消圆角效果"按钮
    myImageView.setClipToOutline(false);
} }
```

上面这段代码在 MyCode \ MySample558 \ app \ src \ main \ java \ com \ bin \ luo \ mysample \ MainActivity.java 文件中。在这段代码中，myProvider = new ViewOutlineProvider(){}用于创建一个圆角半径为 60 像素的 ViewOutlineProvider。myImageView.setOutlineProvider(myProvider)表示 myImageView 控件使用 myProvider 的圆角形状。myImageView.setClipToOutline(true)表示使用 myProvider 裁剪 myImageView 控件的外形。此实例的完整项目在 MyCode\MySample558 文件夹中。

061　使用 AnalogClock 创建自定义时钟

此实例主要通过设置 AnalogClock 控件的 android：dial 和 android：hand_minute 属性，并使用 Timer 中的 schedule()方法定时刷新时间，创建一个动态走动的时钟。当实例运行之后，当时间走到 11：01，时钟的显示效果如图 061.1 的左图所示；当时间走到 11：04，时钟的显示效果如图 061.1 的右图所示。

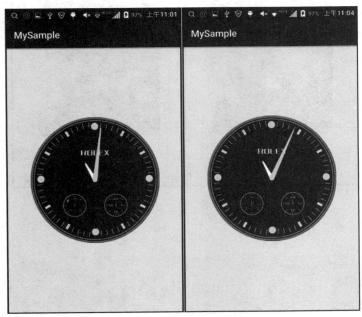

图　061.1

主要代码如下:

```xml
<AnalogClock android:id="@+id/myAnalogClock"
             android:layout_width="wrap_content"
             android:layout_height="wrap_content"
             android:dial="@mipmap/mywatch"
             android:hand_minute="@mipmap/myhand"/>
```

上面这段代码在 MyCode\MySample055\app\src\main\res\layout\activity_main.xml 文件中。在这段代码中,android:dial="@mipmap/mywatch" 表示 AnalogClock 控件的表盘 dial 采用 mywatch 资源进行配置。android:hand_minute="@mipmap/myhand" 表示 AnalogClock 控件的分针 hand_minute 采用 myhand 资源进行配置。AnalogClock 控件的时间走动效果则通过 Java 代码实现,主要代码如下:

```java
public class MainActivity extends Activity {
  private AnalogClock myAnalogClock;
  @Override
  protected void onCreate(Bundle savedInstanceState) {
    super.onCreate(savedInstanceState);
    setContentView(R.layout.activity_main);
    myAnalogClock = (AnalogClock) findViewById(R.id.myAnalogClock);
    new Timer().schedule(new TimerTask() {            //定义周期性地改变时钟的指针
      @Override
      public void run() { myAnalogClock.postInvalidate(); }
    }, 0, 1000); } }
```

上面这段代码在 MyCode\MySample055\app\src\main\java\com\bin\luo\mysample\MainActivity.java 文件中。在这段代码中,new Timer().schedule(new TimerTask() { @Override public void run() { myAnalogClock.postInvalidate(); } }, 0, 1000) 表示每隔 1000ms 调用 postInvalidate() 方法更新 myAnalogClock 一次,中间无延迟(0 表示无延迟)。此实例的完整项目在 MyCode\MySample055 文件夹中。

062 在 TextClock 中定制日期格式

此实例主要通过设置 TextClock 控件的 format12Hour 属性,并使用 Timer 的 schedule() 方法定时刷新时间,实现以指定的格式动态显示时间信息。当实例运行之后,将以指定的格式动态显示当前系统的时间信息,如图 062.1 的左图和右图所示。

主要代码如下:

```xml
<TextClock android:id="@+id/myTextClock"
           android:layout_width="wrap_content"
           android:layout_height="wrap_content"
           android:textSize="18dp"
           android:textColor="#00f"
           android:format12Hour="yyyy年MM月dd日 a hh:mm:ss EEEE" />
```

上面这段代码在 MyCode\MySample056\app\src\main\res\layout\activity_main.xml 文件中。在这段代码中,android:format12Hour="yyyy年MM月dd日 a hh:mm:ss EEEE"中的 yyyy 表示

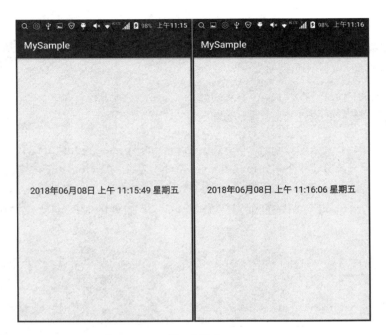

图　062.1

年数、MM 表示月数、dd 表示日数、hh 表示小时数、mm 表示分钟数、ss 表示秒数、EEEE 表示星期数、a 表示上/下午。TextClock 控件的时间动态显示效果则通过 Java 代码实现，主要代码如下：

```java
public class MainActivity extends Activity {
  private TextClock myTextClock;
  @Override
  protected void onCreate(Bundle savedInstanceState) {
    super.onCreate(savedInstanceState);
    setContentView(R.layout.activity_main);
    myTextClock = (TextClock) findViewById(R.id.myTextClock);
    new Timer().schedule(new TimerTask() {            //周期性地改变时钟的指针
      @Override
      public void run() { myTextClock.postInvalidate(); } }, 0, 1000);
} }
```

上面这段代码在 MyCode \ MySample056 \ app \ src \ main \ java \ com \ bin \ luo \ mysample \ MainActivity.java 文件中。此实例的完整项目在 MyCode\MySample056 文件夹中。

063　使用 RatingBar 实现星级评分

此实例主要通过使用 RatingBar 控件，实现星级评分的功能。当实例运行之后，当在 RatingBar 控件中选择不同位置的五星时，效果分别如图 063.1 的左图和右图所示。

主要代码如下：

```java
public class MainActivity extends Activity {
  @Override
  protected void onCreate(Bundle savedInstanceState) {
```

```
super.onCreate(savedInstanceState);
setContentView(R.layout.activity_main);
final ImageView myImageView = (ImageView) findViewById(R.id.myImageView);
RatingBar myRatingBar = (RatingBar) findViewById(R.id.myRatingBar);
myRatingBar.setOnRatingBarChangeListener(
                new RatingBar.OnRatingBarChangeListener() {
    @Override                              //当选择不同位置的五星时触发该方法
    public void onRatingChanged(RatingBar arg0, float rating, boolean fromUser) {
        //动态改变图像透明度,255 表示星级评分条最大值,5 个星星就代表最大值 255
        myImageView.setAlpha((int) (rating * 255 / 5));
}});}}
```

图　063.1

上面这段代码在 MyCode\MySample284\app\src\main\java\com\bin\luo\mysample\MainActivity.java 文件中。在这段代码中,setOnRatingBarChangeListener()事件响应方法用于响应用户在 RatingBar 控件上选择不同位置的五星,RatingBar 控件的用法和功能与 SeekBar 控件十分相似,它们都允许用户通过拖动来改变进度,RatingBar 与 SeekBar 的最大区别在于：RatingBar 使用五星表示进度。此实例的完整项目在 MyCode\MySample284 文件夹中。

064　在登录窗口中使用 SeekBar 实现手动校验

此实例主要通过比较 SeekBar 滑块的最大值,实现在登录窗口强制手动操作校验。当实例运行之后,在"账户名称："和"账户密码："输入框中输入内容,然后单击"登录"按钮,此时"登录"按钮没有任何反应,如图 064.1 的左图所示；如果根据提示"请按住滑块,拖动到最右边"进行操作,则显示"验证成功",然后自动消失,如图 064.1 的右图所示,此时单击"登录"按钮,则会弹出 Toast。

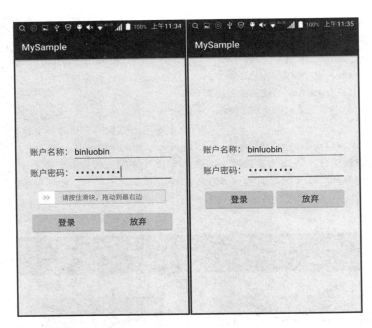

图 064.1

主要代码如下:

```java
public class MainActivity extends Activity implements SeekBar.OnSeekBarChangeListener {
 public Handler myHandler = new Handler() {
  @Override
  public void handleMessage(Message msg) {
   super.handleMessage(msg);
   if (msg.what == 1) {
    mySeekBar.setVisibility(View.GONE);
    myTextView.setVisibility(View.GONE);
   }
   myBtnLogin.setOnClickListener(new View.OnClickListener() {
    @Override
    public void onClick(View view) {
     Toast.makeText(MainActivity.this, "登录成功!", Toast.LENGTH_SHORT).show();
     finish(); } }); } };
 private TextView myTextView;
 private SeekBar mySeekBar;
 private Button myBtnLogin;
 @Override
 protected void onCreate(Bundle savedInstanceState) {
  super.onCreate(savedInstanceState);
  setContentView(R.layout.activity_main);
  myTextView = (TextView) findViewById(R.id.myTextView);
  mySeekBar = (SeekBar) findViewById(R.id.mySeekBar);
  mySeekBar.setOnSeekBarChangeListener(this);
  myBtnLogin = (Button) findViewById(R.id.myBtnLogin);
  myBtnLogin.setClickable(false);
 }
 @Override
 public void onProgressChanged(SeekBar seekBar, int i, boolean b) {
```

```java
//如果滑块抵达右端最大值,则设置"登录"按钮为可用状态(之前该按钮是不可用的)
if (seekBar.getProgress() == seekBar.getMax()) {
  myTextView.setVisibility(View.VISIBLE);
  myTextView.setTextColor(Color.WHITE);
  myTextView.setText("验证成功!");
  myBtnLogin.setClickable(true);
  new Thread() {
   @Override
   public void run() {
    super.run();
    try {
     Thread.sleep(1000);
     myHandler.sendEmptyMessage(1);
    } catch (InterruptedException e) { e.printStackTrace(); }
   } }.start();
  } else { myTextView.setVisibility(View.INVISIBLE); }
 }
 @Override
 public void onStartTrackingTouch(SeekBar seekBar) { }
 @Override
 public void onStopTrackingTouch(SeekBar seekBar) {
  //如果停止滑动滑块,并且没有抵达最大值,则重置为最小值
  if (seekBar.getProgress() != seekBar.getMax()) {
   seekBar.setProgress(0);
   myTextView.setVisibility(View.VISIBLE);
   myTextView.setTextColor(Color.GRAY);
   myTextView.setText("向右滑动,完成验证");
} } }
```

上面这段代码在 MyCode\MySample706\app\src\main\java\com\bin\luo\mysample\MainActivity.java 文件中。此实例的完整项目在 MyCode\MySample706 文件夹中。

第3章 文字

065 使用 ScaleXSpan 创建扁平风格的文字

此实例主要通过使用 ScaleXSpan，实现在 TextView 中以扁平化的风格显示文本。当实例运行之后，在 TextView 中以扁平化的风格显示文本的效果如图 065.1 所示。

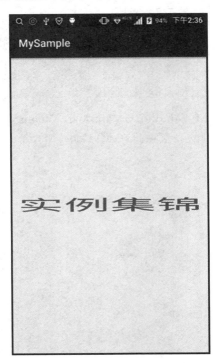

图　065.1

主要代码如下：

```
public class MainActivity extends Activity {
    @Override
    protected void onCreate(Bundle savedInstanceState) {
        super.onCreate(savedInstanceState);
        setContentView(R.layout.activity_main);
```

```
TextView myTextView = (TextView) findViewById(R.id.myTextView);
//创建一个 SpannableString 对象
SpannableString mySpannableString = new SpannableString("实例集锦");
//创建水平拉伸 3 倍的 ScaleXSpan
mySpannableString.setSpan(new ScaleXSpan(3), 0,
        mySpannableString.length(), Spanned.SPAN_EXCLUSIVE_EXCLUSIVE);
myTextView.setText(mySpannableString);
} }
```

上面这段代码在 MyCode \ MySample040 \ app \ src \ main \ java \ com \ bin \ luo \ mysample \ MainActivity.java 文件中。在这段代码中，mySpannableString.setSpan（new ScaleXSpan（3），0，mySpannableString.length（），Spanned.SPAN_EXCLUSIVE_EXCLUSIVE）表示将指定的文字在水平方向拉伸 3 倍，0 和 mySpannableString.length（）表示 ScaleXSpan 作用范围的起止位置。此实例的完整项目在 MyCode\MySample040 文件夹中。

066　使用 MaskFilterSpan 实现文字边缘模糊

此实例主要通过定制模糊滤镜 BlurMaskFilter 的模糊方式为 BlurMaskFilter.Blur.SOLID，在 TextView 中实现文本的背景模糊扩散。当实例运行之后，在 TextView 中文本的背景模糊扩散的效果如图 066.1 所示。

图　066.1

主要代码如下：

```
public class MainActivity extends Activity {
    @Override
    protected void onCreate(Bundle savedInstanceState) {
```

```
        super.onCreate(savedInstanceState);
        setContentView(R.layout.activity_main);
        TextView myTextView = (TextView) findViewById(R.id.myTextView);
        SpannableStringBuilder mySpannableStringBuilder =
                                        new SpannableStringBuilder("炫酷");
        //定制滤镜的方式是BlurMaskFilter.Blur.SOLID
        MaskFilterSpan myMaskFilterSpan = new MaskFilterSpan(
                    new BlurMaskFilter(21, BlurMaskFilter.Blur.SOLID));
        mySpannableStringBuilder.setSpan(myMaskFilterSpan, 0, 2,
                    Spanned.SPAN_INCLUSIVE_INCLUSIVE);          //设置滤镜的作用范围
        myTextView.setText(mySpannableStringBuilder);
    }}
```

上面这段代码在 MyCode\MySample037\app\src\main\java\com\bin\luo\mysample\MainActivity.java 文件中。在这段代码中，myMaskFilterSpan = new MaskFilterSpan(new BlurMaskFilter(21, BlurMaskFilter.Blur.SOLID))用于创建一个模糊距离是21，模糊方式是BlurMaskFilter.Blur.SOLID 的模糊滤镜。mySpannableStringBuilder.setSpan(myMaskFilterSpan, 0, 2, Spanned.SPAN_INCLUSIVE_INCLUSIVE)中的0表示模糊滤镜在字符串中发生作用的开始位置索引，2表示模糊滤镜在字符串中发生作用的结束位置索引。此实例的完整项目在 MyCode\MySample037 文件夹中。

067 使用 MaskFilterSpan 实现文字中心镂空

此实例主要通过定制模糊滤镜 BlurMaskFilter 的模糊方式为 BlurMaskFilter.Blur.OUTER，实现在 TextView 中显示线条描边的镂空文本。当实例运行之后，在 TextView 中显示的线条描边的镂空文本的效果如图 067.1 所示。

图 067.1

主要代码如下：

```java
public class MainActivity extends Activity {
 @Override
 protected void onCreate(Bundle savedInstanceState) {
  super.onCreate(savedInstanceState);
  setContentView(R.layout.activity_main);
  TextView myTextView = (TextView) findViewById(R.id.myTextView);
  SpannableStringBuilder mySpannableStringBuilder =
                        new SpannableStringBuilder("炫酷");
  //定制滤镜的方式是BlurMaskFilter.Blur.OUTER
  MaskFilterSpan myMaskFilterSpan = new MaskFilterSpan(
              new BlurMaskFilter(1, BlurMaskFilter.Blur.OUTER));
  mySpannableStringBuilder.setSpan(myMaskFilterSpan, 0, 2,
         Spanned.SPAN_INCLUSIVE_INCLUSIVE);          //设置滤镜的作用范围
  myTextView.setText(mySpannableStringBuilder);
 }}
```

上面这段代码在 MyCode\MySample036\app\src\main\java\com\bin\luo\mysample\ MainActivity. java 文件中。在这段代码中，myMaskFilterSpan = new MaskFilterSpan（new BlurMaskFilter（1，BlurMaskFilter.Blur.OUTER））用于创建一个模糊距离是1，模糊方式是BlurMaskFilter.Blur.OUTER 的模糊滤镜。mySpannableStringBuilder.setSpan（myMaskFilterSpan，0，2，Spanned.SPAN_INCLUSIVE_INCLUSIVE）中的 0 表示模糊滤镜在字符串中发生作用的开始位置索引，2 表示模糊滤镜在字符串中发生作用的结束位置索引。此实例的完整项目在 MyCode\MySample036 文件夹中。

068　使用 MaskFilterSpan 实现文字整体模糊

此实例主要通过使用 BlurMaskFilter，实现在 TextView 中模糊显示文本。当实例运行之后，在 TextView 中模糊显示文本的效果如图 068.1 所示。

图　068.1

主要代码如下：

```
public class MainActivity extends Activity {
 @Override
 protected void onCreate(Bundle savedInstanceState) {
  super.onCreate(savedInstanceState);
  setContentView(R.layout.activity_main);
  TextView myTextView = (TextView) findViewById(R.id.myTextView);
  SpannableStringBuilder mySpannableStringBuilder =
                          new SpannableStringBuilder("炫酷");
  MaskFilterSpan myMaskFilterSpan = new MaskFilterSpan(
            new BlurMaskFilter(15, BlurMaskFilter.Blur.NORMAL));
  mySpannableStringBuilder.setSpan(myMaskFilterSpan,
                    0, 2, Spanned.SPAN_INCLUSIVE_INCLUSIVE);
  myTextView.setText(mySpannableStringBuilder);
 } }
```

上面这段代码在 MyCode \ MySample035 \ app \ src \ main \ java \ com \ bin \ luo \ mysample \ MainActivity.java 文件中。在这段代码中，myMaskFilterSpan = new MaskFilterSpan（new BlurMaskFilter(15，BlurMaskFilter.Blur.NORMAL)）用于设置模糊滤镜的模糊距离是15，模糊方式是 BlurMaskFilter.Blur.NORMAL，除了 NORMAL 之外，常用的模糊方式还有：OUTER、INNER、SOLID。mySpannableStringBuilder.setSpan（myMaskFilterSpan，0，2，Spanned.SPAN_INCLUSIVE_INCLUSIVE）中的 0 表示模糊滤镜在字符串中发生作用的开始位置索引，2 表示模糊滤镜在字符串中发生作用的结束位置索引。此实例的完整项目在 MyCode\ MySample035 文件夹中。

069　使用 MaskFilterSpan 模糊多个字符串

此实例主要通过使用 MaskFilterSpan，实现非连续的多个字符串以模糊效果显示。当实例运行之后，在"关键词："输入框中输入"软件"，然后单击"开始搜索"按钮，则在下面的文本中，所有非"软件"的文本都将以模糊效果显示，如图 069.1 的左图所示。在"关键词："输入框中输入"的"，然后单击"开始搜索"按钮，则在下面的文本中，所有非"的"的文本都将以模糊效果显示，如图 069.1 的右图所示。测试其他关键词将取得类似的效果。

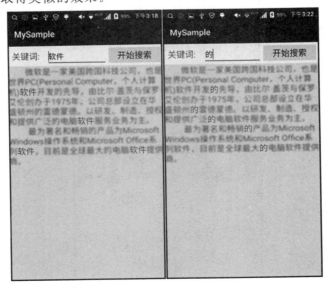

图　069.1

主要代码如下：

```java
public class MainActivity extends Activity {
    TextView myTextView;
    EditText myEditText;
    String myText;
    @Override
    protected void onCreate(Bundle savedInstanceState) {
        super.onCreate(savedInstanceState);
        getWindow().getDecorView().setLayerType(View.LAYER_TYPE_SOFTWARE, null);
        setContentView(R.layout.activity_main);
        myEditText = (EditText) findViewById(R.id.myEditText);
        myTextView = (TextView) findViewById(R.id.myTextView);
        myText = myTextView.getText().toString();              //记录原始文本内容
    }
    public void onClickButton(View v) {                        //响应单击"开始搜索"按钮
        //获取搜索字符串(关键词)
        String mySearch = myEditText.getText().toString();
        SpannableString mySpannableString = new SpannableString(myText);
        for (int i = 0; i < myText.length(); i++) {
            int start = i;                                     //记录起始索引值
            i = myText.indexOf(mySearch, i);                   //获取符合搜索结果的索引值
            if (i < 0) {                                       //未搜索到
                mySpannableString.setSpan(new MaskFilterSpan(new BlurMaskFilter(8,
                        BlurMaskFilter.Blur.NORMAL)), start + mySearch.length() - 1, myText.
                        length(), Spannable.SPAN_EXCLUSIVE_EXCLUSIVE);
                break;
            }
            //模糊显示不符合搜索结果的部分,以默认样式显示符合搜索结果的部分
            mySpannableString.setSpan(new MaskFilterSpan(new BlurMaskFilter(8,
                    BlurMaskFilter.Blur.NORMAL)), start + mySearch.length() - 1,
                    i, Spannable.SPAN_EXCLUSIVE_EXCLUSIVE);
        }
        myTextView.setText(mySpannableString);                 //应用 SpannableString 对象
    }
}
```

上面这段代码在 MyCode\MySample799\app\src\main\java\com\bin\luo\mysample\MainActivity.java 文件中。在这段代码中，mySpannableString.setSpan(new MaskFilterSpan(new BlurMaskFilter(8，BlurMaskFilter.Blur.NORMAL))，start ＋ mySearch.length()-1，myText.length()，Spannable.SPAN_EXCLUSIVE_EXCLUSIVE)表示对指定的文本进行模糊，范围从 start＋mySearch.length()-1 到 myText.length()，模糊半径为 8 像素。此实例的完整项目在 MyCode\MySample799 文件夹中。

070 使用 BulletSpan 在文本首字前添加小圆点

此实例主要通过使用 BulletSpan，实现在 TextView 文本的首字前添加小圆点符号。当实例运行之后，在 TextView 文本的首字前添加小圆点符号的效果如图 070.1 所示。

图　070.1

主要代码如下：

```java
public class MainActivity extends Activity {
 @Override
 protected void onCreate(Bundle savedInstanceState) {
  super.onCreate(savedInstanceState);
  setContentView(R.layout.activity_main);
  TextView myTextView = (TextView) findViewById(R.id.myTextView);
  SpannableString mySpannableString =
                      new SpannableString("炫酷应用实例集锦");
  //创建指定颜色的符号,并指定符号与后面文字之间宽度的BulletSpan
  mySpannableString.setSpan(new BulletSpan(50,
              Color.RED), 0, 1, Spanned.SPAN_EXCLUSIVE_EXCLUSIVE);
  myTextView.setText(mySpannableString);
 }
}
```

上面这段代码在 MyCode\MySample039\app\src\main\java\com\bin\luo\mysample\MainActivity.java 文件中。在这段代码中,mySpannableString.setSpan(new BulletSpan(50,Color.RED),0,1,Spanned.SPAN_EXCLUSIVE_EXCLUSIVE)用于创建一个距离文本首字50px的红色小圆点。如果省略颜色值,则创建的小圆点的颜色与文本的颜色相同。此实例的完整项目在 MyCode\MySample039 文件夹中。

071　使用 StrikethroughSpan 添加文字删除线

此实例主要通过使用 StrikethroughSpan,实现在 TextView 中的部分文字上添加删除线。当实例运行之后,"便捷的"三个字的中间有一条删除线,如图 071.1 所示。

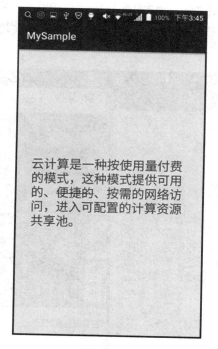

图 071.1

主要代码如下：

```
public class MainActivity extends Activity {
 @Override
 protected void onCreate(Bundle savedInstanceState) {
  super.onCreate(savedInstanceState);
  setContentView(R.layout.activity_main);
  TextView myTextView = (TextView) findViewById(R.id.myTextView);
  SpannableString mySpannableString = new SpannableString("云计算是一种按使用量付费的模式,这种模式提供可用的、便捷的、按需的网络访问,进入可配置的计算资源共享池.");
  mySpannableString.setSpan(new StrikethroughSpan(), 26, 29,
     Spanned.SPAN_EXCLUSIVE_EXCLUSIVE);         // 在"便捷的"上设置删除线
  myTextView.setText(mySpannableString);
 }
}
```

上面这段代码在 MyCode\MySample028\app\src\main\java\com\bin\luo\mysample\MainActivity.java 文件中。在这段代码中，mySpannableString.setSpan(new StrikethroughSpan()，26，29，Spanned.SPAN_EXCLUSIVE_EXCLUSIVE)中的 26 和 29 指示删除线的范围，即"便捷的"三个字在字符串中的索引位置。如果需要为"便捷的"三个字添加下画线，则可以把代码修改为 myText.setSpan(new UnderlineSpan()，26，29，Spanned.SPAN_EXCLUSIVE_EXCLUSIVE)。此实例的完整项目在 MyCode\MySample028 文件夹中。

072　使用 URLSpan 为部分内容添加超链接

此实例主要通过使用 URLSpan，实现在 TextView 中为部分内容添加超链接。当实例运行之后，"电话""公司网站"和"公司地址"将以超链接的形式显示，如图 072.1 的左图所示；单击"电话"超链接，则弹出手机的电话拨号盘；单击"公司网站"超链接，则打开网址指定的网站，如果手机上有多个浏览器，则显示浏览器选择窗口；单击"公司地址"超链接，则通过百度地图（默认地图工具）定位指定的地址，如图 072.1 的右图所示。

图　072.1

主要代码如下：

```java
public class MainActivity extends Activity {
    @Override
    protected void onCreate(Bundle savedInstanceState) {
        super.onCreate(savedInstanceState);
        setContentView(R.layout.activity_main);
        TextView myTextView = (TextView) findViewById(R.id.myTextView);
        SpannableString mySpannableString = new SpannableString("我公司以国际化和专业化为目标,集聚了一批复合型、开拓型的营销和管理等各方面的人才,你可以通过电话与相关人员进行联系,也可以通过公司网站了解更多信息,还可以在这里查看公司地址.");
        //查找"电话"在字符串中的索引位置
        int myStartPos = mySpannableString.toString().indexOf("电话");
        int myEndPos = myStartPos + 2;
        mySpannableString.setSpan(new URLSpan("tel:13996060872"), myStartPos,
            myEndPos, Spanned.SPAN_EXCLUSIVE_EXCLUSIVE);          //响应单击"电话"超链接
        //查找"公司网站"在字符串中的索引位置
        myStartPos = mySpannableString.toString().indexOf("公司网站");
        myEndPos = myStartPos + 4;
        mySpannableString.setSpan(new URLSpan("http://www.baidu.com"), myStartPos,
            myEndPos, Spanned.SPAN_EXCLUSIVE_EXCLUSIVE);          //响应单击"公司网站"超链接
        //查找"公司地址"在字符串中的索引位置
        myStartPos = mySpannableString.toString().indexOf("公司地址");
        myEndPos = myStartPos + 4;
        //响应单击"公司地址"超链接
        mySpannableString.setSpan(new URLSpan("geo:38.899533,-77.036476"),
                myStartPos, myEndPos, Spanned.SPAN_EXCLUSIVE_EXCLUSIVE);
```

```
    myTextView.setText(mySpannableString);
    //实现单击超链接的功能
    myTextView.setMovementMethod(LinkMovementMethod.getInstance());
  }}
```

上面这段代码在 MyCode \ MySample033 \ app \ src \ main \ java \ com \ bin \ luo \ mysample \ MainActivity.java 文件中。在这段代码中，mySpannableString.setSpan（new URLSpan（"tel：13996060872"），myStartPos，myEndPos，Spanned.SPAN_EXCLUSIVE_EXCLUSIVE）用于设置电话超链接，URLSpan("tel：13996060872")表示自动根据参数内容调用手机缺省应用，如果该参数是"tel：13996060872"，则调用电话拨号盘；如果该参数是"http：//www.baidu.com"网址格式，则调用浏览器；如果该参数是"geo：38.899533，-77.036476"地址格式，则调用地图工具；myStartPos 表示超链接作用范围的开始位置，myEndPos 表示超链接作用范围的结束位置。但是仅设置了 URLSpan("tel：13996060872")，超链接不会起作用，必须调用 setMovementMethod(LinkMovementMethod.getInstance())方法，超链接才会发生作用。此实例的完整项目在 MyCode\MySample033 文件夹中。

073　使用 ImageSpan 同时显示 QQ 表情和文字

此实例主要通过使用 ImageSpan，实现在 TextView 中混合显示 QQ 表情和文字。当实例运行之后，在 TextView 中混合显示 QQ 表情和文字的效果如图 073.1 所示。

图　073.1

主要代码如下：

```
public class MainActivity extends Activity {
  @Override
  protected void onCreate(Bundle savedInstanceState) {
```

```
super.onCreate(savedInstanceState);
setContentView(R.layout.activity_main);
TextView myTextView = (TextView) findViewById(R.id.myTextView);
SpannableString mySpannableString =
        new SpannableString("大爷,伤不起啊!我伤不起啊!伤不起啊!");
//查找"我伤不起啊!"在字符串中的索引位置
int myStartPos = mySpannableString.toString().indexOf("我伤不起啊!");
int myEndPos = myStartPos + "我伤不起啊!".toString().length();
//从 Drawable 获取 QQ 表情资源
Drawable myDrawable = getResources().getDrawable(R.drawable.myqq);
myDrawable.setBounds(0, 0, myDrawable.getIntrinsicWidth()/5,
        myDrawable.getIntrinsicHeight()/5);           //设置 QQ 表情图像的大小
//使用 QQ 表情图像替换"我伤不起啊!"
mySpannableString.setSpan(new ImageSpan(myDrawable),
            myStartPos, myEndPos, Spanned.SPAN_EXCLUSIVE_EXCLUSIVE);
myTextView.setText(mySpannableString);
}}
```

上面这段代码在 MyCode\MySample034\app\src\main\java\com\bin\luo\mysample\MainActivity.java 文件中。在这段代码中，mySpannableString.setSpan（new ImageSpan（myDrawable），myStartPos，myEndPos，Spanned.SPAN_EXCLUSIVE_EXCLUSIVE）表示使用 QQ 表情图像替换文字"我伤不起啊!"，myStartPos 和 myEndPos 表示"我伤不起啊!"在整个字符串中的起止位置。此实例的完整项目在 MyCode\MySample034 文件夹中。

074　使用 StyleSpan 实现以粗斜体显示文字

此实例主要通过使用 StyleSpan，实现以粗斜体风格显示 TextView 的部分文字。当实例运行之后，"便捷的"三个字将以粗斜体风格显示，如图 074.1 所示。

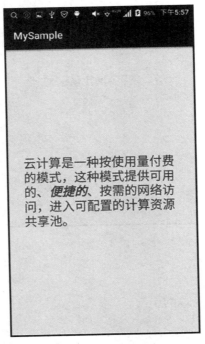

图　074.1

主要代码如下:

```java
public class MainActivity extends Activity {
    @Override
    protected void onCreate(Bundle savedInstanceState) {
        super.onCreate(savedInstanceState);
        setContentView(R.layout.activity_main);
        TextView myTextView = (TextView) findViewById(R.id.myTextView);
        SpannableString mySpannableString = new SpannableString("云计算是一种按使用量付费的模式,这种模式提供可用的、便捷的、按需的网络访问,进入可配置的计算资源共享池.");
        //以粗斜体风格显示"便捷的"三个字
        mySpannableString.setSpan(
                    new StyleSpan(android.graphics.Typeface.BOLD_ITALIC),
                    26, 29, Spanned.SPAN_EXCLUSIVE_EXCLUSIVE);
        myTextView.setText(mySpannableString);
    }
}
```

上面这段代码在 MyCode\MySample029\app\src\main\java\com\bin\luo\mysample\MainActivity.java 文件中。在这段代码中,mySpannableString.setSpan(new StyleSpan(android.graphics.Typeface.BOLD_ITALIC),26,29,Spanned.SPAN_EXCLUSIVE_EXCLUSIVE)中的 26 和 29 指示粗斜体的作用范围,即"便捷的"三个字在字符串中的索引位置;android.graphics.Typeface.BOLD_ITALIC 表示字体是粗斜体;如果此参数为 android.graphics.Typeface.ITALIC,则指定文字仅以斜体显示;如果此参数为 android.graphics.Typeface.BOLD,则指定文字仅以粗体显示。此实例的完整项目在 MyCode\MySample029 文件夹中。

075 使用 SuperscriptSpan 绘制勾股定理公式

此实例主要通过使用 SuperscriptSpan 和 RelativeSizeSpan,实现在 TextView 中以标准格式显示勾股定理公式。当实例运行之后,以标准格式显示勾股定理的效果如图 075.1 所示。

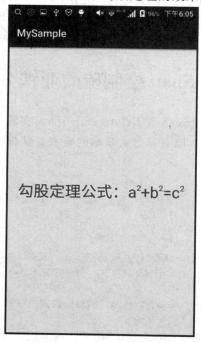

图 075.1

主要代码如下：

```java
public class MainActivity extends Activity {
 @Override
 protected void onCreate(Bundle savedInstanceState) {
  super.onCreate(savedInstanceState);
  setContentView(R.layout.activity_main);
  TextView myTextView = (TextView) findViewById(R.id.myTextView);
  SpannableString mySpannableString =
                   new SpannableString("勾股定理公式:a2 + b2 = c2");
  //以上标的形式显示指数
  mySpannableString.setSpan(new SuperscriptSpan(),
                   8, 9, Spanned.SPAN_EXCLUSIVE_EXCLUSIVE);
  mySpannableString.setSpan(new SuperscriptSpan(),
                   11, 12, Spanned.SPAN_EXCLUSIVE_EXCLUSIVE);
  mySpannableString.setSpan(new SuperscriptSpan(),
                   14, 15, Spanned.SPAN_EXCLUSIVE_EXCLUSIVE);
  // 0.5f 表示指数字体大小只有底数字体大小的一半
  mySpannableString.setSpan(new RelativeSizeSpan(0.5f),
                   8, 9, Spanned.SPAN_EXCLUSIVE_EXCLUSIVE);
  mySpannableString.setSpan(new RelativeSizeSpan(0.5f),
                   11, 12, Spanned.SPAN_EXCLUSIVE_EXCLUSIVE);
  mySpannableString.setSpan(new RelativeSizeSpan(0.5f),
                   14, 15, Spanned.SPAN_EXCLUSIVE_EXCLUSIVE);
  myTextView.setText(mySpannableString);
 }
}
```

上面这段代码在 MyCode\MySample030\app\src\main\java\com\bin\luo\mysample\MainActivity.java 文件中。在这段代码中，mySpannableString.setSpan(new SuperscriptSpan(),8,9,Spanned.SPAN_EXCLUSIVE_EXCLUSIVE)用于将指数 2 放在 a 的右上角，此时指数 2 的字体大小仍然与底数 a 相同；mySpannableString.setSpan(new RelativeSizeSpan(0.5f),8,9,Spanned.SPAN_EXCLUSIVE_EXCLUSIVE)用于将指数 2 的字体大小缩小一半。在 Android 中，SuperscriptSpan 用于以上标的形式重置文本的显示位置，它一般用于数学公式中，RelativeSizeSpan 用于根据指定的参数缩放字体大小。此实例的完整项目在 MyCode\MySample030 文件夹中。

076　使用 SubscriptSpan 绘制硫酸亚铁分子式

此实例主要通过使用 SubscriptSpan 和 RelativeSizeSpan，实现在 TextView 中以下标形式显示硫酸亚铁的分子式。当实例运行之后，以下标形式显示的硫酸亚铁分子式的效果如图 076.1 所示。

主要代码如下：

```java
public class MainActivity extends Activity {
 @Override
 protected void onCreate(Bundle savedInstanceState) {
  super.onCreate(savedInstanceState);
  setContentView(R.layout.activity_main);
  TextView myTextView = (TextView) findViewById(R.id.myTextView);
  SpannableString mySpannableString =
                   new SpannableString("硫酸亚铁的分子式:FeSO4");
  //查找 4 在字符串中的索引位置
  int myFePos = mySpannableString.toString().indexOf("4");
```

```
//以下标的形式表示 FeSO4 中的 4
mySpannableString.setSpan(new SubscriptSpan(),
                myFePos, myFePos + 1, Spanned.SPAN_EXCLUSIVE_EXCLUSIVE);
//缩小下标字体的尺寸
mySpannableString.setSpan(new RelativeSizeSpan(0.8f),
                myFePos, myFePos + 1, Spanned.SPAN_EXCLUSIVE_EXCLUSIVE);
myTextView.setText(mySpannableString);
} }
```

图 076.1

上面这段代码在 MyCode\MySample031\app\src\main\java\com\bin\luo\mysample\MainActivity.java 文件中。在这段代码中，mySpannableString.setSpan（new SubscriptSpan()，myFePos，myFePos + 1，Spanned.SPAN_EXCLUSIVE_EXCLUSIVE）用于将 4 以下标的形式放在右下角，此时 4 的字体大小不变。myFePos 和 myFePos + 1 表示 SubscriptSpan()作用范围的开始和结束索引值。mySpannableString.setSpan（new RelativeSizeSpan(0.8f)，myFePos，myFePos + 1，Spanned.SPAN_EXCLUSIVE_EXCLUSIVE）用于将 4 的字体大小缩小至原来的 4/5。在 Android 中，SubscriptSpan 用于以下标的形式重置文本的显示位置，它一般用于化学分子式中，RelativeSizeSpan 用于根据指定的参数缩放字体大小。此实例的完整项目在 MyCode\MySample031 文件夹中。

077　使用 TypefaceSpan 定制文本的部分内容

此实例主要通过使用 TypefaceSpan，实现对字符串中的部分文本定制个性化的字体。当实例运行之后，"微软风格的字体好看呀！"中的"微软"二字将以自定义字体显示，效果如图 077.1 所示。

图 077.1

主要代码如下：

```java
public class MainActivity extends Activity {
    @Override
    protected void onCreate(Bundle savedInstanceState) {
        super.onCreate(savedInstanceState);
        setContentView(R.layout.activity_main);
        TextView myTextView = (TextView) findViewById(R.id.myTextView);
        myTextView.setText("微软风格的字体好看呀!");
        Typeface myTypeface = Typeface.createFromAsset(getAssets(),"myfont.ttf"); //加载指定字体
        myTextView.setTypeface(myTypeface);                                       //在 TextView 控件上应用该字体

        SpannableString mySpannableString = new SpannableString(myTextView.getText());
        //通过反向选取区域并应用字体,使其以默认字体样式显示
        mySpannableString.setSpan(new TypefaceSpan("myfont"), 2,
                myTextView.getText().length(), Spanned.SPAN_EXCLUSIVE_EXCLUSIVE);
        myTextView.setText(mySpannableString);                                    //生成效果并显示
    }
}
```

上面这段代码在 MyCode \ MySample800 \ app \ src \ main \ java \ com \ bin \ luo \ mysample \ MainActivity.java 文件中。在这段代码中,mySpannableString.setSpan（new TypefaceSpan("myfont"),2,myTextView.getText().length(),Spanned.SPAN_EXCLUSIVE _EXCLUSIVE) 表示对 mySpannableString 字符串中的前 2 个字符使用字体"myfont"。需要说明的是,自定义字体文件 myfont.ttf 通常放在 MyCode\MySample800\app\src\main\ assets 目录中。此实例的完整项目在 MyCode\MySample800 文件夹中。

078　使用 ForegroundColorSpan 创建光照文字

此实例主要通过同时使用 ForegroundColorSpan 和 MaskFilterSpan，使字符串的部分文本呈现光照的浮雕特效。当实例运行之后，同时使用 ForegroundColorSpan 和 MaskFilterSpan 产生的部分文本（Google）浮雕特效如图 078.1 所示。

图　078.1

主要代码如下：

```
public class MainActivity extends Activity {
  @Override
  protected void onCreate(Bundle savedInstanceState) {
    super.onCreate(savedInstanceState);
    setContentView(R.layout.activity_main);
    TextView myTextView = (TextView) findViewById(R.id.myTextView);
    myTextView.setText("Google 谷歌");
    SpannableString mySpannableString = new SpannableString(myTextView.getText());
    mySpannableString.setSpan(new ForegroundColorSpan(Color.RED),
        0, 6, Spanned.SPAN_EXCLUSIVE_EXCLUSIVE);          //应用红色前景
    mySpannableString.setSpan(new MaskFilterSpan(
        new EmbossMaskFilter(new float[]{1, 1, 1}, 0.25f, 6, 3.5f)),
        0, 6, Spannable.SPAN_EXCLUSIVE_EXCLUSIVE);        //应用光照特效
    myTextView.setText(mySpannableString);
} }
```

上面这段代码在 MyCode\MySample801\app\src\main\java\com\bin\luo\mysample\MainActivity.java 文件中。在这段代码中，mySpannableString.setSpan（new ForegroundColorSpan（Color.RED），0，6，Spanned.SPAN_EXCLUSIVE_EXCLUSIVE）表示对 mySpannableString 字符

串的前6个字符("Google")应用红色前景。mySpannableString.setSpan(new MaskFilterSpan(new EmbossMaskFilter(new float[]{1,1,1},0.25f,6,3.5f)),0,6,Spannable.SPAN_EXCLUSIVE _EXCLUSIVE)表示对mySpannableString字符串的前6个字符应用EmbossMaskFilter产生的光照特效。在测试此实例时,通常应该关闭硬件加速,即设置android:hardwareAccelerated="false",该代码在MyCode\MySample801\app\src\main\AndroidManifest.xml文件中。此实例的完整项目在MyCode\MySample801文件夹中。

079 使用BlurMaskFilter创建阴影扩散文字

此实例主要通过在自定义View中创建BlurMaskFilter模糊滤镜,并在画笔Paint中使用setMaskFilter()方法设置模糊滤镜,实现绘制阴影扩散的文字。当实例运行之后,在自定义View中使用BlurMaskFilter模糊滤镜绘制阴影扩散的文字的效果如图079.1所示。

主要代码如下:

```java
public class MainActivity extends Activity {
    @Override
    protected void onCreate(Bundle savedInstanceState) {
        setContentView(new MyView(this));
        super.onCreate(savedInstanceState);
    }
class MyView extends View {
    private Paint myPaint;
    public MyView(Context context) {
        super(context);
        myPaint = new Paint();
        myPaint.setFlags(Paint.ANTI_ALIAS_FLAG);
        myPaint.setAntiAlias(true);
        myPaint.setColor(Color.BLACK);
        myPaint.setTextSize(220);
        myPaint.setStyle(Paint.Style.FILL_AND_STROKE);
        myPaint.setStrokeWidth(2);
        BlurMaskFilter myBlurMaskFilter =
                new BlurMaskFilter(40, BlurMaskFilter.Blur.SOLID);
                                                        //创建模糊滤镜
        myPaint.setMaskFilter(myBlurMaskFilter);     //设置模糊滤镜画笔
    }
    @Override
    protected void onDraw(Canvas myCanvas) {
        super.onDraw(myCanvas);
        Display myDisplay = getWindowManager().getDefaultDisplay();
        int myWidth = myDisplay.getWidth();
        int myHeight = myDisplay.getHeight();
        myCanvas.drawText("炫酷实例", myWidth/10,
                myHeight/2 - 150, myPaint);  //显示扩散阴影文本
}}}
```

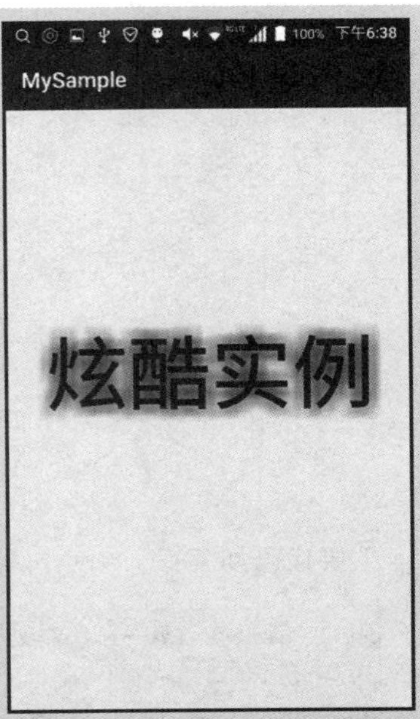

图 079.1

上面这段代码在MyCode\MySample087\app\src\main\java\com\bin\luo\mysample\MainActivity.java文件中。在这段代码中,myBlurMaskFilter = new BlurMaskFilter(40, BlurMaskFilter.Blur.SOLID)用于创建模糊滤镜,40表示模糊半径,BlurMaskFilter.Blur.SOLID表示模糊模式是SOLID。BlurMaskFilter支持下列4种模糊模式。

（1）NORMAL，表示同时绘制图形本身内容＋内阴影＋外阴影，正常阴影效果。
（2）INNER，表示绘制图形内容本身＋内阴影，不绘制外阴影。
（3）OUTER，表示不绘制图形内容以及内阴影，只绘制外阴影。
（4）SOLID，表示只绘制外阴影和图形内容本身，不绘制内阴影。
此实例的完整项目在 MyCode\MySample087 文件夹中。

080　使用 EmbossMaskFilter 创建浮雕文字

此实例主要通过使用浮雕滤镜 EmbossMaskFilter，实现在 TextView 中创建浮雕风格的文本。当实例运行之后，在 TextView 中显示的浮雕文本的效果如图 080.1 所示。

图　080.1

主要代码如下：

```
public class MainActivity extends Activity {
    @Override
    protected void onCreate(Bundle savedInstanceState) {
        super.onCreate(savedInstanceState);
        setContentView(R.layout.activity_main);
        TextView myTextView = (TextView) findViewById(R.id.myTextView);
        SpannableStringBuilder mySpannableStringBuilder =
                new SpannableStringBuilder("炫酷");
        float[] myDirection = new float[]{ 10, 10, 10 };     //设置光源的方向
        float myLight = 0.1f;                                 //环境光亮度
        float mySpecular = 5;                                 //反射等级
        float myBlur = 5;                                     //模糊半径
        //根据指定的参数创建浮雕滤镜 EmbossMaskFilter
        EmbossMaskFilter myEmbossMaskFilter =
```

```
        new EmbossMaskFilter(myDirection,myLight,mySpecular,myBlur);
//根据浮雕滤镜创建 MaskFilterSpan
 MaskFilterSpan myMaskFilterSpan = new MaskFilterSpan(myEmbossMaskFilter);
 mySpannableStringBuilder.setSpan(myMaskFilterSpan,0, 2,
     Spanned.SPAN_INCLUSIVE_INCLUSIVE);            //设置 MaskFilterSpan 的作用范围
 myTextView.setText(mySpannableStringBuilder);
} }
```

上面这段代码在 MyCode \ MySample038 \ app \ src \ main \ java \ com \ bin \ luo \ mysample \ MainActivity.java 文件中。在这段代码中，myEmbossMaskFilter = new EmbossMaskFilter(myDirection，myLight，mySpecular，myBlur)用于根据指定的参数 myDirection，myLight，mySpecular，myBlur 创建浮雕滤镜；当浮雕滤镜创建后,再使用该浮雕滤镜创建 MaskFilterSpan,即 myMaskFilterSpan = new MaskFilterSpan（myEmbossMaskFilter）。mySpannableStringBuilder. setSpan(myMaskFilterSpan，0，2，Spanned.SPAN_INCLUSIVE_ INCLUSIVE)中的 0 表示浮雕滤镜发生作用的开始位置索引,2 表示浮雕滤镜发生作用的结束位置索引。在测试此实例时,通常应该关闭硬件加速,即在 MyCode\ MySample038\app\src \main\ AndroidManifest.xml 文件中设置 android：hardwareAccelerated＝"false"。此实例的完整项目在 MyCode\MySample038 文件夹中。

081　通过自定义 View 在半圆弧上绘制文字

此实例主要通过使用 Path 的 addArc()方法和 Canvas 的 drawTextOnPath()方法,在自定义 View 中实现在半圆弧路径上绘制文字。当实例运行之后,在半圆弧路径上绘制的文字"Android 炫酷应用实例集锦"的效果如图 081.1 所示。

图　081.1

主要代码如下:

```
<com.bin.luo.mysample.CustomTextView android:layout_width = "match_parent"
                                     android:layout_height = "wrap_content"/>
```

上面这段代码在 MyCode\MySample775\app\src\main\res\layout\activity_main.xml 文件中。在这段代码中,com.bin.luo.mysample.CustomTextView 是自定义 View 类(控件)在布局文件中的应用,com.bin.luo.mysample 是包名,在实际应用中通常应该修改为自己的包名,自定义类 CustomTextView 的主要代码如下:

```
public class CustomTextView extends View {
  String myText = "Android炫酷应用实例集锦";
  Paint myPaint = null;
  Path myPath = null;
  public CustomTextView(Context context, AttributeSet attrs) {
    super(context, attrs);
    Init();
  }
  public void Init() {                                            //初始化画笔和路径对象
    myPaint = new Paint();
    myPaint.setAntiAlias(true);
    myPaint.setTextSize(96);
    LinearGradient myGradient = new LinearGradient(100,100, 800, 800,
        Color.GREEN, Color.BLUE, Shader.TileMode.MIRROR);         //初始化线性渐变对象
    myPaint.setShader(myGradient);                                //使用渐变色填充画笔颜色
    if (myPath == null) {                                         //若未设置路径对象,则默认将以半圆路径显
                                                                  //示文本
      myPath = new Path();
      myPath.addArc(100,100, 800, 800,180, 180);                  //绘制半圆路径
    } }
  @Override
  protected void onDraw(Canvas canvas) {
    super.onDraw(canvas);
    canvas.translate(100,490);                                    //平移画布至指定位置
    canvas.drawTextOnPath(myText, myPath, 3,
                          -10, myPaint);                          //在指定路径上绘制指定文本
} }
```

上面这段代码在 MyCode\MySample775\app\src\main\java\com\bin\luo\mysample\CustomTextView.java 文件中。在这段代码中,myGradient = new LinearGradient(100,100,800,800,Color.GREEN,Color.BLUE,Shader.TileMode.MIRROR)用于创建一个填充 Paint 的线性渐变色对象,LinearGradient()构造函数的语法声明如下:

```
public LinearGradient(float x0, float y0, float x1,
                      float y1, int color0, int color1, TileMode tile)
```

其中,参数 float x0 表示渐变起点的 x 坐标;参数 float y0 表示渐变起点的 y 坐标;参数 float x1 表示渐变终点的 x 坐标;参数 float y1 表示渐变终点的 y 坐标;参数 int color0 表示渐变开始颜色;参数 int color1 表示渐变结束颜色;参数 TileMode tile 表示平铺方式。TileMode 有 3 种参数可供选择,分别为 CLAMP、REPEAT 和 MIRROR;CLAMP 的作用是如果渲染器超出原始边界范围,则会复制边缘颜色对超出范围的区域进行着色,REPEAT 的作用是在横向和纵向上以平铺的形式重复渲染,MIRROR 的作用是在横向和纵向上以镜像的方式重复渲染。

myPath.addArc(100,100,800,800,180,180)的作用是绘制一个半圆弧路径。addArc()方法的语法声明如下：

```
public void addArc(float left, float top,
        float right, float bottom, float startAngle, float sweepAngle)
```

其中，参数 float left 表示圆弧外接矩形左上角的 x 坐标，参数 float top 表示圆弧外接矩形左上角的 y 坐标，参数 float right 表示圆弧外接矩形右下角的 x 坐标，参数 float bottom 表示圆弧外接矩形右下角的 y 坐标，float startAngle 表示圆弧起始角，此角度不好理解，它以水平线右端为 0°，以顺时针方向为增量，此实例所用的 180°即在水平线的左端，float sweepAngle 表示圆弧的度数范围，在此实例中，startAngle 为 180，sweepAngle 为 180，因此终止角为 360°，即水平线的右端。如果 myPath.addArc(9,9,350,350,90,180)，则圆弧文字表现为左端，而不是上端。

canvas.drawTextOnPath(myText，myPath，3，−10，myPaint)表示在指定的圆弧路径上绘制渐变色文字。drawTextOnPath()方法的语法声明如下：

```
public void drawTextOnPath(@NonNull String text, @NonNull Path path,
        float hOffset, float vOffset, @NonNull Paint paint)
```

其中，参数 String text 表示显示的文字，参数 Path path 表示文字的显示路径，参数 float hOffset 表示水平偏移量，参数 float vOffset 表示垂直偏移量，参数 Paint paint 表示定制的画笔。

此实例的完整项目在 MyCode\MySample775 文件夹中。

082　通过自定义 View 在圆弧上滚动文字

此实例主要通过使用 Timer 定时修改 drawTextOnPath()方法的 hOffset 参数（水平偏移量），实现使文字沿着自定义圆弧路径滚动显示。当实例运行之后，单击"开始滚动"按钮，则演示文本"人生得意须尽欢"将沿着自定义圆弧路径以顺时针方向滚动显示；单击"暂停滚动"按钮，则滚动显示停止；再次单击"开始滚动"按钮，则演示文本将从上次暂停的位置继续滚动显示，效果分别如图 082.1 的左图和右图所示。

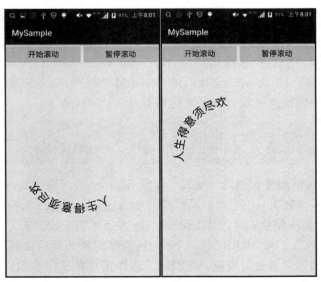

图　082.1

主要代码如下：

```
//响应单击"开始滚动"按钮
public void onClickBtn1(View v){ myCustomTextView.startMarqueeAnim(); }
 //响应单击"暂停滚动"按钮
public void onClickBtn2(View v){ myCustomTextView.pauseMaruqeeAnim(); }
```

上面这段代码在 MyCode \ MySample776 \ app \ src \ main \ java \ com \ bin \ luo \ mysample \ MainActivity.java 文件中。在这段代码中，myCustomTextView =（CustomTextView）findViewById(R.id.myCustomTextView)中的 CustomTextView 是自定义控件（类），该控件（类）的主要代码如下：

```
public class CustomTextView extends View {
 String myText = "人生得意须尽欢";
 Paint myPaint = null;
 Path myPath = null;
 Timer myTimer;
 TimerTask myTask;
 float myPositionX = 0;
 public CustomTextView(Context context, AttributeSet attrs) {
  super(context, attrs);
  Init();
 }
 public void Init() {                      //初始化画笔和路径对象
  myPaint = new Paint();
  myPaint.setAntiAlias(true);
  myPaint.setTextSize(80);
  if (myPath == null) {                    //若未设置路径对象,则默认将以半圆路径显示文本
   myPath = new Path();
   myPath.addArc(10, 10, 700, 700, 0, 360); //绘制半圆路径
  } }
 @Override
 protected void onDraw(Canvas canvas) {
  super.onDraw(canvas);
  canvas.translate(150, 300);              //平移画布至指定位置
  canvas.drawTextOnPath(myText, myPath,
   myPositionX, -10, myPaint);             //在指定路径上绘制指定文本
 }
  public void startMarqueeAnim() {         //创建定时任务
  if (myTask != null) { myTask.cancel(); }
  myTimer = new Timer();
  myTask = new CustomTimerTask();          //初始化自定义定时任务类
  myTimer.schedule(myTask, 0, 100);        //提交任务至定时器并开始执行
 }
 //取消定时任务
 public void pauseMaruqeeAnim() { myTask.cancel(); }
 private class CustomTimerTask extends TimerTask {
  @Override
  public void run() {
   myPositionX += 10;
   if (myPositionX == 1500) { myPositionX = 0; }
   postInvalidate();
} } }
```

上面这段代码在 MyCode \ MySample776 \ app \ src \ main \ java \ com \ bin \ luo \ mysample \

CustomTextView.java 文件中。在这段代码中，myTimer.schedule(myTask，0，100)表示间隔执行滚动显示文本任务，schedule()方法的语法声明如下：

```
public void schedule(TimerTask task, long delay, long period)
```

其中，参数 TimerTask task 是一个 TimerTask 派生类的实例，该派生类通常需要实现 public void run()方法，因为 TimerTask 类实现了 Runnable 接口。参数 long delay 用于设置 timer 定时器第一次调用 run 方法的时间，0 表示不设置时间，立刻执行任务。参数 long period 表示第一次执行任务之后，从第二次开始每隔多长的时间调用一次 run()方法，以 ms 为单位。

此实例的完整项目在 MyCode\MySample776 文件夹中。

083　通过自定义 View 绘制渐变色的文字

此实例主要通过使用线性渐变 LinearGradient 创建着色器 Shader，并在画笔 Paint 中使用 setShader()方法设置 Shader，实现绘制颜色渐变的文字。当实例运行之后，在自定义 View 中绘制的颜色渐变文字的效果如图 083.1 所示。

图　083.1

主要代码如下：

```
public class MainActivity extends Activity {
  @Override
  protected void onCreate(Bundle savedInstanceState) {
    setContentView(new MyView(this));
    super.onCreate(savedInstanceState);
  }
}
```

```
class MyView extends View {
  public MyView(Context context) { super(context); }
  @Override
  protected void onDraw(Canvas myCanvas) {
    Display myDisplay = getWindowManager().getDefaultDisplay();
    int myWidth = myDisplay.getWidth();                    //获取屏幕宽度
    int myHeight = myDisplay.getHeight();                  //获取屏幕高度
    Shader myShader = new LinearGradient(0, 0, 160, 160,
            new int[]{Color.RED, Color.GREEN, Color.BLUE, Color.YELLOW},
            null, Shader.TileMode.REPEAT);                 //创建渐变色的着色器
    Paint myPaint = new Paint();
    myPaint.setShader(myShader);                           //使用渐变色着色器设置画笔
    myPaint.setTextSize(160);                              //设置文字大小
    myCanvas.drawColor(Color.WHITE);                       //设置背景颜色
    myCanvas.drawText("炫酷应用实例", 50,
            myHeight / 2 - 200, myPaint);                  //绘制渐变色文字
  }
}
```

上面这段代码在 MyCode\MySample085\app\src\main\java\com\bin\luo\mysample\MainActivity.java 文件中。在这段代码中，myShader = new LinearGradient(0，0，160，160，new int[]{Color.RED，Color.GREEN，Color.BLUE，Color.YELLOW}，null，Shader. TileMode.REPEAT)用于创建渐变色的着色器。myCanvas.drawText("炫酷应用实例"，50，myHeight / 2 − 200，myPaint)中的 50 表示绘制文字开始位置的水平坐标，myHeight / 2 − 200 表示绘制文字开始位置的垂直坐标。此实例的完整项目在 MyCode\MySample085 文件夹中。

084 通过自定义 View 绘制线条描边文字

此实例主要通过设置画笔的样式为 STROKE，实现在自定义 View 中绘制线条描边的空心文字。当实例运行之后，在自定义 View 中绘制的线条描边的空心文字的效果如图 084.1 所示。

主要代码如下：

```
public class MainActivity extends Activity {
  @Override
  protected void onCreate(Bundle savedInstanceState) {
    setContentView(new MyView(this));
    super.onCreate(savedInstanceState);
  }
  class MyView extends View {
    private Paint myPaint;
    public MyView(Context context) {
      super(context);
      myPaint = new Paint(Paint.ANTI_ALIAS_FLAG);
      myPaint.setColor(Color.BLUE);                        //设置文字颜色
      myPaint.setStyle(Paint.Style.STROKE);                //创建空心文字
      myPaint.setStrokeWidth(1);                           //设置文字线条宽度
      myPaint.setTextSize(220);                            //设置字体大小
    }
    @Override
    protected void onDraw(Canvas myCanvas) {
      super.onDraw(myCanvas);
      Display myDisplay = getWindowManager().getDefaultDisplay();
      int myWidth = myDisplay.getWidth();
```

图 084.1

```
        int myHeight = myDisplay.getHeight();
        myCanvas.scale(1.0f,1.5f);                    //在垂直方向拉伸1.5倍
        myCanvas.drawText("炫酷实例", myWidth/10, myHeight * 3/10, myPaint);
    } }
}
```

上面这段代码在 MyCode \ MySample088 \ app \ src \ main \ java \ com \ bin \ luo \ mysample \ MainActivity.java 文件中。在这段代码中，myPaint.setStyle(Paint.Style.STROKE)用于设置画笔的样式为空心。在 Android 中，Paint.Style 用于设置画笔的样式，取值如下。

（1）FILL，该样式用于创建实心画笔。

（2）FILL_AND_STROKE，该样式用于使画笔同时实现实心和空心效果，该样式在某些场合会带来不可预期的显示效果。

（3）STROKE，该样式用于创建空心画笔。

此实例的完整项目在 MyCode\MySample088 文件夹中。

085　通过自定义 View 绘制阴影扩散文字

此实例主要通过使用 setShadowLayer()方法，实现在自定义 View 中绘制阴影扩散的文字。当实例运行之后，在自定义 View 中绘制的阴影扩散文字的效果如图 085.1 所示。

主要代码如下：

```
public class MainActivity extends Activity {
    @Override
    protected void onCreate(Bundle savedInstanceState) {
        setContentView(new MyView(this));
        super.onCreate(savedInstanceState);
    }
    class MyView extends View {
        private Paint myPaint;
        public MyView(Context context) {
            super(context);
            myPaint = new Paint();
            myPaint.setFlags(Paint.ANTI_ALIAS_FLAG);
            myPaint.setAntiAlias(true);
            myPaint.setColor(Color.BLUE);
            myPaint.setTextSize(200);
            myPaint.setStyle(Paint.Style.FILL_AND_STROKE);
            myPaint.setStrokeWidth(2);
            myPaint.setShadowLayer(15f, 10f, 10f, Color.GRAY);     //设置阴影画笔
        }
        @Override
        protected void onDraw(Canvas myCanvas) {
            super.onDraw(myCanvas);
            Display myDisplay = getWindowManager().getDefaultDisplay();
            int myWidth = myDisplay.getWidth();
            int myHeight = myDisplay.getHeight();
            myCanvas.drawText("炫酷实例", myWidth / 10,
                              myHeight * 2 / 5, myPaint);  //绘制阴影文字
    } } }
```

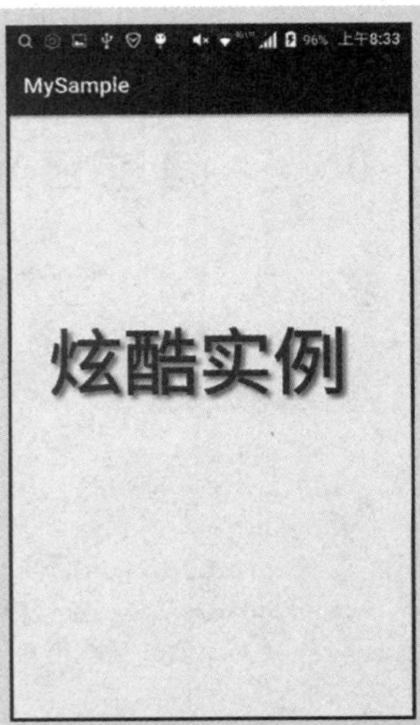

图 085.1

上面这段代码在 MyCode\MySample091\app\src\main\java\com\bin\luo\mysample\MainActivity.java 文件中。在这段代码中，myPaint.setShadowLayer(15f,10f,10f,Color.GRAY)用于设置阴影画笔，15f 表示扩散半径，10f 表示水平方向的偏移量，10f 表示垂直方向的偏移量，Color.GRAY 表示阴影颜色。此实例的完整项目在 MyCode\MySample091 文件夹中。

086　加载字库文件显示自定义草书字体

此实例主要通过使用 Typeface 的 createFromAsset()方法根据指定的草体字库文件创建自定义字体，实现以草体字显示文本。当实例运行之后，单击"以默认字体显示"按钮，则唐诗将以默认的黑体字显示，如图 086.1 的左图所示。单击"以草体显示"按钮，则唐诗将以自定义的草体字显示，如图 086.1 的右图所示。

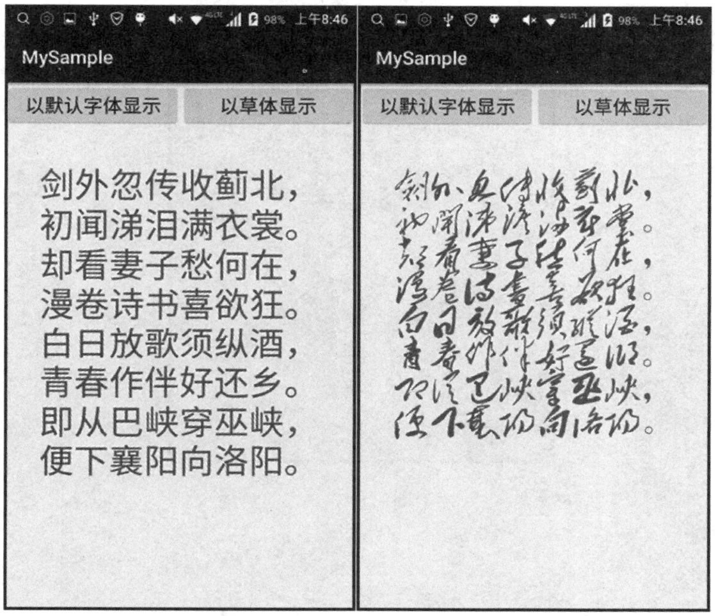

图　086.1

主要代码如下：

```java
//响应单击"以默认字体显示"按钮
public void onClickmyBtn1(View v){ myTextView.setTypeface(Typeface.DEFAULT);}
//响应单击"以草体显示"按钮
public void onClickmyBtn2(View v) {
    Typeface myFont = Typeface.createFromAsset(this.getAssets(), "myfont.ttf");
    myTextView.setTypeface(myFont);          //加载自定义草体字
}
```

上面这段代码在 MyCode\MySample644\app\src\main\java\com\bin\luo\mysample\MainActivity.java 文件中。在这段代码中，myFont = Typeface.createFromAsset（this.getAssets()，"myfont.ttf"）用于根据草体字库文件 myfont.ttf 创建草体字。需要说明的是，如果当前项目不存在 assets 子目录，应该首先在 app\src\main 目录下创建 assets 子目录，然后在 assets 子目录中添加草体字库文件 myfont.ttf。此实例的完整项目在 MyCode\MySample644 文件夹中。

087 加载字库文件显示自定义液晶字体

此实例主要通过在 assets 目录中添加自定义液晶字体,并使用 createFromAsset()方法获取此字体,实现在 TextView 中显示自定义的液晶字体文字。当实例运行之后,"HTML AND CSS"以液晶字体显示的效果如图 087.1 所示。

图 087.1

主要代码如下:

```java
public class MainActivity extends Activity {
    @Override
    protected void onCreate(Bundle savedInstanceState) {
        super.onCreate(savedInstanceState);
        setContentView(R.layout.activity_main);
        TextView myTextView = (TextView) findViewById(R.id.myTextView);
        AssetManager myAssetManager = getAssets();                          //获取 AssetManager
        Typeface myTypeface = Typeface.createFromAsset(myAssetManager,
                        "myLed.TTF");                                       //根据液晶字体路径获取 Typeface
        myTextView.setTypeface(myTypeface);                                 //设置当前字体为液晶字体
    }
}
```

上面这段代码在 MyCode \ MySample019 \ app \ src \ main \ java \ com \ bin \ luo \ mysample \ MainActivity.java 文件中。在这段代码中,setTypeface(myTypeface)用于设置 myTypeface 字体为当前文字的字体,myTypeface = Typeface.createFromAsset(myAssetManager, "myLed.TTF")则用于将 myLed.TTF 字体文件以资源的形式导入字体库。此实例的完整项目在 MyCode \ MySample019 文件夹中。

088　判断在一个字符串中是否包含汉字

此实例主要通过使用Pattern和Matcher的成员方法,实现根据正则表达式"[\u4E00－\u9FA5\uF900－\uFA2D]"判断在一个字符串中是否包含汉字。当实例运行之后,如果在"测试内容:"输入框中输入"China 是一个伟大的国家",然后单击"检测该字符串是否包含汉字"按钮,则在弹出的Toast中提示"该字符串有汉字!",如图088.1的左图所示。如果在"测试内容:"输入框中输入"China is a great country",然后单击"检测该字符串是否包含汉字"按钮,则在弹出的Toast中提示"该字符串没有汉字!",如图088.1的右图所示。

图　088.1

主要代码如下:

```java
public void onClickmyBtn1(View v) {              //响应单击"检测该字符串是否包含汉字"按钮
    String myText = myEditText.getText().toString();
    final String format = "[\\u4E00 - \\u9FA5\\uF900 - \\uFA2D]";
    Pattern myPattern = Pattern.compile(format);
    Matcher myMatcher = myPattern.matcher(myText);
    boolean myResult = myMatcher.find();
    if(myResult){
     Toast.makeText(getApplicationContext(),
                "该字符串有汉字!", Toast.LENGTH_SHORT). show();
    }else{
     Toast.makeText(getApplicationContext(),
                "该字符串没有汉字!", Toast.LENGTH_SHORT). show();
    }}
```

上面这段代码在 MyCode\MySample539\app\src\main\java\com\bin\luo\mysample\MainActivity.java文件中。在这段代码中,myMatcher.find()的返回值如果为true,表示测试内容与正则表达式匹配;如果为false,表示测试内容与正则表达式不匹配。此实例的完整项目在MyCode\MySample539文件夹中。

第4章 图形和图像

089 在自定义View中绘制径向渐变的图形

此实例主要通过使用RadialGradient类,实现在自定义View中绘制径向渐变的图形。当实例运行之后,在自定义View中绘制的径向渐变图形如图089.1所示。

主要代码如下:

```java
public class MainActivity extends Activity {
    @Override
    protected void onCreate(Bundle savedInstanceState) {
        setContentView(new MyView(this));
        super.onCreate(savedInstanceState);
    }
    class MyView extends View {
        public MyView(Context context) { super(context); }
        @Override
        protected void onDraw(Canvas myCanvas) {
            Display myDisplay = getWindowManager().getDefaultDisplay();
            int myWidth = myDisplay.getWidth();
            int myHeight = myDisplay.getHeight();
            Paint myPaint = new Paint();
            myPaint.setFlags(Paint.ANTI_ALIAS_FLAG);
            myPaint.setAntiAlias(true);
            RadialGradient myGradient = new RadialGradient(myWidth / 2,
                myHeight / 2 - 160, myWidth / 2 - 20, new int[]{Color.YELLOW, Color.GREEN,
                Color.TRANSPARENT, Color.RED}, null, Shader.TileMode.REPEAT);
            myPaint.setShader(myGradient);                    //设置径向渐变画笔
            myCanvas.drawCircle(myWidth / 2, myHeight / 2 - 160,
                            myWidth / 2 - 20, myPaint);       //绘制径向渐变的圆形
        }
    }
}
```

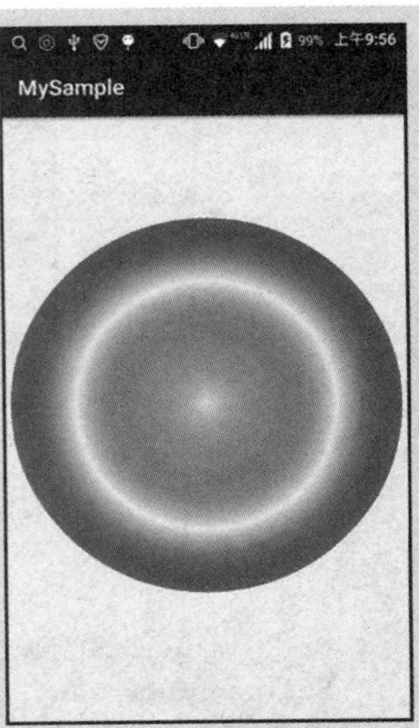

图 089.1

上面这段代码在 MyCode\MySample092\app\src\main\java\com\bin\luo\mysample\MainActivity.java 文件中。在这段代码中,myGradient = new RadialGradient(myWidth / 2, myHeight / 2 - 160, myWidth / 2 - 20, new int[]{Color.YELLOW, Color.GREEN, Color.TRANSPARENT, Color.RED}, null, Shader.TileMode.REPEAT)用于创建一个径向渐变RadialGradient,该构造函数的语法声明如下:

```
public RadialGradient(float x, float y, float radius,
                 int[] colors, float[] positions,Shader.TileMode tile)
```

其中,参数 float x 表示圆心 x 坐标,参数 float y 表示圆心 y 坐标,参数 float radius 表示径向渐变的半径,参数 int[] colors 表示渲染颜色数组,参数 float[] positions 表示相对位置数组,可为 null,即颜色沿渐变方向均匀分布,参数 Shader.TileMode tile 表示渲染平铺模式,包括 CLAMP、MIRROR 和 REPEAT 3 种模式。

此实例的完整项目在 MyCode\MySample092 文件夹中。

090 在自定义 View 中实现图像波纹起伏效果

此实例主要通过使用 drawBitmapMesh()方法绘制扭曲图像,实现图像的波纹起伏效果。当实例运行之后,图像将会像波浪一样上下起伏,效果分别如图 090.1 的左图和右图所示。

图 090.1

主要代码如下:

```
< LinearLayout   android:layout_width = "match_parent"
                 android:layout_height = "match_parent"
                 android:orientation = "vertical">
    < com.bin.luo.mysample.myImageView  android:layout_width = "wrap_content"
                                         android:layout_height = "wrap_content"
                                         android:scaleType = "center"
                                         android:visibility = "visible"/>
</LinearLayout >
```

上面这段代码在 MyCode\MySample146\app\src\main\res\layout\activity_main.xml 文件中。在这段代码中,com.bin.luo.mysample.myImageView 即是用于实现图像波浪起伏的自定义控件,myImageView 类(控件)的主要代码如下:

```java
public class myImageView extends View {
    //定义两个常量表示需要将图像划分成多少个小方格
    private int meshWidth = 40;                              //横向划分的方格数目
    private int meshHeight = 40;                             //纵向划分的方格数目
    private float FREQUENCY = 0.1f;                          //三角函数的频率大小
    private int AMPLITUDE = 80;                              //三角函数的振幅大小
    private int myCount = (meshWidth + 1) * (meshHeight + 1);
    private Bitmap myBmp =
            BitmapFactory.decodeResource(getResources(), R.mipmap.myimage);
    //保存每个小方格的交叉点坐标
    private float[] myPoints = new float[myCount * 2];
    private float[] myVerts = new float[myCount * 2];
    private float k;
    public myImageView(Context context, AttributeSet attrs, int defStyleAttr) {
      super(context, attrs, defStyleAttr);
      initData();
    }
    public myImageView(Context context, AttributeSet attrs){this(context, attrs, 0);}
    @Override
    protected void onDraw(Canvas myCanvas) {
      flagWave();
      k += FREQUENCY;
      myCanvas.translate(0,150);                             //纵向平移画布
      myCanvas.drawBitmapMesh(myBmp,
                  meshWidth, meshHeight, myVerts, 0, null, 0, null);
      invalidate();
    }
    private void initData() {
      float myBmpWidth = myBmp.getWidth();
      float myBmpHeight = myBmp.getHeight();
      int index = 0;
      //通过遍历所有小方格,得到原图每个交叉点坐标,并把它们保存在数组中
      for (int i = 0; i <= meshHeight; i++) {
        float fy = myBmpHeight * i / meshHeight;
        for (int j = 0; j <= meshWidth; j++) {
          float fx = myBmpHeight * j / meshWidth;
          myPoints[index * 2 + 0] = myVerts[index * 2 + 0] = fx;
          myPoints[index * 2 + 1] = myVerts[index * 2 + 1] = fy + AMPLITUDE * 1.2f;
          index++;
    } } }
    //加入正弦函数算法,修改交叉点坐标得到波纹效果,这里只修改 y 坐标,x 坐标不变
    public void flagWave() {
      for (int i = 0; i <= meshHeight; i++) {
        for (int j = 0; j < meshWidth; j++) {
          myVerts[(i * (meshWidth + 1) + j) * 2 + 0] += 0;
          float offSetY =
              (float) Math.sin((float) j / meshWidth * 2 * Math.PI + Math.PI * k);
          myVerts[(i * (meshWidth + 1) + j) * 2 + 1] =
                  myPoints[(i * (meshWidth + 1) + j) * 2 + 1] + offSetY * AMPLITUDE;
    } } } }
```

上面这段代码在 MyCode\MySample146\app\src\main\java\com\bin\luo\mysample\myImageView.java 文件中。在这段代码中,drawBitmapMesh()方法用于绘制扭曲图像,该方法的语法声明如下:

```
drawBitmapMesh(@NonNull Bitmap bitmap, int meshWidth, int meshHeight, @NonNull float[] verts, int
vertOffset, @Nullable int[] colors, int colorOffset,@Nullable Paint paint)
```

其中，参数 Bitmap bitmap 表示需要扭曲的图像。参数 int meshWidth 表示在横向上把该图像划成多少格。参数 int meshHeight 表示在纵向上把该图像划成多少格。参数 float[] verts 表示长度为 (meshWidth + 1) * (meshHeight + 1) * 2 的数组，它记录了扭曲后的图像各顶点位置。参数 int vertOffset 表示 verts 数组中从第几个数组元素开始才对图像进行扭曲。参数 int[] colors 指定每个顶点之间的颜色。参数 int colorOffset 表示绘制前需要跳过的颜色数，即偏移量。参数 Paint paint 表示定制的画笔。

此实例的完整项目在 MyCode\MySample146 文件夹中。

091　在自定义 View 中使用椭圆裁剪图像

此实例主要通过使用 drawOval()方法绘制椭圆，并在 setXfermode()方法的参数中使用 android.graphics.PorterDuff.Mode.ADD 模式创建 PorterDuffXfermode 实例，实现将图像裁剪成椭圆。当实例运行之后，在自定义 View 中使用椭圆裁剪图像的效果如图 091.1 所示。

主要代码如下：

```
public class MainActivity extends Activity {
    @Override
    protected void onCreate(Bundle savedInstanceState) {
        setContentView(new MyView(this));
        super.onCreate(savedInstanceState);
    }
}
class MyView extends View {
    public MyView(Context context) { super(context); }
    @Override
    protected void onDraw(Canvas myCanvas) {
        Display myDisplay = getWindowManager().getDefaultDisplay();
        int myWidth = myDisplay.getWidth();
        int myHeight = myDisplay.getHeight();
        Bitmap myBitmap =
                BitmapFactory.decodeResource(getResources(), R.mipmap.myimage);
        Paint myPaint = new Paint();
        myCanvas.drawOval(10, myHeight / 5, myWidth - 10,
                        myHeight * 3 / 5, myPaint);          //以椭圆形裁剪图像
        myPaint.setXfermode(new PorterDuffXfermode(PorterDuff.Mode.ADD));
        myCanvas.drawBitmap(myBitmap, 0, 0, myPaint);        //绘制图像
} } }
```

图 091.1

上面这段代码在 MyCode\MySample081\app\src\main\java\com\bin\luo\mysample\MainActivity.java 文件中。在这段代码中，BitmapFactory.decodeResource（getResources（），R.mipmap.myimage）用于从 R.mipmap.myimage 资源中获取图像。myCanvas.drawOval（10，myHeight/5，myWidth－10，myHeight * 3/5，myPaint）用于绘制椭圆，10 表示左上角的水平坐标（因为椭圆内切于矩形），myHeight/5 表示左上角的垂直坐标，myWidth－10 表示右下角的水平坐标，myHeight * 3/5 表示右下角的垂直坐标，myPaint 表示画笔。myPaint.setXfermode（new

PorterDuffXfermode（PorterDuff.Mode.ADD））表示以 PorterDuff.Mode.ADD 模式定制画笔，当在 myCanvas.drawBitmap(myBitmap,0,0,myPaint)中使用此画笔绘制图像时，则仅绘制椭圆部分的对应内容。此实例的完整项目在 MyCode\MySample081 文件夹中。

092　通过 PorterDuff 模式增暗显示两幅图像

此实例主要通过设置 Paint 的 PorterDuffXfermode 为 DARKEN，实现仅显示两幅图像颜色较深的部分。当实例运行之后，单击"特效图像"按钮，则两幅图像在叠加之后仅显示颜色较深部分的效果如图 092.1 的左图所示。单击"原始图像"按钮，则显示原始图像，如图 092.1 的右图所示。

图　092.1

主要代码如下：

```java
public void onClickBtn1(View v) {                    //响应单击"特效图像"按钮
    Bitmap myBitmap1 =
            BitmapFactory.decodeResource(getResources(),R.mipmap.myimage1);
    Bitmap myBitmap2 =
            BitmapFactory.decodeResource(getResources(),R.mipmap.myimage2);
    Bitmap myBitmapCopy = myBitmap2.copy(Bitmap.Config.ARGB_8888,true);
    Canvas myCanvas = new Canvas(myBitmapCopy);
    Paint myPaint = new Paint();
    //获得每个位置上两幅图像中最暗的像素并显示
    myPaint.setXfermode(new PorterDuffXfermode(PorterDuff.Mode.DARKEN));
    myCanvas.drawBitmap(myBitmap1,0,0,myPaint);
    myImageView.setImageBitmap(myBitmapCopy);
}
```

上面这段代码在 MyCode\MySample796\app\src\main\java\com\bin\luo\mysample\MainActivity.java 文件中。在这段代码中，myPaint.setXfermode（new PorterDuffXfermode（PorterDuff.Mode.DARKEN））中的 DARKEN 表示在两幅图像叠加之后，仅显示颜色较深的部分。此实例的完整项目在 MyCode\MySample796 文件夹中。

093 通过 PorterDuff 模式将图像裁剪成五角星

此实例主要通过使用 PorterDuff.Mode.SRC_IN 相交模式叠加图像和五角星,实现将图像裁剪成五角星形状。当实例运行之后,单击"显示原图"按钮,则在屏幕上显示原始图像,如图 093.1 的左图所示;单击"显示五星图"按钮,则将以五角星形状在屏幕上显示裁剪后的图像,如图 093.1 的右图所示。

图 093.1

主要代码如下:

```
public void onClickmyBtn2(View v) {                   //响应单击"显示五星图"按钮
    myImageView.setImageBitmap(getNewImage(myBmp));
}
public Bitmap getNewImage(Bitmap myBitmap) {
    Bitmap myStarBmp = Bitmap.createBitmap(myWidth,myHeight,
                Bitmap.Config.ARGB_8888);             //创建新位图
    Canvas myCanvas = new Canvas(myStarBmp);          //创建带有新位图的画布
    Paint myPaint = new Paint();
    Path myPath = new Path();
    myPath.moveTo(myWidth * 0.5f, myWidth * 0);
    myPath.lineTo(myWidth * 0.63f, myWidth * 0.38f);
    myPath.lineTo(myWidth, myWidth * 0.38f);
    myPath.lineTo(myWidth * 0.69f, myWidth * 0.59f);
    myPath.lineTo(myWidth * 0.82f, myWidth);
    myPath.lineTo(myWidth * 0.5f, myWidth * 0.75f);
    myPath.lineTo(myWidth * 0.18f, myWidth);
    myPath.lineTo(myWidth * 0.31f, myWidth * 0.59f);
    myPath.lineTo(0, myWidth * 0.38f);
    myPath.lineTo(myWidth * 0.37f, myWidth * 0.38f);
    myPath.close();
    myCanvas.translate(0,360);
```

```
myCanvas.drawPath(myPath, myPaint);                //绘制五角星
//设置相交模式
myPaint.setXfermode(new PorterDuffXfermode(PorterDuff.Mode.SRC_IN));
Rect myRect = new Rect(0, 0, myWidth, myWidth);    //把图像画到五星
myCanvas.drawBitmap(myBitmap, null, myRect, myPaint);
return myStarBmp;
}
```

上面这段代码在 MyCode\MySample108\app\src\main\java\com\bin\luo\mysample\MainActivity.java 文件中。在这段代码中，myPath 表示使用五角星的 10 个点的坐标位置用直线连接而成的封闭路径。myCanvas.drawPath（myPath，myPaint）则是根据封闭路径绘制五角星。myPaint.setXfermode（new PorterDuffXfermode（PorterDuff.Mode.SRC_IN））用于设置画笔的 PorterDuffXfermode 模式为 SRC_IN，即相交模式，该模式规定只在源图像和目标图像相交的地方绘制源图像。myCanvas.drawBitmap（myBitmap，null，myRect，myPaint）则表示在五角星中绘制图像。此实例的完整项目在 MyCode\MySample108 文件夹中。

094 通过 PorterDuff 模式改变 tint 属性叠加效果

此实例主要通过使用 PorterDuff 的多种图像处理模式，使 tint 属性指定的颜色与图像叠加产生不同的效果。当实例运行之后，单击"设置 ADD 模式"按钮，则 ImageView 控件的图像显示效果如图 094.1 的左图所示；单击"设置 SRC_OUT 模式"按钮，则 ImageView 控件的图像显示效果如图 094.1 的右图所示。

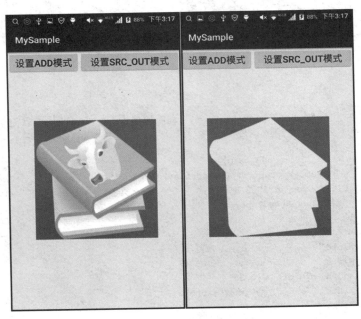

图　094.1

主要代码如下：

```
< ImageView android:id = "@ + id/myImageView"
            android:layout_width = "wrap_content"
            android:layout_height = "wrap_content"
```

```
            android:layout_centerInParent = "true"
            android:tint = "@android:color/holo_green_dark"
            android:tintMode = "multiply"
            android:src = "@mipmap/myimage1"/>
```

上面这段代码在 MyCode\MySample561\app\src\main\res\layout\activity_main.xml 文件中。在这段代码中，android：tint＝"@android：color/holo_green_dark"表示使用深绿色与图像进行叠加。android：tintMode＝"multiply"表示叠加模式是 PorterDuff.Mode.MULTIPLY。在 Java 代码中，则使用 setImageTintMode()方法设置叠加模式，主要代码如下：

```
public void onClickBtn1(View v) {         //响应单击"设置 ADD 模式"按钮
    myImageView.setImageTintMode(PorterDuff.Mode.ADD);
}
public void onClickBtn2(View v) {         //响应单击"设置 SRC_OUT 模式"按钮
    myImageView.setImageTintMode(PorterDuff.Mode.SRC_OUT);
}
```

上面这段代码在 MyCode \ MySample561 \ app \ src \ main \ java \ com \ bin \ luo \ mysample \ MainActivity.java 文件中。在这段代码中，myImageView.setImageTintMode（PorterDuff.Mode.ADD）用于设置 ImageView 控件的 tintMode 属性。tintMode 属性通常支持 PorterDuff 列举的模式，如 CLEAR、SRC、DST、SRC_OVER、DST_OVER、SRC_IN、DST_IN、SRC_OUT、DST_OUT、SRC_ATOP、DST_ATOP、XOR、DARKEN、LIGHTEN、MULTIPLY、SCREEN、ADD、OVERLAY 等。此实例的完整项目在 MyCode\MySample561 文件夹中。

095　使用 Region 的 DIFFERENCE 实现抠图功能

此实例主要通过使用 Region.Op.DIFFERENCE 裁剪模式，实现在图像上抠出一个六边形的窟窿。当实例运行之后，单击"显示原始图像"按钮，则在屏幕上显示原始图像，如图 095.1 的左图所示；单击"显示抠图结果"按钮，则原始图像根据六边形抠图之后的结果如图 095.1 的右图所示。

图　095.1

主要代码如下：

```java
public void onClickBtn2(View v) {                           //响应单击"显示抠图结果"按钮
    Bitmap myBitmap = BitmapFactory.decodeResource(getResources(),
                    R.mipmap.myimage1).copy(Bitmap.Config.ARGB_8888,true);
    int myWidth = myBitmap.getWidth();
    int myYoffset = 250;
    Path myPath = new Path();
    myPath.moveTo(0, myWidth * 0.5f + myYoffset);
    myPath.lineTo(myWidth * 0.25f, 0 + myYoffset);
    myPath.lineTo(myWidth * 0.75f, 0 + myYoffset);
    myPath.lineTo(myWidth , myWidth * 0.5f + myYoffset);
    myPath.lineTo(myWidth * 0.75f, myWidth + myYoffset);
    myPath.lineTo(myWidth * 0.25f, myWidth + myYoffset);
    myPath.close();
    Region myRegion = new Region();
    myRegion.setPath(myPath,new Region(new Rect(0,0,
            myBitmap.getWidth(),
            myBitmap.getHeight())));            //根据封闭的六边形创建 Region 对象
    Canvas myCanvas = new Canvas(myBitmap);
    myCanvas.drawColor(Color.WHITE);
    //myCanvas.clipRegion(myRegion,Region.Op.XOR);
    //Region.Op.DIFFERENCE 表示获取图像减去 Region 的剩余部分,即抠图
    myCanvas.clipRegion(myRegion,Region.Op.DIFFERENCE);
    myCanvas.drawBitmap(BitmapFactory.decodeResource(getResources(),
                    R.mipmap.myimage1),0,0,null);
    //在 ImageView 上显示抠图(不相交的部分)结果
    myImageView.setImageBitmap(myBitmap);
}
```

上面这段代码在 MyCode \ MySample828 \ app \ src \ main \ java \ com \ bin \ luo \ mysample \ MainActivity.java 文件中。在这段代码中，myCanvas.clipRegion(myRegion，Region.Op.DIFFERENCE)表示使用 Region.Op.DIFFERENCE 模式裁剪图像，Region.Op.DIFFERENCE 模式表示裁剪结果是全集减去交集的剩余部分（即差值）。此实例的完整项目在 MyCode \ MySample828 文件夹中。

096　使用 ShapeDrawable 裁剪三角形图像

此实例主要以三角形 PathShape 创建 ShapeDrawable，并使用图像填充三角形风格的 ShapeDrawable，实现裁剪三角形图像的效果。当实例运行之后，单击"显示原始图像"按钮，则在屏幕上显示原始图像，如图 096.1 的左图所示；单击"显示三角形图像"按钮，则将以三角形在屏幕上显示裁剪的图像，如图 096.1 的右图所示。

主要代码如下：

```java
public void onClickBtn2(View v) {                           //响应单击"显示三角形图像"按钮
    Bitmap myBitmap =
            BitmapFactory.decodeResource(getResources(), R.mipmap.myimage1);
```

```
    Path myPath = new Path();
    myPath.moveTo(myBitmap.getWidth()/2,0);
    myPath.lineTo(myImageDrawable.getIntrinsicWidth(),
            myImageDrawable.getIntrinsicHeight());
    myPath.lineTo(0,myImageDrawable.getIntrinsicHeight());
    myPath.close();                              //根据三角形的3个顶点创建封闭三角形路径
    ShapeDrawable myShapeDrawable = new ShapeDrawable(new PathShape(myPath,
    myImageDrawable.getIntrinsicWidth(),myImageDrawable.getIntrinsicHeight()));
    BitmapShader myBitmapShader = new BitmapShader(myBitmap, Shader.TileMode.CLAMP,
            Shader.TileMode.CLAMP);              //根据图像设置对应BitmapShader对象
    myShapeDrawable.getPaint().setShader(myBitmapShader); //应用该BitmapShader
    myImageView.setBackground(myShapeDrawable);
}
```

图 096.1

上面这段代码在 MyCode\MySample821\app\src\main\java\com\bin\luo\mysample\MainActivity.java 文件中。在这段代码中，myPath 主要用于创建封闭的三角形路径。myShapeDrawable 主要根据 myPath 创建三角形 ShapeDrawable。myBitmapShader 则根据图像创建着色器。myShapeDrawable.getPaint().setShader(myBitmapShader) 则使用该图像着色器填充（绘制）三角形的内部，实现裁剪三角形图像的效果。此实例的完整项目在 MyCode\MySample821 文件夹中。

097 使用 ClipDrawable 裁剪图像实现星级评分

此实例主要通过使用 ClipDrawable 控制裁剪区域，实现星级评分功能。当实例运行之后，单击"增加评分"按钮，则一次增加一颗星；单击"减少评分"按钮，则一次减少一颗星，效果分别如图 097.1 的左图和右图所示。

图 097.1

主要代码如下：

```java
public class MainActivity extends Activity {
    ImageView myImageView;
    ClipDrawable myClipDrawable;
    int myLevel = 10000;
    Bitmap myBitmap;
    @Override
    protected void onCreate(Bundle savedInstanceState) {
        super.onCreate(savedInstanceState);
        setContentView(R.layout.activity_main);
        myImageView = (ImageView) findViewById(R.id.myImageView);
        myBitmap = BitmapFactory.decodeResource(getResources(), R.mipmap.myscorestar);
        myClipDrawable = new ClipDrawable(new BitmapDrawable(myBitmap),
                                Gravity.LEFT, ClipDrawable.HORIZONTAL);
        myImageView.setImageDrawable(myClipDrawable);
        myClipDrawable.setLevel(myLevel);
    }
    public void onClickBtn1(View v) {                //响应单击"增加评分"按钮
        if(myLevel < 10000){ myClipDrawable.setLevel(myLevel += 2000); }
    }
    public void onClickBtn2(View v) {                //响应单击"减少评分"按钮
        if(myLevel > 0){ myClipDrawable.setLevel(myLevel -= 2000); }
    }
}
```

上面这段代码在 MyCode \ MySample782 \ app \ src \ main \ java \ com \ bin \ luo \ mysample \ MainActivity.java 文件中。在这段代码中，myDrawable.setLevel(myLevel += 2000)用于改变裁剪区域，setLevel()方法的参数取值范围从 0 到 10000，为 0 时完全不显示，为 10000 时完全显示；10000/2000＝5，由于 5 颗星是一幅图像（灰色的五星也是一幅图像，被设置为背景），因此每增加 2000，即增加一颗。myDrawable ＝ new ClipDrawable(new BitmapDrawable (myBitmap),Gravity.

LEFT,ClipDrawable.HORIZONTAL)用于创建指定规则的ClipDrawable对象,此实例在裁剪图像时,裁剪规则是将对象(图像)放在容器的左端,在水平方向上从右端开始裁剪。此实例的完整项目在MyCode\MySample782 文件夹中。

098　使用自定义 Drawable 实现对图像进行圆角

此实例主要通过使用 Drawable 为基类创建自定义类 RoundImageDrawable,实现对 ImageView 控件中的图像进行圆角。当实例运行之后,两个 ImageView 控件的图像经过圆角之后的效果如图 098.1 所示。

主要代码如下:

```
public class MainActivity extends Activity {
 ImageView myImageView1;
 ImageView myImageView2;
 @Override
 protected void onCreate(Bundle savedInstanceState) {
  super.onCreate(savedInstanceState);
  setContentView(R.layout.activity_main);
  myImageView1 = (ImageView) findViewById(R.id.myImageView1);
  Bitmap myBmp1 =
      BitmapFactory.decodeResource(getResources(),R.mipmap.myimage1);
  myImageView1.setImageDrawable(new RoundImageDrawable(myBmp1,30));
  myImageView2 = (ImageView) findViewById(R.id.myImageView2);
  Bitmap myBmp2 =
      BitmapFactory.decodeResource(getResources(),R.mipmap.myimage2);
  myImageView2.setImageDrawable(new RoundImageDrawable(myBmp2,30));
 }
 public class RoundImageDrawable extends Drawable{
  private Paint myPaint;
  private Bitmap myBitmap;
  private RectF myRect;
  private float myRadius;
  public RoundImageDrawable(Bitmap bitmap,float radius) {
   myRadius = radius;
   myBitmap = bitmap;
   BitmapShader myBitmapShader = new BitmapShader(bitmap,
                     Shader.TileMode.CLAMP,Shader.TileMode.CLAMP);
   myPaint = new Paint();
   myPaint.setAntiAlias(true);
   myPaint.setShader(myBitmapShader);
  }
  @Override
  public void setBounds(int left, int top, int right, int bottom){
   super.setBounds(left, top, right, bottom);
   myRect = new RectF(left, top, right, bottom);
  }
  @Override
  public void draw(Canvas canvas){              //绘制圆角矩形
   canvas.drawRoundRect(myRect, myRadius, myRadius, myPaint);
  }
  @Override
  public int getIntrinsicWidth(){ return myBitmap.getWidth(); }
  @Override
```

图　098.1

```
    public int getIntrinsicHeight(){ return myBitmap.getHeight(); }
    @Override
    public void setAlpha(int alpha){ myPaint.setAlpha(alpha); }
    @Override
    public void setColorFilter(ColorFilter cf){ myPaint.setColorFilter(cf); }
    @Override
    public int getOpacity(){ return PixelFormat.TRANSLUCENT; }
    }
}
```

上面这段代码在 MyCode\MySample673\app\src\main\java\com\bin\luo\mysample\MainActivity.java 文件中。在这段代码中，canvas.drawRoundRect(myRect, myRadius, myRadius, myPaint)用于根据传递的圆角半径 myRadius 绘制圆角矩形，即对图像进行圆角。此实例的完整项目在 MyCode\MySample673 文件夹中。

099 使用 Matrix 实现按照指定方向倾斜图像

此实例主要通过使用 Matrix 的 postSkew()方法倾斜矩阵，实现按照一定的方向倾斜图像。当实例运行之后，单击"显示原图"按钮，则在屏幕上显示原始图像，如图 099.1 的左图所示；单击"倾斜图像"按钮，则图像将会按照指定的方向进行倾斜，如图 099.1 的右图所示。

图　099.1

主要代码如下：

```
public void onClickmyBtn2(View v) {              //响应单击"倾斜图像"按钮
    myImageView.setImageBitmap(getNewImage(myBitmap));
}
public Bitmap getNewImage(Bitmap oldBmp) {
    int myWidth = oldBmp.getWidth();
    int myHeight = oldBmp.getHeight();
```

```
    Matrix myMatrix = new Matrix();
    //myMatrix.preSkew(0.1f, 0.2f);
    myMatrix.postSkew(0.1f, 0.2f);                //倾斜图像
    Bitmap newBmp = Bitmap.createBitmap(oldBmp, 0, 0, myWidth,
        myHeight, myMatrix, true); //根据原始图像和倾斜之后的矩阵创建新图像
    return newBmp;
}
```

上面这段代码在 MyCode\MySample127\app\src\main\java\com\bin\luo\mysample\MainActivity.java 文件中。在这段代码中，myMatrix.postSkew(0.1f, 0.2f)表示将 myMatrix 矩阵相对于 x 轴倾斜 0.1f，相对于 y 轴倾斜 0.2f，此处的倾斜理解为扭曲可能更合适些。Bitmap.createBitmap(oldBmp, 0, 0, myWidth, myHeight, myMatrix, true)用于根据 oldBmp 图像创建与倾斜之后的 myMatrix 相匹配的新图像。此实例的完整项目在 MyCode\MySample127 文件夹中。

100　使用 ColorMatrix 为图像添加泛紫效果

此实例主要通过直接使用 ColorMatrix 颜色矩阵配置画笔，实现为图像添加泛紫效果。当实例运行之后，单击"显示原图"按钮，则在屏幕上显示原始图像，如图 100.1 的左图所示；单击"泛紫图像"按钮，则显示泛紫效果的图像，如图 100.1 的右图所示。

图　100.1

主要代码如下：

```
public void onClickmyBtn2(View v) {                    //响应单击"泛紫图像"按钮
    Bitmap newImage = getNewImage(myBitmap);
    myImageView.setImageBitmap(newImage);
}
public Bitmap getNewImage(Bitmap oldBmp) {
    int myWidth = oldBmp.getWidth();
```

```
        int myHeight = oldBmp.getHeight();
        float[] myColorArray = {1, 0, 11, 0, 10,
                                0, 1, 0, 0, 100,
                                0, 100, 1, 0, 0,
                                0, 0, 0, 1, 0};              //泛紫颜色矩阵
        Bitmap newBmp = Bitmap.createBitmap(myWidth,
                                    myHeight, Bitmap.Config.ARGB_8888);
        Canvas myCanvas = new Canvas(newBmp);
        Paint myPaint = new Paint();
        ColorMatrix myColorMatrix = new ColorMatrix();       //新建颜色矩阵
        myColorMatrix.set(myColorArray);                     //设置泛紫颜色矩阵
        //通过颜色过滤器配置泛紫画笔
        myPaint.setColorFilter(new ColorMatrixColorFilter(myColorMatrix));
        myCanvas.drawBitmap(oldBmp, 0, 0, myPaint);          //绘制泛紫图像
        return newBmp;
    }
```

上面这段代码在 MyCode \ MySample133 \ app \ src \ main \ java \ com \ bin \ luo \ mysample \ MainActivity.java 文件中。在这段代码中，float[ ] myColorArray = {1，0，11，0，10，0，1，0，0，100，0，100，1，0，0，0，0，0，1，0}是泛紫颜色矩阵的数组表达式，myColorMatrix.set(myColorArray)用于实现在使用该数组构建颜色矩阵与图像在画布上进行绘制时（即 myCanvas.drawBitmap(oldBmp，0，0，myPaint)），将使图像产生泛紫效果。此实例的完整项目在 MyCode\MySample133 文件夹中。

101 使用 ColorMatrix 实现图像的加暗效果

此实例主要通过直接使用 ColorMatrix 颜色矩阵配置画笔，实现为图像添加加暗效果。当实例运行之后，单击"显示原图"按钮，则在屏幕上显示原始图像，如图 101.1 的左图所示；单击"加暗图像"按钮，则显示加暗效果的图像，如图 101.1 的右图所示。

图 101.1

主要代码如下：

```
public void onClickmyBtn2(View v) {              //响应单击"加暗图像"按钮
    Bitmap newImage = getNewImage(myBitmap);
    myImageView.setImageBitmap(newImage);
}
public Bitmap getNewImage(Bitmap oldBmp) {
    int myWidth = oldBmp.getWidth();
    int myHeight = oldBmp.getHeight();
    float[] myColorArray = { 1,0,0,0,-100,
                             0,1,0,0,-100,
                             0,0,1,0,-100,
                             0,0,0,1,0};       //加暗颜色矩阵
    Bitmap newBmp = Bitmap.createBitmap(myWidth,
                              myHeight, Bitmap.Config.ARGB_8888);
    Canvas myCanvas = new Canvas(newBmp);
    Paint myPaint = new Paint();
    ColorMatrix myColorMatrix = new ColorMatrix();   //新建颜色矩阵
    myColorMatrix.set(myColorArray);                 //设置加暗颜色矩阵
    //通过颜色过滤器配置加暗画笔
    myPaint.setColorFilter(new ColorMatrixColorFilter(myColorMatrix));
    myCanvas.drawBitmap(oldBmp, 0, 0, myPaint);      //绘制加暗图像
    return newBmp;
}
```

上面这段代码在 MyCode\MySample141\app\src\main\java\com\bin\luo\mysample\MainActivity.java 文件中。在这段代码中，float[] myColorArray = {1,0,0,0,-100,0,1,0,0,-100,0,0,1,0,-100,0,0,0,1,0}是加暗图像的颜色矩阵的数组表达式，myColorMatrix.set(myColorArray)用于实现在使用该数组构建颜色矩阵与图像在画布上进行绘制时（即 myCanvas.drawBitmap(oldBmp,0,0,myPaint)），将使图像产生加暗效果。此实例的完整项目在 MyCode\MySample141 文件夹中。

102 通过自定义 ColorMatrix 调整图像蓝色色调

此实例主要通过直接使用 ColorMatrix 颜色矩阵配置画笔，实现使用蓝色通道过滤图像。当实例运行之后，单击"显示原图"按钮，则在屏幕上显示原始图像，如图 102.1 的左图所示；单击"使用蓝色通道过滤图像"按钮，则图像在经过蓝色通道过滤之后的效果如图 102.1 的右图所示。

主要代码如下：

```
//响应单击"使用蓝色通道过滤图像"按钮
public void onClickmyBtn2(View v) {
    Bitmap newImage = getNewImage(myBitmap);
    myImageView.setImageBitmap(newImage);
}
public Bitmap getNewImage(Bitmap oldBmp) {
    int myWidth = oldBmp.getWidth();
    int myHeight = oldBmp.getHeight();
    //定义蓝色通道颜色矩阵数组值
    float[] myColorArray = {
            0, 0, 0, 0, 0,
```

```
            0, 0, 0, 0, 0,
            0, 0, 1, 0, 0,
            0, 0, 0, 1, 0 };
Bitmap newBmp = Bitmap.createBitmap(myWidth,
                                myHeight, Bitmap.Config.ARGB_8888);
Canvas myCanvas = new Canvas(newBmp);
Paint myPaint = new Paint();
//新建颜色矩阵
ColorMatrix myColorMatrix = new ColorMatrix();
//设置蓝色通道颜色矩阵
myColorMatrix.set(myColorArray);
//通过颜色过滤器配置画笔
myPaint.setColorFilter(new ColorMatrixColorFilter(myColorMatrix));
//绘制蓝色通道过滤的图像
myCanvas.drawBitmap(oldBmp, 0, 0, myPaint);
return newBmp;
}
```

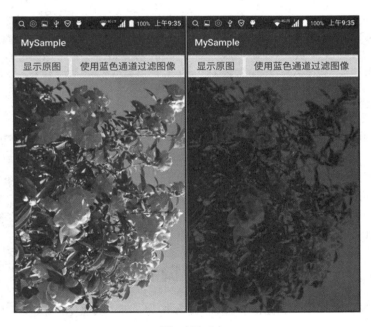

图 102.1

上面这段代码在 MyCode\MySample417\app\src\main\java\com\bin\luo\mysample\MainActivity.java 文件中。在这段代码中，float[] myColorArray = {0,0,0,0,0,0,0,0,0,0,0,0,1,0,0,0,0,0,1,0}是蓝色通道颜色矩阵的数组表达式，myColorMatrix.set(myColorArray)用于在使用该数组构建颜色矩阵与图像在画布上进行绘制时（即 myCanvas.drawBitmap(oldBmp,0,0,myPaint)），将使图像仅显示蓝色通道支持的部分。此实例的完整项目在 MyCode\MySample417 文件夹中。

103　使用 RenderScript 实现高斯算法模糊图像

此实例主要通过使用 RenderScript 和 ScriptIntrinsicBlur，实现以高斯模式模糊图像。当实例运行之后，单击"显示原始图像"按钮，则在屏幕上显示原始图像，如图 103.1 的左图所示；单击"高斯模糊图像"按钮,则显示经过模糊处理之后的图像,如图 103.1 的右图所示。

图 103.1

主要代码如下:

```java
//响应单击"高斯模糊图像"按钮
public void onClickmyBtn2(View v) {
    myBitmap = BitmapFactory.decodeResource(getResources(), R.mipmap.myimage1);
    Bitmap myBitmap = blurBitmap(this.myBitmap, this);
    myImageView.setImageBitmap(myBitmap);
}
public static Bitmap blurBitmap(Bitmap bitmap, Context context) {
    Bitmap outBitmap = Bitmap.createBitmap(bitmap.getWidth(),
                        bitmap.getHeight(),
                        Bitmap.Config.ARGB_8888);
    //初始化 Renderscript,这个类提供了 RenderScript 上下文,在创建其他 RS 类
    //之前必须要先创建这个类,以控制 RenderScript 的初始化、资源管理、释放
    RenderScript myRenderScript = RenderScript.create(context);
    ScriptIntrinsicBlur myBlurScript = ScriptIntrinsicBlur.create(
        myRenderScript,Element.U8_4(myRenderScript));         //创建高斯模糊对象
    //创建 Allocations,此类是将数据传递给 RenderScript 内核的主要方法,
    //并制定后备类型存储给定类型
    Allocation myIn = Allocation.createFromBitmap(myRenderScript, bitmap);
    Allocation myOut = Allocation.createFromBitmap(myRenderScript, outBitmap);
    //设定模糊度
    myBlurScript.setRadius(15.f);
    //执行 Renderscript
    myBlurScript.setInput(myIn);
    myBlurScript.forEach(myOut);
    myOut.copyTo(outBitmap);
    bitmap.recycle();
    myRenderScript.destroy();
    return outBitmap;
}
```

上面这段代码在 MyCode \ MySample661 \ app \ src \ main \ java \ com \ bin \ luo \ mysample \ MainActivity.java 文件中。此实例的完整项目在 MyCode\MySample661 文件夹中。

104 使用拉普拉斯模板实现图像的锐化特效

此实例主要实现了使用拉普拉斯模板对图像进行锐化特效处理。锐化图像实质上就是要突出图像中有关形体的边缘。所谓形体的边缘就是图像像素点的颜色值发生显著变化的地方，在图像的平淡区，这种颜色值的变化比较平缓，而在图像的边缘区域这种变化则相当明显。当实例运行之后，单击"显示原图"按钮，则在屏幕上显示原始图像，如图 104.1 的左图所示；单击"显示锐化图像"按钮，则将显示经过锐化处理之后的图像，如图 104.1 的右图所示。

图　104.1

主要代码如下：

```java
//响应单击"显示锐化图像"按钮
public void onClickmyBtn2(View v) {
    //第一次锐化图像
    Bitmap myShowBmp = getNewImage(myBitmap);
    //第二次锐化图像
    myShowBmp = getNewImage(myShowBmp);
    //显示锐化之后的图像
    myImageView.setImageBitmap(myShowBmp);
}
public Bitmap getNewImage(Bitmap myBitmap) {
    //定义拉普拉斯矩阵数组值
    int[] myLaplacian = new int[]{-1, -1, -1,
                                  -1,  9, -1,
                                  -1, -1, -1};
    int myWidth = myBitmap.getWidth();
    int myHeight = myBitmap.getHeight();
```

```
        int oldR = 0,oldG = 0,oldB = 0;
        int[] myPixels = new int[myWidth * myHeight];
        myBitmap.getPixels(myPixels, 0, myWidth, 0, 0, myWidth, myHeight);
        for (int i = 1; i < myHeight - 1; i++) {
         for (int j = 1; j < myWidth - 1; j++) {
          int myPixel = 0;
          int newR = 0;
          int newG = 0;
          int newB = 0;
          int myIndex = 0;
          for (int m = -1; m <= 1; m++) {
           for (int n = -1; n <= 1; n++) {
            myPixel = myPixels[(i + n) * myWidth + (j + m)];
            oldR = Color.red(myPixel);
            oldG = Color.green(myPixel);
            oldB = Color.blue(myPixel);
            newR = newR + oldR * myLaplacian[myIndex];
            newG = newG + oldG * myLaplacian[myIndex];
            newB = newB + oldB * myLaplacian[myIndex];
            myIndex++;
           } }
          newR = Math.min(255, Math.max(0, newR));
          newG = Math.min(255, Math.max(0, newG));
          newB = Math.min(255, Math.max(0, newB));
          myPixels[(i-1) * myWidth + (j-1)] = Color.argb(255,newR, newG, newB);
        } }
        //根据原始图像创建与其尺寸完全相同的新图像
        Bitmap newBmp = Bitmap.createBitmap(myWidth,
                              myHeight, Bitmap.Config.RGB_565);
        //根据新的像素值填充新图像
        newBmp.setPixels(myPixels, 0, myWidth, 0, 0, myWidth, myHeight);
        return newBmp;
    }
```

上面这段代码在 MyCode\MySample122\app\src\main\java\com\bin\luo\mysample\MainActivity.java 文件中。在这段代码中，int[] myLaplacian = new int[]{-1, -1, -1, -1, 9, -1, -1, -1, -1}即是拉普拉斯模板的数组表示方式。myBitmap.getPixels(myPixels, 0, myWidth, 0, 0, myWidth, myHeight)用于获取原始图像的像素值，并保存在数组 myPixels 中。newR = newR + oldR * myLaplacian[myIndex]、newG = newG + oldG * myLaplacian[myIndex]和 newB = newB + oldB * myLaplacian[myIndex]即是根据拉普拉斯模板对原像素的 RGB 分量进行数学运算从而产生新的像素以形成整幅图像的锐化效果。newR = Math.min(255, Math.max(0, newR))、newG = Math.min(255, Math.max(0, newG))和 newB = Math.min(255, Math.max(0, newB))用于处理 R、G、B 分量溢出的情况，即禁止 R、G、B 分量出现小于 0 或大于 255 的情况。newBmp.setPixels(myPixels, 0, myWidth, 0, 0, myWidth, myHeight)用于根据处理之后的像素值在空白图像中填充像素，即生成锐化图像。此实例的完整项目在 MyCode\MySample122 文件夹中。

105 通过像素操作实现在图像上添加光照效果

此实例主要通过使用 getPixels() 方法和 setPixels() 方法操作图像的像素,实现在图像上添加光照效果。当实例运行之后,单击"显示原图"按钮,则在屏幕上显示原始图像,如图 105.1 的左图所示;单击"显示光照图像"按钮,则将显示中心光照的图像,如图 105.1 的右图所示。

图 105.1

主要代码如下:

```
public void onClickmyBtn2(View v) {              //响应单击"显示光照图像"按钮
    myImageView.setImageBitmap(getNewImage(myBitmap));
}
public Bitmap getNewImage(Bitmap oldBmp) {
    final int myWidth = oldBmp.getWidth();
    final int myHeight = oldBmp.getHeight();
    int centerX = myWidth / 2;
    int centerY = myHeight / 2;
    int myRadius = Math.min(centerX, centerY);   //设置光照半径
    final float myLight = 150F;                  //设置光照强度
    int[] myPixels = new int[myWidth * myHeight];
    oldBmp.getPixels(myPixels, 0, myWidth, 0, 0, myWidth, myHeight);
    int myPos = 0;
    for (int i = 1, length = myHeight - 1; i < length; i++) {
        for (int j = 1, len = myWidth - 1; j < len; j++) {
            myPos = i * myWidth + j;
            int oldR = 0, oldG = 0, oldB = 0, myPixel = 0;
            int newR = 0, newG = 0, newB = 0;
            myPixel = myPixels[myPos];
            oldR = Color.red(myPixel);
            oldG = Color.green(myPixel);
```

```
            oldB = Color.blue(myPixel);
            newR = oldR;
            newG = oldG;
            newB = oldB;
            //计算当前点到中心的距离,在平面坐标系中求两点之间的距离
            int myLength = (int)(Math.pow((centerY - i), 2) + Math.pow(centerX - j, 2));
            if (myLength < myRadius * myRadius) {
             //按照距离大小计算增加的光照强度
             int myDifference = (int)(myLight * (1.0 - Math.sqrt(myLength) / myRadius));
             newR = oldR + myDifference;
             newG = oldG + myDifference;
             newB = oldB + myDifference;
            }
            newR = Math.min(255, Math.max(0, newR));
            newG = Math.min(255, Math.max(0, newG));
            newB = Math.min(255, Math.max(0, newB));
            myPixels[myPos] = Color.argb(255, newR, newG, newB);
        }}
        //根据原始图像创建与其尺寸完全相同的新图像
        Bitmap newBmp = Bitmap.createBitmap(myWidth, myHeight, Bitmap.Config.RGB_565);
        //根据新的像素值填充新图像
        newBmp.setPixels(myPixels, 0, myWidth, 0, 0, myWidth, myHeight);
        return newBmp;
    }
```

上面这段代码在 MyCode \ MySample123 \ app \ src \ main \ java \ com \ bin \ luo \ mysample \ MainActivity.java 文件中。在这段代码中,oldBmp.getPixels(myPixels,0,myWidth,0,0,myWidth,myHeight)用于获取原始图像的像素值,并保存在数组 myPixels 中。newBmp.setPixels(myPixels,0,myWidth,0,0,myWidth,myHeight)用于根据添加中心光照之后的像素值在空白图像中填充像素。newR = oldR + myDifference、newG = oldG + myDifference 和 newB = oldB + myDifference 用于给指定范围的像素添加光照。newR = Math.min(255,Math.max(0,newR))、newG = Math.min(255,Math.max(0,newG))和 newB = Math.min(255,Math.max(0,newB))用于处理 R、G、B 分量溢出。因为当应用上面的数学算法重置 R、G、B 分量时,很可能产生 R、G、B 分量大于 255 的情况;这种情况在像素值中是不可能存在的,所以一旦发生这种情况,则将其强行设置为最大值 255。此实例的完整项目在 MyCode\MySample123 文件夹中。

106 通过像素操作使彩色图像呈现浮雕特效

此实例主要通过使用 getPixels()方法和 setPixels()方法操作图像的像素,实现将彩色图像转化为浮雕图像。当实例运行之后,单击"显示原图"按钮,则在屏幕上显示原始图像,如图 106.1 的左图所示;单击"显示浮雕图像"按钮,则将显示浮雕效果的图像,如图 106.1 的右图所示。

主要代码如下:

```
public void onClickmyBtn2(View v) {                    //响应单击"显示浮雕图像"按钮
    int myWidth = myBitmap.getWidth();
    int myHeight = myBitmap.getHeight();
    int myPixel = 0, prePixel = 0, a, r, g, b;
    int r1, g1, b1;
```

```
int[] oldPixels = new int[myWidth * myHeight];
int[] newPixels = new int[myWidth * myHeight];
//获取原始图像每个点的像素值
myBitmap.getPixels(oldPixels, 0, myWidth, 0, 0, myWidth, myHeight);
for (int i = 1; i < oldPixels.length; i++) {
 prePixel = oldPixels[i - 1];
 a = Color.alpha(prePixel);
 r = Color.red(prePixel);
 g = Color.green(prePixel);
 b = Color.blue(prePixel);
 myPixel = oldPixels[i];
 r1 = Color.red(myPixel);
 g1 = Color.green(myPixel);
 b1 = Color.blue(myPixel);
 r = r1 - r + 127;
 g = g1 - g + 127;
 b = b1 - b + 127;                   //处理相邻两个像素点的RGB分量,使之产生浮雕效果
 newPixels[i] = Color.argb(a, r > 255 ? 255 : r,
     g > 255 ? 255 : g, b > 255 ? 255 : b);    //处理R、G、B值超过255情形
}
Bitmap newBmp = Bitmap.createBitmap(myWidth, myHeight,
    Bitmap.Config.ARGB_8888);              //根据原始图像创建与其尺寸完全相同的新图像
newBmp.setPixels(newPixels, 0, myWidth, 0, 0,
                 myWidth, myHeight);        //根据新的像素值填充新图像
myImageView.setImageBitmap(newBmp);
}
```

图 106.1

上面这段代码在 MyCode\MySample121\app\src\main\java\com\bin\luo\mysample\MainActivity.java 文件中。在这段代码中,myBitmap.getPixels(oldPixels, 0, myWidth, 0, 0, myWidth, myHeight)用于获取原始图像的像素值,并保存在数组中。r = r1 - r + 127、g = g1 - g + 127 和 b = b1 - b + 127 表示将相邻像素点的 R、G、B 分量相减后加 127,从而使整个图像产生浮雕效果。newPixels[i] = Color.argb(a, r > 255 ? 255 : r, g > 255 ? 255 : g, b > 255 ? 255 : b)

用于处理 R、G、B 分量溢出的情况。newBmp.setPixels(newPixels，0，myWidth，0，0，myWidth，myHeight)用于根据处理之后的像素值在空白图像中填充像素,即生成的浮雕图像。此实例的完整项目在 MyCode\MySample121 文件夹中。

107　使用 BitmapShader 实现文字线条图像化

此实例主要通过使用 BitmapShader 自定义 Paint 对象,实现使用图像填充文字线条。当实例运行之后,单击"图像文字"按钮,则使用图像填充文字线条,效果如图 107.1 的左图所示。单击"渐变文字"按钮,则使用线性渐变色填充文字线条的效果,如图 107.1 的右图所示。

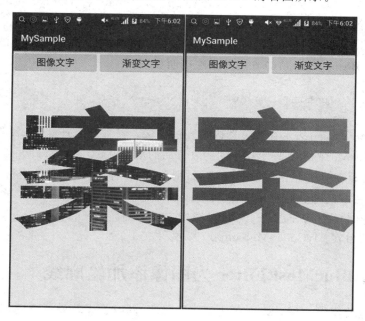

图　107.1

主要代码如下:

```
public void onClickBtn1(View v) {                    //响应单击"图像文字"按钮
    Bitmap myBackground =
            BitmapFactory.decodeResource(getResources(), R.mipmap.myimage1);
    drawTextWithEffect("案", myBackground, null);
}
public void onClickBtn2(View v) {                    //响应单击"渐变文字"按钮
    LinearGradient myGradient = new LinearGradient(0, 0, 1080, 0, Color.RED,
        Color.GREEN, Shader.TileMode.REPEAT);        //创建从红色至绿色的渐变对象
    drawTextWithEffect("案", null, myGradient);
}
public void drawTextWithEffect(String text,
                                Bitmap bitmap, LinearGradient gradient){
    Bitmap myBitmap = Bitmap.createBitmap(
            getWindowManager().getDefaultDisplay().getWidth(),
            getWindowManager().getDefaultDisplay().getHeight(),
            Bitmap.Config.ARGB_8888);
    Canvas myCanvas = new Canvas(myBitmap);
    Paint myPaint = new Paint();
    myPaint.setTextSize(1080);                       //设置字体大小
```

```
myPaint.setTypeface(Typeface.DEFAULT_BOLD);        //加粗文字
if (bitmap == null) {                              //使用渐变色填充文字
 myPaint.setShader(gradient);
} else if (gradient == null) {                     //使用图像填充文字
 myPaint.setShader(new BitmapShader(bitmap,
                   Shader.TileMode.REPEAT, Shader.TileMode.REPEAT));
}
myCanvas.drawText(text, 0, 1200, myPaint);
myImageView.setImageBitmap(myBitmap);
}
```

上面这段代码在 MyCode \ MySample790 \ app \ src \ main \ java \ com \ bin \ luo \ mysample \ MainActivity.java 文件中。在这段代码中，new BitmapShader（bitmap，Shader.TileMode.REPEAT，Shader.TileMode.REPEAT）用于创建位图渲染器，在 Paint 中使用该渲染器来实现使用图像填充文字线条。BitmapShader()方法的语法声明如下：

```
public BitmapShader(@NonNull Bitmap bitmap, TileMode tileX, TileMode tileY)
```

其中，参数 Bitmap bitmap 表示在渲染器内呈现的图像。参数 TileMode tileX 表示在位图的 x 方向的平铺模式。参数 TileMode tileY 表示在位图的 y 方向的平铺模式。TileMode 一共有如下 3 种类型。

（1）CLAMP，如果渲染器超出原始边界范围，则复制范围内边缘的颜色进行渲染。
（2）REPEAT，在横向和纵向重复图像，平铺。
（3）MIRROR，在横向和纵向重复图像，以镜像方式平铺。

此实例的完整项目在 MyCode\MySample790 文件夹中。

108　使用 BlurMaskFilter 为图像添加轮廓线

此实例主要通过使用 BlurMaskFilter 定制阴影画笔，同时为画笔设置颜色，实现为 png 格式的图像添加轮廓线。当实例运行之后，单击"添加特效"按钮，则在图像的轮廓上出现一条红色的线条，效果如图 108.1 的左图所示。单击"移除特效"按钮，则将显示原始图像，如图 108.1 的右图所示。

图　108.1

主要代码如下：

```java
public void onClickBtn1(View v) {                      //响应单击"添加特效"按钮
    Bitmap myBitmap =
            BitmapFactory.decodeResource(getResources(),R.mipmap.myimage1);
    //初始化外阴影遮罩过滤器
    BlurMaskFilter myFilter = new BlurMaskFilter(10,BlurMaskFilter.Blur.OUTER);
    Paint myPaint = new Paint();
    myPaint.setColor(Color.RED);                       //设置红色画笔
    myPaint.setMaskFilter(myFilter);                   //在画笔中应用模糊过滤器
    int[] myOffset = new int[2];
    Bitmap myImage = myBitmap.extractAlpha();          //生成图像底图
    Bitmap myImageCopy =
            myImage.copy(Bitmap.Config.ARGB_8888,true); //生成底图副本
    Canvas myCanvas = new Canvas(myImageCopy);         //根据该副本初始化画布
    myCanvas.drawBitmap(myImage,0,0,myPaint);          //绘制底图和阴影
    myCanvas.drawBitmap(myBitmap, - myOffset[0],
                                  - myOffset[1],null); //在底图上覆盖绘制原图
    myImageView.setImageBitmap(myImageCopy);           //显示特效
}
public void onClickBtn2(View v) {                      //响应单击"移除特效"按钮
    Bitmap myBitmap =
            BitmapFactory.decodeResource(getResources(), R.mipmap.myimage1);
    myImageView.setImageBitmap(myBitmap);
}
```

上面这段代码在 MyCode\MySample789\app\src\main\java\com\bin\luo\mysample\MainActivity.java 文件中。在这段代码中，BlurMaskFilter myFilter = new BlurMaskFilter（10，BlurMaskFilter.Blur.OUTER）用于创建半径为 10 的外阴影过滤器，由于外阴影半径较小，它实际看起来就像是一根线条。myPaint.setColor(Color.RED)用于设置画笔颜色，该颜色将作用于外阴影过滤器，因此可以据此设置不同的阴影颜色。此实例的完整项目在 MyCode\MySample789 文件夹中。

109　使用 PathDashPathEffect 实现椭圆线条

此实例主要通过使用 PathDashPathEffect 创建路径特效，实现使用椭圆填充线条。当实例运行之后，使用椭圆线条绘制的椭圆效果如图 109.1 所示。

主要代码如下：

```java
public class MainActivity extends Activity {
    @Override
    protected void onCreate(Bundle savedInstanceState) {
        setContentView(new MyView(this));
        super.onCreate(savedInstanceState);
    }
}
class MyView extends View {
    public MyView(Context context) { super(context); }
    @Override
    protected void onDraw(Canvas myCanvas) {
        Display myDisplay = getWindowManager().getDefaultDisplay();
        int myWidth  = myDisplay.getWidth();
        int myHeight = myDisplay.getHeight();
        Path myOval = new Path();
        myOval.addOval(0, 0, 46, 28, Path.Direction.CCW);
        PathEffect myEffect = new PathDashPathEffect(myOval,
```

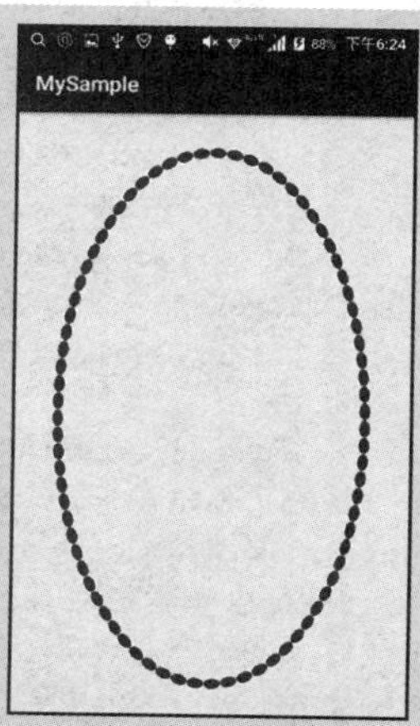

图　109.1

```
                        48,0,PathDashPathEffect.Style.MORPH);
    Paint myPaint = new Paint();
    myPaint.setColor(Color.BLUE);
    myPaint.setAntiAlias(true);
     myPaint.setPathEffect(myEffect);                             //设置椭圆路径特效
    myCanvas.drawOval(myWidth/10,myHeight/20,
          myWidth - myWidth/10,myHeight - myHeight/6,myPaint);    //绘制大椭圆
}}}
```

上面这段代码在 MyCode \ MySample103 \ app \ src \ main \ java \ com \ bin \ luo \ mysample \ MainActivity.java 文件中。在这段代码中，PathDashPathEffect（myOval，48，0，PathDashPathEffect.Style.MORPH）用于创建一个路径特效，myOval 表示路径中的图形（如椭圆、矩形等），48 表示路径中图形与图形之间的间距，0 表示首个图形的偏移量，PathDashPathEffect.Style.MORPH 表示图形的摆放风格，除了 MORPH 风格之外，还有 ROTATE 和 TRANSLATE 风格。myOval.addOval（0，0，46，28，Path.Direction.CCW）用于在路径中添加椭圆图形，（0，0，46，28）用于定义椭圆图形的大小，Path.Direction.CCW 表示逆时针方向，如果是 Path.Direction.CW，则表示顺时针方向。此实例的完整项目在 MyCode\MySample103 文件夹中。

110 使用 SumPathEffect 叠加多种路径特效

此实例主要通过使用 SumPathEffect 叠加两种路径特效，使图像的四周产生散射的毛刺。当实例运行之后，图像的四周产生散射的毛刺效果如图 110.1 所示。

主要代码如下：

```
public class MainActivity extends Activity {
 @Override
 protected void onCreate(Bundle savedInstanceState) {
  setContentView(new MyView(this));
  super.onCreate(savedInstanceState);
 }
 class MyView extends View {
  public MyView(Context context) { super(context); }
  @Override
  protected void onDraw(Canvas myCanvas) {
   Display myDisplay = getWindowManager().getDefaultDisplay();
   int myWidth = myDisplay.getWidth();
   int myHeight = myDisplay.getHeight();
   Bitmap myBitmap =
            BitmapFactory.decodeResource(getResources(), R.mipmap.
myimage);
   PathEffect myEffect1 = new DiscretePathEffect(5.0f, 70.0f);
   PathEffect myEffect2 = new DiscretePathEffect(1.0f, 50.0f);
   PathEffect myEffect = new SumPathEffect(myEffect1, myEffect2);
   Paint myPaint = new Paint();
   myPaint.setPathEffect(myEffect);
   myCanvas.drawOval(myWidth/10,myHeight/20,
       myWidth - myWidth/10, myHeight - myHeight/7,myPaint);    //绘制椭圆
   //以椭圆形裁剪图像
   myPaint.setXfermode(new PorterDuffXfermode(PorterDuff.Mode.ADD));
   myCanvas.drawBitmap(myBitmap, 0, 30, myPaint);               //绘制图像
}}}
```

图 110.1

上面这段代码在 MyCode\MySample104\app\src\main\java\com\bin\luo\mysample\MainActivity.java 文件中。在这段代码中,myEffect1 = new DiscretePathEffect(5.0f,70.0f)用于生成散列路径特效,5.0f 表示碎片长度,70.0f 表示偏移量。myEffect2 = new DiscretePathEffect(1.0f,50.0f)也用于生成散列路径特效,1.0f 表示碎片长度,50.0f 表示偏移量。myEffect = new SumPathEffect(myEffect1,myEffect2)用于叠加两种路径特效,myEffect1 表示第一种路径特效,myEffect2 表示第二种路径特效。myPaint.setPathEffect(myEffect)则表示在画笔中应用叠加的路径特效 myEffect。此实例的完整项目在 MyCode\MySample104 文件夹中。

111　通过 BitmapShader 实现以图像填充椭圆

此实例主要通过使用 BitmapShader 类,实现使用水平和垂直镜像图像填充椭圆内部。当实例运行之后,使用水平和垂直镜像图像填充椭圆内部的效果如图 111.1 所示。

图　111.1

主要代码如下:

```
public class MainActivity extends Activity {
 @Override
 protected void onCreate(Bundle savedInstanceState) {
  setContentView(new MyView(this));
  super.onCreate(savedInstanceState);
 }
 class MyView extends View {
  private BitmapShader myShader = null;
  private Bitmap myBitmap = null;
  private ShapeDrawable myDrawable = null;
  public MyView(Context context) {
   super(context);
```

```
    myBitmap = ((BitmapDrawable) getResources().getDrawable(
                    R.mipmap.myimage)).getBitmap();           //获取资源图像
    myShader = new BitmapShader(myBitmap,Shader.TileMode.MIRROR,
                    Shader.TileMode.MIRROR);                  //构造渲染器 BitmapShader
  }
  protected void onDraw(Canvas myCanvas) {
    super.onDraw(myCanvas);
    int myWidth = 1080,myHeight = 1820;
    myCanvas.drawColor(Color.BLACK);                          //设置背景颜色
    myDrawable = new ShapeDrawable(new OvalShape());          //构建椭圆 ShapeDrawable
    myDrawable.getPaint().setShader(myShader);                //获取画笔并设置渲染器
    myDrawable.setBounds(20, 20, myWidth - 20, myHeight - 150); //设置显示区域
    myDrawable.draw(myCanvas);                                //绘制 shapeDrawable
} } }
```

上面这段代码在 MyCode \ MySample093 \ app \ src \ main \ java \ com \ bin \ luo \ mysample \ MainActivity.java 文件中。在这段代码中，BitmapShader(myBitmap,Shader.TileMode.MIRROR, Shader.TileMode.MIRROR)用于创建一个位图渲染器 BitmapShader，该构造函数的语法声明如下：

```
public BitmapShader(Bitmap bitmap,Shader.TileMode tileX,Shader.TileMode tileY)
```

其中，参数 Bitmap bitmap 表示在渲染器内使用的位图。参数 Shader.TileMode tileX 表示位图在 x 方向的堆放模式。参数 Shader.TileMode tileY 表示位图在 y 方向的堆放模式。Shader.TileMode 堆放模式有下列 3 种：CLAMP、REPEAT、MIRROR。

此实例的完整项目在 MyCode\MySample093 文件夹中。

112　使用 ComposeShader 创建渐变图像

此实例主要通过使用 ComposeShader 叠加 BitmapShader 和 LinearGradient 两种着色器，实现以线性渐变透明的效果显示图像。当实例运行之后，单击"绿色渐变"按钮，则在图像的上面蒙上一层从透明至绿色渐变的遮罩层，如图 112.1 的左图所示；单击"白色渐变"按钮，则在图像的上面蒙上一层从透明至白色渐变的遮罩层，如图 112.1 的右图所示。单击其他按钮将会实现按钮标题所示的效果。

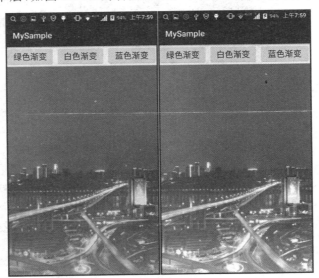

图　112.1

主要代码如下：

```java
public class MainActivity extends Activity {
    MyView myView;
    Bitmap myBitmap;
    Shader myBmpShader;
    @Override
    protected void onCreate(Bundle savedInstanceState) {
        super.onCreate(savedInstanceState);
        setContentView(R.layout.activity_main);
        myView = (MyView) findViewById(R.id.myView);
        Display myDisplay = getWindowManager().getDefaultDisplay();
        myView.myWidth = myDisplay.getWidth();
        myView.myHeight = myDisplay.getHeight();
        myBitmap = ((BitmapDrawable) getResources().getDrawable(
            R.mipmap.myimage)).getBitmap();                    //获取图像资源
        myBmpShader = new BitmapShader(myBitmap,
            Shader.TileMode.REPEAT, Shader.TileMode.REPEAT);   //创建 BitmapShader
        myView.myShader = myBmpShader;
    }
    public void onClickBtnGreen(View v) {                      //响应单击"绿色渐变"按钮
        //创建沿着透明色至绿色的方向进行线性渐变的 LinearGradient
        Shader myLinearGradientShader = new LinearGradient(0, 0, 0, myView.myHeight,
            new int[] {Color.TRANSPARENT, Color.GREEN}, null, Shader.TileMode.REPEAT);
        myView.myShader = new ComposeShader(myBmpShader, myLinearGradientShader,
            PorterDuff.Mode.ADD);                              //混合渲染,将线性透明渐变和图像叠加
        myView.postInvalidate();
    }
    public void onClickBtnWhite(View v) {                      //响应单击"白色渐变"按钮
        //创建沿着透明色至白色的方向进行线性渐变的 LinearGradient
        Shader myLinearGradientShader = new LinearGradient(0, 0, 0, myView.myHeight,
            new int[] {Color.TRANSPARENT, Color.WHITE}, null, Shader.TileMode.REPEAT);
        myView.myShader = new ComposeShader(myBmpShader,
            myLinearGradientShader, PorterDuff.Mode.ADD);
        myView.postInvalidate();
    }
    public void onClickBtnBlue(View v) {                       //响应单击"蓝色渐变"按钮
        //创建沿着透明色至蓝色的方向进行线性渐变的 LinearGradient
        Shader myLinearGradientShader = new LinearGradient(0, 0, 0, myView.myHeight,
            new int[] {Color.TRANSPARENT, Color.BLUE}, null, Shader.TileMode.REPEAT);
        myView.myShader = new ComposeShader(myBmpShader,
            myLinearGradientShader, PorterDuff.Mode.ADD);
        myView.postInvalidate();
    }
}
```

上面这段代码在 MyCode \ MySample097 \ app \ src \ main \ java \ com \ bin \ luo \ mysample \ MainActivity.java 文件中。在这段代码中，myBmpShader = new BitmapShader(myBitmap, Shader.TileMode.REPEAT, Shader.TileMode.REPEAT)用于根据图像资源创建 BitmapShader 图像着色器，myBitmap 表示图像，Shader.TileMode.REPEAT(前一个)表示图像在水平方向上的平铺模式，Shader.TileMode.REPEAT(后一个)表示图像在垂直方向上的平铺模式。myLinearGradientShader = new LinearGradient(0, 0, 0, myView.myHeight, new int[] {Color.TRANSPARENT, Color.BLUE}, null, Shader.TileMode.REPEAT)用于创建蓝色渐变透明的着色器，(0, 0)表示渐变开始位置的坐标，(0, myView.myHeight)表示渐变结束位置的坐标，{Color.TRANSPARENT, Color.BLUE}表示渐变的开始颜色是透明色、结束颜色是蓝色，null 表示渐变过程均匀分布，Shader.

TileMode. REPEAT 表示如果渐变不能填充目标对象，则用重复模式平铺渐变效果。myView. myShader = new ComposeShader(myBmpShader, myLinearGradientShader, PorterDuff. Mode. ADD)表示使用 myBmpShader 图像着色器和 myLinearGradientShader 线性渐变着色器以 PorterDuff. Mode. ADD 叠加的模式创建一个组合着色器。myView. postInvalidate()用于刷新 myView。myView 是自定义（控件）类 MyView 的实例，MyView 类的主要代码如下：

```
public class MyView extends View {
  public Shader myShader = null;
  public Paint myPaint = null;
  public int myWidth,myHeight;
  public MyView(Context context, AttributeSet attrs) {
    super(context, attrs);
    myPaint = new Paint();
  }
  @Override
  protected void onDraw(Canvas myCanvas) {
    myPaint.setShader(myShader);
    myCanvas.drawRect(0,0,myWidth,myHeight, myPaint);
  } }
```

上面这段代码在 MyCode\MySample097\app\src\main\java\com\bin\luo\mysample\ MyView.java 文件中。myPaint.setShader(myShader)用于在画笔中设置包括图像着色器和渐变透明着色器的组合着色器。myCanvas.drawRect(0,0,myWidth,myHeight,myPaint)表示使用组合着色器绘制矩形图像。此实例的完整项目在 MyCode\MySample097 文件夹中。

113　使用 ImageView 显示 XML 实现的矢量图形

此实例主要通过设置 ImageView 控件的 src 属性为 XML 格式的矢量图形文件，实现在 ImageView 控件中显示以矢量形式创建的桃心图形。当实例运行之后，矢量桃心图形的显示效果如图 113.1 所示。

图　113.1

主要代码如下：

```
<ImageView android:layout_width="wrap_content"
           android:layout_height="wrap_content"
           android:src="@drawable/myvectordata"
           android:layout_centerInParent="true"/>
```

上面这段代码在 MyCode\MySample238\app\src\main\res\layout\activity_main.xml 文件中。在这段代码中，android：src="@drawable/myvectordata"的 myvectordata 是一个 XML 格式的文件，该文件包含了创建矢量桃心图形的数据。myvectordata.xml 文件的主要内容如下：

```
<vector xmlns:android="http://schemas.android.com/apk/res/android"
        android:width="480dp"
        android:height="480dp"
        android:viewportWidth="480.0"
        android:viewportHeight="480.0">
<path android:strokeColor="#FF0000"
      android:strokeWidth="1"
      android:fillColor="#FF0000"
      android:pathData="M233.642 288 C249.9 265.728 318 231.266 318 195.613 C318 133.896 247.069 148.581 232.993 172.957 C218.041 147.832 149 137.805 149 195.613 C149 232.135 219.429 265.911 233.642 288 z"/>
</vector>
```

上面这段代码在 MyCode\MySample238\app\src\main\res\drawable\myvectordata.xml 文件中。在这段代码中，矢量桃心图形数据主要在<path>标签中定义，线条粗细、颜色等属性也在此设置。此实例的完整项目在 MyCode\MySample238 文件夹中。

114 使用 BitmapFactory 压缩图像的大小

此实例主要通过在 BitmapFactory 类的 decodeFile()方法中设置 RGB_565 参数属性，实现使用 RGB_565 法压缩图像大小。当实例运行之后，在"图像文件："输入框中输入图像文件的路径，单击"显示压缩前图像"按钮，则在弹出的 Toast 中显示压缩前的图像大小和宽高，如图 114.1 的左图所示。单击"显示压缩后图像"按钮，则在弹出的 Toast 中显示压缩后的图像大小和宽高，如图 114.1 的右图所示。很明显，压缩前后的图像质量基本不变，但是大小几乎减小一半。需要注意的是，在测试时，请确保在 SD 卡上有 myimg.png 图像文件。

主要代码如下：

```
public void onClickmyBtn1(View v) {                    //响应单击"显示压缩前图像"按钮
    BitmapFactory.Options myOptions = new BitmapFactory.Options();
    myOptions.inPreferredConfig = Bitmap.Config.ARGB_8888;
    Bitmap myBmp =
            BitmapFactory.decodeFile(myFile.getText().toString(), myOptions);
    myImage.setImageBitmap(myBmp);
    String myInfo = "压缩前的图像大小" + (myBmp.getByteCount() / 1024 / 1024)
            + "M,\n 宽度为" + myBmp.getWidth() + ",\n 高度为" + myBmp.getHeight();
    Toast.makeText(MainActivity.this, myInfo, Toast.LENGTH_SHORT).show();
}
public void onClickmyBtn2(View v) {                    //响应单击"显示压缩后图像"按钮
```

```
BitmapFactory.Options myOptions = new BitmapFactory.Options();
myOptions.inPreferredConfig = Bitmap.Config.RGB_565;
Bitmap myBmp = 
        BitmapFactory.decodeFile(myFile.getText().toString(), myOptions);
myImage.setImageBitmap(myBmp);
String myInfo = "压缩后的图像大小" + (myBmp.getByteCount() / 1024 / 1024)
        + "M,\n 宽度为" + myBmp.getWidth() + ",\n 高度为" + myBmp.getHeight();
Toast.makeText(MainActivity.this, myInfo, Toast.LENGTH_SHORT).show();
}
```

图 114.1

上面这段代码在 MyCode \ MySample234 \ app \ src \ main \ java \ com \ bin \ luo \ mysample \ MainActivity.java 文件中。在这段代码中，myBmp = BitmapFactory.decodeFile(myFile.getText().toString()，myOptions)用于将指定路径的文件根据指定的参数 myOptions 解码为一个位图对象，在参数 myOptions 的 inPreferredConfig 属性中，如果设置该属性值为 Bitmap.Config.ARGB_8888，则将保持原始图像，如果设置该属性值为 Bitmap.Config.RGB_565，则按照 RGB_565 压缩图像。此外，在 SD 卡读写文件需要在 AndroidManifest.xml 文件中添加权限< uses－permission android：name＝"android.permission.READ_EXTERNAL_STORAGE"/>和< uses－permission android：name＝"android.permission.WRITE_EXTERNAL_STORAGE"/>。此实例的完整项目在 MyCode \MySample234 文件夹中。

115　在自定义类中使用 Movie 显示动态图像

此实例主要通过在自定义类 MyGifView 中使用 Movie 类的相关方法，实现显示 GIF 动态图像。当实例运行之后，单击"开始播放"按钮，则显示 GIF 动态图像；单击"停止播放"按钮，则显示 GIF 静态图像，效果分别如图 115.1 的左图和右图所示。

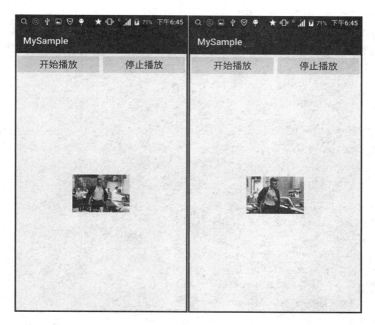

图 115.1

主要代码如下：

```
public void onClickmyBtn1(View v) {              //响应单击"开始播放"按钮
    myImageView.setVisibility(View.GONE);
    myGIFImage.setVisibility(View.VISIBLE);
}
public void onClickmyBtn2(View v) {              //响应单击"停止播放"按钮
    myImageView.setVisibility(View.VISIBLE);
    myGIFImage.setVisibility(View.GONE);
}
```

上面这段代码在 MyCode \ MySample687 \ app \ src \ main \ java \ com \ bin \ luo \ mysample \ MainActivity.java 文件中。在这段代码中，myImageView.setVisibility(View.VISIBLE)用于显示 GIF 静态图像，该静态图像通常是 GIF 动态图像的首幅图像。myGIFImage.setVisibility(View.GONE)用于隐藏静态图像。myGIFImage 是自定义类 MyGifView 的实例，自定义类 MyGifView 的主要代码如下：

```
public class MyGifView extends ImageView {
    private Movie myGif;
    private long myGifStartTime;
    private int myWidth;
    private int myHeight;
    public MyGifView(Context context) { super(context); }
    public MyGifView(Context context, AttributeSet attrs) {
        super(context, attrs);
        loadGifImage(R.raw.myimage1);
    }
    private void loadGifImage(int resourceId) {
        InputStream myInputStream = getResources().openRawResource(resourceId);
        myGif = Movie.decodeStream(myInputStream);
```

```java
    if (myGif != null) {
     Bitmap myBitmap = BitmapFactory.decodeStream(myInputStream);
     myWidth = myBitmap.getWidth();
     myHeight = myBitmap.getHeight();
     myBitmap.recycle();
    } }
    @Override
    protected void onDraw(Canvas canvas) {
     showGifImage(canvas);
     invalidate();
    }
    @Override
    protected void onMeasure(int widthMeasureSpec, int heightMeasureSpec) {
     super.onMeasure(widthMeasureSpec, heightMeasureSpec);
    }
    private boolean showGifImage(Canvas canvas) {
     long now = SystemClock.uptimeMillis();
     if (myGifStartTime == 0) { myGifStartTime = now; }
     int myDuration = myGif.duration();
     if (myDuration == 0) { myDuration = 100; }
     int myRelativeTime = (int) ((now - myGifStartTime) % myDuration);
     myGif.setTime(myRelativeTime);
     myGif.draw(canvas, (getWidth() - myWidth) / 2, (getHeight() - myHeight) / 2);
     if ((now - myGifStartTime) >= myDuration) {
      myGifStartTime = 0;
      return true;
     }
     return false;
    } }
```

上面这段代码在 MyCode\MySample687\app\src\main\java\com\bin\luo\mysample\MyGifView.java 文件中。需要说明的是,如果在编译安装时出现错误,问题可能出在硬件加速上,因此需要在 AndroidManifest.xml 文件中关闭硬件加速,即设置<activity android：name=".MainActivity" android：hardwareAccelerated="false">。此实例的完整项目在 MyCode\MySample687 文件夹中。

116　通过使用图像作为画布创建带水印图像

此实例主要通过使用 Bitmap 对象创建画布,实现在图像上添加水印文字。当实例运行之后,显示的图像没有水印文字,如图 116.1 的左图所示。在"水印文字："输入框中输入内容,如"纽约之夜",单击"添加水印文字"按钮,则在图像上显示添加的文字,如图 116.1 的右图所示。

主要代码如下：

```java
public void onClickBtn1(View v) {                    //响应单击"添加水印文字"按钮
    Bitmap myOldBmp = ((BitmapDrawable) myImageView.getDrawable()).getBitmap();
    Bitmap myNewBmp = drawTextToCenter(this,
            myOldBmp, myEditText.getText().toString(),240,Color.YELLOW);
    myImageView.setImageBitmap(myNewBmp);
}
```

```
public static Bitmap drawTextToCenter(Context context,
                    Bitmap bitmap, String text,int size, int color) {
    Paint myPaint = new Paint(Paint.ANTI_ALIAS_FLAG);
    myPaint.setColor(color);
    myPaint.setTextSize(size);
    Rect myRect = new Rect();
    myPaint.getTextBounds(text, 0, text.length(), myRect);
    return drawTextToBitmap(context, bitmap, text, myPaint, myRect,
                    (bitmap.getWidth() - myRect.width()) / 2,
                    (bitmap.getHeight() + myRect.height()) / 2);
}
private static Bitmap drawTextToBitmap(Context context, Bitmap bitmap,
        String text,Paint paint, Rect bounds, int paddingLeft, int paddingTop) {
    Bitmap.Config myConfig = bitmap.getConfig();
    paint.setDither(true);
    paint.setFilterBitmap(true);
    bitmap = bitmap.copy(myConfig, true);
    Canvas canvas = new Canvas(bitmap);
    paint.setStyle(Paint.Style.STROKE);            //创建空心文字
    paint.setStrokeWidth(1);                       //设置文字线条宽度
    canvas.drawText(text, paddingLeft, paddingTop, paint);
    return bitmap;
}
```

图 116.1

上面这段代码在 MyCode \ MySample457 \ app \ src \ main \ java \ com \ bin \ luo \ mysample \ MainActivity.java 文件中。在这段代码中，drawTextToCenter()方法用于在图像中央绘制水印文字。drawTextToBitmap()方法用于在图像的指定位置绘制水印文字。canvas = new Canvas(bitmap)表示使用 Bitmap 创建 Canvas。paint.setStyle(Paint.Style.STROKE)用于创建空心风格的 Paint。此实例的完整项目在 MyCode\MySample457 文件夹中。

117　通过操作根布局实现将屏幕内容保存为图像

此实例主要通过使用 View 的 draw() 方法,实现将屏幕内容保存为一幅图像。当实例运行之后,单击"将屏幕内容保存为一幅图像"按钮,则将把当前屏幕内容(通知栏除外)作为一幅图像保存在 SD 卡的根目录下,如图 117.1 的左图所示。在 SD 卡根目录下的截屏图像文件名是"ScreenImage.png",如图 117.1 的右图所示。

图　117.1

主要代码如下:

```
public void viewSaveToImage(View view) {                    //将图像保存为文件
    Bitmap myBmp = loadBitmapFromView(view);                // 把一个 View 转换成图像
    FileOutputStream myFileStream;
    try {
        boolean ismySDCard = Environment.getExternalStorageState().equals
(android.os.Environment.MEDIA_MOUNTED);                     //判断手机是否已经安装了 SD 卡
        if (ismySDCard) {
            File myRoot =
                    Environment.getExternalStorageDirectory(); //获取 SD 卡根目录
            File myFile = new File(myRoot, "ScreenImage.png");
            myFileStream = new FileOutputStream(myFile);
        } else throw new Exception("创建文件失败!");
        myBmp.compress(Bitmap.CompressFormat.PNG, 90, myFileStream);
        myFileStream.flush();
        myFileStream.close();
    } catch (Exception e) { e.printStackTrace(); }
    view.destroyDrawingCache();
}
private Bitmap loadBitmapFromView(View myView) {            //将 View 转换成 Bitmap
    int myWidth = myView.getWidth();
```

```
    int myHeight = myView.getHeight();
    Bitmap myBmp = Bitmap.createBitmap(myWidth, myHeight, Bitmap.Config.ARGB_8888);
    Canvas myCanvas = new Canvas(myBmp);
    myCanvas.drawColor(Color.WHITE);
    myView.layout(0, 0, myWidth, myHeight);
    myView.draw(myCanvas);
    return myBmp;
}
public void onClickmyBtn1(View v) {                    //响应单击"将屏幕内容保存为一幅图像"按钮
    View myView = myLayout.getRootView();
    viewSaveToImage(myView);
    Toast.makeText(getApplicationContext(), "成功将屏幕内容保存为一幅图像,请查看在存储卡上的
ScreenImage.png 文件",Toast.LENGTH_SHORT).show();
}
```

上面这段代码在 MyCode\MySample459\app\src\main\java\com\bin\luo\mysample\MainActivity.java 文件中。在这段代码中,myView = myLayout.getRootView()用于取得 RelativeLayout 布局控件的 View 对象。myView.draw(myCanvas)用于将 RelativeLayout 布局控件上的所有内容绘制到 myCanvas 中,这里实际上是图像;当 RelativeLayout 充满整个屏幕的时候,就相当于除通知栏之外的截屏。此外,在 SD 卡上写入文件需要在 AndroidManifest.xml 文件中添加<uses-permission android:name="android.permission.WRITE_EXTERNAL_STORAGE"/>权限。此实例的完整项目在 MyCode\MySample459 文件夹中。

118　通过手势变化实现平移旋转缩放图像

此实例主要通过设置 ImageView 控件的 android:scaleType="matrix",然后在 ImageView 控件的 setOnTouchListener()方法中监听手势的平移、旋转、缩放等动作,并将其应用于 Matrix 矩阵,即通过平移、旋转、缩放矩阵的方式控制 ImageView 控件的图像实现相应的变化。当实例运行之后,在 ImageView 控件的图像上使用两个手指执行平移、旋转、缩放等动作,即可平移、旋转、缩放图像,效果分别如图 118.1 的左图和右图所示。

图　118.1

主要代码如下：

```java
public class MainActivity extends Activity {
    //初始化起始位置,两指间距,旋转角度等参数
    float startX = 0, startY = 0, startDistance = 1f, startRotation = 0;
    PointF middlePoint = new PointF();
    Matrix myMatrix = new Matrix(), myCurrentMatrix =
            new Matrix(), myPrepareMatrix = new Matrix();    //初始化矩阵对象
    private static final int NONE = 0, DRAG = 1, ZOOM = 2;    //手势模式
    int MODE = NONE;
    int myWidth, myHeight;
    Bitmap myBitmap;
    @Override
    protected void onCreate(Bundle savedInstanceState) {
        super.onCreate(savedInstanceState);
        setContentView(R.layout.activity_main);
        final ImageView myImageView = (ImageView) findViewById(R.id.myImageView);
        myBitmap = BitmapFactory.decodeResource(getResources(), R.mipmap.myimage);
        DisplayMetrics myMetrics = new DisplayMetrics();
        getWindowManager().getDefaultDisplay().getMetrics(myMetrics);
        myWidth = myMetrics.widthPixels;
        myHeight = myMetrics.heightPixels;
        myImageView.setOnTouchListener(new View.OnTouchListener() {
            @Override
            public boolean onTouch(View view, MotionEvent motionEvent) {
                if ((motionEvent.getAction() &
                        MotionEvent.ACTION_MASK) == MotionEvent.ACTION_DOWN) {
                                                            //拖曳操作,如果取消,则无法平移图像
                    MODE = DRAG;
                    startX = motionEvent.getX();
                    startY = motionEvent.getY();            //记录手势起始位置
                    myPrepareMatrix.set(myMatrix);
                } else if ((motionEvent.getAction() &
                        MotionEvent.ACTION_MASK) == MotionEvent.ACTION_MOVE) {
                    if (MODE == ZOOM) {                     //缩放操作
                        myCurrentMatrix.set(myPrepareMatrix);
                        float myRotation = calc_Rotation(motionEvent) - startRotation;
                        float myDistance = calc_Distance(motionEvent);
                        float myScale = myDistance / startDistance;
                        myCurrentMatrix.postScale(myScale, myScale,
                                middlePoint.x, middlePoint.y);    //缩放图像(矩阵)
                        myCurrentMatrix.postRotate(myRotation,
                                middlePoint.x, middlePoint.y);    //旋转图像(矩阵)
                        myMatrix.set(myCurrentMatrix);
                    } else if (MODE == DRAG) {              //执行拖曳平移操作
                        myCurrentMatrix.set(myPrepareMatrix);
                        myCurrentMatrix.postTranslate(motionEvent.getX() - startX,
                                motionEvent.getY() - startY);
                        myMatrix.set(myCurrentMatrix);
                    }
                } else if ((motionEvent.getAction() &
                        MotionEvent.ACTION_MASK) == MotionEvent.ACTION_UP) {
                    MODE = NONE;
                } else if ((motionEvent.getAction() &
```

```
                MotionEvent.ACTION_MASK) == MotionEvent.ACTION_POINTER_DOWN) {
            MODE = ZOOM;
            startDistance = calc_Distance(motionEvent);        //重新计算两手指间距离
            startRotation = calc_Rotation(motionEvent);        //重新计算旋转角度
            myPrepareMatrix.set(myImageView.getImageMatrix());
            calc_MiddlePoint(middlePoint, motionEvent);        //重新计算中心点
        } else if ((motionEvent.getAction() &
                MotionEvent.ACTION_MASK) == MotionEvent.ACTION_POINTER_UP) {
            MODE = NONE;
        }
        //根据平移、旋转、缩放之后的矩阵来平移、旋转、缩放图像
        myImageView.setImageMatrix(myMatrix);
        return true;
    } }); }
    private float calc_Distance(MotionEvent event) {
        float x = event.getX(0) - event.getX(1);
        float y = event.getY(0) - event.getY(1);
        return (float) Math.sqrt(x * x + y * y);               //计算两个触点间的距离
    }
    private void calc_MiddlePoint(PointF point, MotionEvent event) {
        float x = event.getX(0) + event.getX(1);
        float y = event.getY(0) + event.getY(1);
        point.set(x / 2, y / 2);                               //计算旋转手势中心点
    }
    private float calc_Rotation(MotionEvent event) {
        double delta_x = (event.getX(0) - event.getX(1));
        double delta_y = (event.getY(0) - event.getY(1));
        double radius = Math.atan2(delta_y, delta_x);
        return (float) Math.toDegrees(radius);                 //计算旋转角度
    }
```

上面这段代码在 MyCode \ MySample733 \ app \ src \ main \ java \ com \ bin \ luo \ mysample \ MainActivity.java 文件中。此实例的完整项目在 MyCode\MySample733 文件夹中。

119 使用 ThumbnailUtils 提取大图像的缩略图

此实例主要通过使用 ThumbnailUtils 的 extractThumbnail()方法,实现从一幅大图像中提取对应的缩略图。当实例运行之后,单击"显示原图"按钮,则将显示大图像,如图 119.1 的左图所示。单击"显示缩略图"按钮,则将显示大图像的缩略图,如图 119.1 的右图所示。

主要代码如下:

```
public void onClickBtn1(View v) {                              //响应单击"显示原图"按钮
    myImageView.setImageBitmap(myBitmap);
}
public void onClickBtn2(View v) {                              //响应单击"显示缩略图"按钮
    Bitmap smallBmp = ThumbnailUtils.extractThumbnail(this.myBitmap, 150, 150);
    myImageView.setImageBitmap(smallBmp);
}
```

上面这段代码在 MyCode \ MySample107 \ app \ src \ main \ java \ com \ bin \ luo \ mysample \ MainActivity.java 文件中。在这段代码中,smallBmp = ThumbnailUtils.extractThumbnail(this.myBitmap,150,150)用于提取缩略图,smallBmp 表示缩略图,myBitmap 表示大图像,150(前一个)

图 119.1

表示缩略图的宽度,150(后一个)表示缩略图的高度。此实例的完整项目在 MyCode\MySample107 文件夹中。

120　通过采用取模的方式实现轮流显示多幅图像

此实例主要通过对数组索引取模,实现轮流显示多幅图像。当实例运行之后,默认将显示第一幅图像,如图 120.1 的左图所示。单击图像,则将显示下一幅图像。图 120.1 右图所示的图像是第四幅图像,它是图库的最后一幅图像,单击此幅图像则将显示第一幅图像,即轮流显示。

图　120.1

主要代码如下：

```java
public class MainActivity extends Activity {
  int[] myImages = new int[]{ R.mipmap.cpp, R.mipmap.java,
                              R.mipmap.spark, R.mipmap.jee};    //定义图像数组
  int myCurrent = 0;
  @Override
  protected void onCreate(Bundle savedInstanceState) {
    super.onCreate(savedInstanceState);
    LinearLayout myLinearLayout = new LinearLayout(this);        //创建线性布局管理器
    super.setContentView(myLinearLayout);                        //加载线性布局管理器
    //宽度和高度匹配父窗口
    ViewGroup.LayoutParams myLayoutParams = myLinearLayout.getLayoutParams();
    myLayoutParams.width = LinearLayout.LayoutParams.MATCH_PARENT;
    myLayoutParams.height = LinearLayout.LayoutParams.MATCH_PARENT;
    myLinearLayout.setLayoutParams(myLayoutParams);
    final ImageView myImageView = new ImageView(this);           //创建 ImageView 控件
    myImageView.setLayoutParams(myLayoutParams);                 //宽度和高度匹配父窗口
    myLinearLayout.addView(myImageView);                         //将 ImageView 控件添加到 LinearLayout
    myImageView.setImageResource(myImages[0]);                   //初始化时显示第一幅图像
    myImageView.setScaleType(ImageView.ScaleType.CENTER);        //居中显示图像
    //添加 Click 单击事件响应显示下一幅图像
    myImageView.setOnClickListener(new View.OnClickListener() {
      @Override
      public void onClick(View v) {                              //轮流显示 myImages 数组的图像
        myImageView.setImageResource(myImages[++myCurrent % myImages.length]);
} }); } }
```

上面这段代码在 MyCode\MySample002\app\src\main\java\com\bin\luo\mysample\MainActivity.java 文件中。在这段代码中，myImages 数组中的元素：R.mipmap.cpp，R.mipmap.java，R.mipmap.spark，R.mipmap.jee 等是在添加图像资源时，Android Studio 自动添加的，需要注意，图像文件名一定不要大写。myImageView.setImageResource（myImages[＋＋myCurrent % myImages.length]）则以取模的方式控制数组索引，从而实现多幅图像周而复始地轮流显示。例如：9％4，余数为1，因此显示 myImages[1] 图像；5％4，余数为1，也显示 myImages[1] 图像；即无论 myCurrent 怎样增加，myImages 数组总有对应的图像显示。大多数情况下，使用条件语句判断是否到达数组的末端，但是通过取模（取余数）运算，则简化和省略了大量的代码，并且运行更加高效。此实例的完整项目在 MyCode\ MySample002 文件夹中。

第5章

动画

121 使用 ObjectAnimator 创建上下振动动画

此实例主要通过使用 PropertyValuesHolder 创建多个平移和旋转等动画,并使用 ObjectAnimator 类的 ofPropertyValuesHolder() 方法组合这些动画,从而产生振动效果。当实例运行之后,单击"开始播放动画"按钮,则图像(电话机)将不停地上下振动,效果分别如图 121.1 的左图和右图所示。

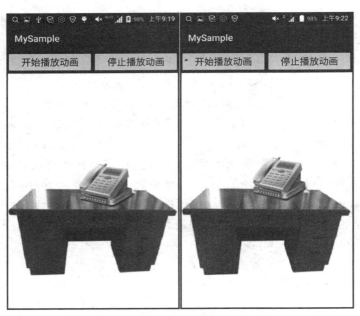

图　121.1

主要代码如下:

```
public class MainActivity extends Activity {
    ImageView myImageView;
    ObjectAnimator myAnimator;
    @Override
    protected void onCreate(Bundle savedInstanceState) {
```

```
super.onCreate(savedInstanceState);
setContentView(R.layout.activity_main);
myImageView = (ImageView) findViewById(R.id.myImageView);
PropertyValuesHolder myHolder1 =
                PropertyValuesHolder.ofFloat("y", 540, 550);
PropertyValuesHolder myHolder2 =
                PropertyValuesHolder.ofFloat("rotationY", 0, 25);
PropertyValuesHolder myHolder3 =
                PropertyValuesHolder.ofFloat("rotationX", 0, 15);
myAnimator = ObjectAnimator.ofPropertyValuesHolder(myImageView,
                        myHolder1,myHolder2,myHolder3);
myAnimator.setRepeatCount(ObjectAnimator.INFINITE);
myAnimator.setRepeatMode(ObjectAnimator.REVERSE);
myAnimator.setDuration(90);
Button myBtn1 = (Button) findViewById(R.id.myBtn1);
//响应单击"开始播放动画"按钮
myBtn1.setOnClickListener(new View.OnClickListener() {
 public void onClick(View arg0) { myAnimator.start(); } });
Button myBtn2 = (Button) findViewById(R.id.myBtn2);
//响应单击"停止播放动画"按钮
myBtn2.setOnClickListener( new View.OnClickListener() {
 public void onClick(View arg0) { myAnimator.cancel(); } });
} }
```

上面这段代码在 MyCode\MySample173\app\src\main\java\com\bin\luo\mysample\MainActivity.java 文件中。在这段代码中，myHolder1 = PropertyValuesHolder.ofFloat("y",540,550)用于创建控件的 y 坐标从 540 改变到 550 的动画。myHolder2 = PropertyValuesHolder.ofFloat("rotationY",0,25)用于创建控件围绕 y 轴从 0 度旋转到 25 度的动画。myHolder3 = PropertyValuesHolder.ofFloat("rotationX",0,15)用于创建控件围绕 x 轴从 0 度旋转到 15 度的动画。myAnimator = ObjectAnimator.ofPropertyValuesHolder(myImageView，myHolder1，myHolder2，myHolder3)用于将上述 3 种动画添加到 myImageView 控件上，并且 3 个动画同时执行。此实例的完整项目在 MyCode\MySample173 文件夹中。

122　使用 ObjectAnimator 实现沿弧线路径平移

此实例主要通过使用 Path 的 arcTo()方法创建弧线路径，并指定该路径为 ObjectAnimator 动画的平移路径，实现 ImageView 控件沿着弧线平移的效果。当实例运行之后，单击"开始播放动画"按钮，则 ImageView 控件(妖怪图像)将沿着细实线指示的弧线路径平移，效果分别如图 122.1 的左图和右图所示。

主要代码如下：

```
//响应单击"开始播放动画"按钮
public void onClickButton(View v) {
  myImageView.setVisibility(View.VISIBLE);
  Path myPath = new Path();
  RectF myRect = new RectF(50, 50, 700, 700);
  myPath.arcTo(myRect, 600, 600);           //创建圆弧路径
  ObjectAnimator myAnimator =
            ObjectAnimator.ofFloat(myImageView, View.X, View.Y, myPath);
```

```
    Path myPathInterpolator = new Path();
    myPathInterpolator.lineTo(0.6f, 0.9f);
    myPathInterpolator.lineTo(0.75f, 0.2f);
    myPathInterpolator.lineTo(1f, 1f);
    myAnimator.setInterpolator(new PathInterpolator(myPathInterpolator));
    myAnimator.setDuration(5000);
    myAnimator.start();
}
```

图 122.1

上面这段代码在 MyCode\MySample566\app\src\main\java\com\bin\luo\mysample\MainActivity.java 文件中。在这段代码中，myAnimator = ObjectAnimator.ofFloat(myImageView，View.X，View.Y，myPath)用于为 myImageView 控件创建一个平移动画，它使 myImageView 控件在指定的时间内在 myPath 指定的弧线路径上平移。此实例的完整项目在 MyCode\MySample566 文件夹中。

123 使用 ObjectAnimator 滚动显示多幅图像

此实例主要通过使用 Handler 定时调用 ObjectAnimator 动画，实现轮播多幅图像。当实例运行之后，将从右向左依次轮播多幅图像（电影海报），效果分别如图 123.1 的左图和右图所示。

主要代码如下：

```
public class MainActivity extends Activity {
    int[] myImages = {R.mipmap.myimage1, R.mipmap.myimage2,
                    R.mipmap.myimage3, R.mipmap.myimage4, R.mipmap.myimage5};
    Handler myHandler = new Handler();
    int myCurrentIndex = 0;
    @Override
    protected void onCreate(Bundle savedInstanceState) {
        super.onCreate(savedInstanceState);
        setContentView(R.layout.activity_main);
        final HorizontalScrollView myScrollView =
```

```
                    (HorizontalScrollView) findViewById(R.id.myScrollView);
final int myWidth = getWindowManager().getDefaultDisplay().getWidth();
int myHeight = getWindowManager().getDefaultDisplay().getHeight();
final LinearLayout myLayout = (LinearLayout) findViewById(R.id.myLayout);
for (int i = 0; i < myImages.length; i++) {
    ImageView myImageView = new ImageView(MainActivity.this);
    myImageView.setScaleType(ImageView.ScaleType.FIT_XY);
    myImageView.setImageResource(myImages[i]);
    myImageView.setLayoutParams(
                    new RelativeLayout.LayoutParams(myWidth, myHeight));
    myLayout.addView(myImageView);
}
myHandler.postDelayed(new Runnable() {
    @Override
    public void run() {
        if (myCurrentIndex >= myImages.length) myCurrentIndex = 0;
        ObjectAnimator myObjectAnimator = ObjectAnimator.ofInt(myLayout,
"scrollX",myLayout.getScrollX(), myWidth * myCurrentIndex++);
        myObjectAnimator.setStartDelay(1000);       //设置图像停滞显示时长
        myObjectAnimator.setDuration(1000);         //设置平移过渡动画时长
        myObjectAnimator.start();                   //启动动画
        myHandler.postDelayed(this, 2000);
                                                    //使用 Handler 发送消息
    } }, 1000);
} }
```

图 123.1

上面这段代码在 MyCode\MySample841\app\src\main\java\com\bin\luo\mysample\ MainActivity.java 文件中。在这段代码中，myObjectAnimator = ObjectAnimator.ofInt（myLayout," scrollX"，myLayout.getScrollX()，myWidth * myCurrentIndex++）表示将 myLayout 的 scrollX 属性值在指定的时间内(1000ms)从 myLayout.getScrollX()改变成 myWidth * myCurrentIndex++，即每次向左移动一个屏幕宽度。ObjectAnimator.ofInt()方法的语法声明如下：

```
public static ObjectAnimator ofInt(Object target,
                    String propertyName, int... values)
```

其中,参数 Object target 指定该动画要操作的是哪个控件。参数 String propertyName 指定该动画要操作该控件的哪个属性。参数 int... values 是可变长参数,与 ValueAnimator 的可变长参数的意义一样,即指这个属性值是从什么改变成什么。

此实例的完整项目在 MyCode\MySample841 文件夹中。

124　使用 ObjectAnimator 实现图形数字形变

此实例主要通过在 ObjectAnimator 动画中改变矢量图形数字的 pathData 属性值,实现矢量图形数字 5 在单击之后执行收缩挤压等动画效果后变成 6。当实例运行之后,将显示矢量图形数字 5,如图 124.1 的左图所示;单击屏幕,则执行收缩挤压等动画变成 6,如图 124.1 的右图所示。

图　124.1

主要代码如下:

```
public void onClickImageView(View v){                    //响应单击 ImageView 控件
  AnimatedVectorDrawable myDrawable =
            (AnimatedVectorDrawable) myImageView.getDrawable();
  if(!bChecked) myDrawable.start();
  else myDrawable.stop();
  bChecked = !bChecked;
}
```

上面这段代码在 MyCode\MySample626\app\src\main\java\com\bin\luo\mysample\MainActivity.java 文件中。在这段代码中,myDrawable.start()表示启动 ImageView 控件的动画,myDrawable.stop()表示停止 ImageView 控件的动画,因此 ImageView 控件的动画不能直接为纯粹的图像资源,而应是 XML 文件。如 android:src="@drawable/ mytransition",表示在 ImageView 控

件上加载的 XML 动画 mytransition。mytransition 动画的主要内容如下：

```xml
<animated-vector xmlns:android="http://schemas.android.com/apk/res/android"
                 android:drawable="@drawable/myvector">
<target android:name="myPathName"
        android:animation="@animator/myanimated"/>
</animated-vector>
```

上面这段代码在 MyCode\MySample626\app\src\main\res\drawable\mytransition.xml 文件中。在这段代码中，android:name="myPathName" 表示自定义的路径（即矢量图形数字 5）名称，android:animation="@animator/myanimated" 中的 myanimated 文件用于指定动画的具体参数，myanimated 文件的主要内容如下：

```xml
<?xml version="1.0" encoding="utf-8"?>
<set xmlns:android="http://schemas.android.com/apk/res/android">
<objectAnimator android:duration="1000"
                android:propertyName="pathData"
                android:valueFrom="M 0.806629834254144, 0.110497237569061 C 0.502762430939227, 0.110497237569061 0.502762430939227, 0.110497237569061 0.502762430939227, 0.110497237569061 C 0.397790055248619, 0.430939226519337 0.397790055248619, 0.430939226519337 0.397790055248619, 0.430939226519337 C 0.535911602209945, 0.364640883977901 0.801104972375691, 0.469613259668508 0.801104972375691, 0.712071823204420 C 0.773480662983425, 1.01104972375691 0.375690607734807, 1.09392265193370 0.248618784530387, 0.850828729281768"
                android:valueTo="M 0.607734806629834, 0.110497237569061 C 0.607734806629834, 0.110497237569061 0.607734806629834, 0.110497237569061 0.607734806629834, 0.110497237569061 C 0.392265193370166, 0.436464088397790 0.265193370165746, 0.508287292817680 0.254143646408840, 0.696132596685083 C 0.287292817679558, 1.13017127071823 0.872928176795580, 1.06077348066298 0.845303867403315, 0.696132596685083 C 0.806629834254144, 0.364640883977901 0.419889502762431, 0.353591160220994 0.295580110497238, 0.552486187845304"
                android:valueType="pathType"/>
</set>
```

上面这段代码在 MyCode\MySample626\app\src\main\res\animator\myanimated.xml 文件中。在这段代码中，android:valueFrom 属性值在此处就是矢量图形数字 5 的数据，android:valueTo 属性值在此处就是矢量图形数字 6 的数据，因此当 objectAnimator 动画开始执行时，就直接从上面两个属性值进行过渡。此外，如果当前工程项目的 res 目录中不存在 animator 子目录，应该先创建此子目录，再在该子目录中添加 myanimated.xml 文件。

在 mytransition.xml 文件中的 android:drawable="@drawable/myvector" 的 myvector 表示使用 pathData 描述的矢量图形数字 5 的文件，myvector 文件的主要内容如下：

```xml
<vector xmlns:android="http://schemas.android.com/apk/res/android"
        android:width="100dp"
        android:height="100dp"
        android:tint="#00f"
        android:viewportHeight="1"
        android:viewportWidth="1">
<group android:translateX="-0.06"
       android:translateY="-0.06">
<path android:name="myPathName"
```

```
                android:pathData = " M 0.806629834254144, 0.110497237569061 C 0.502762430939227, 0.
            110497237569061 0.502762430939227, 0.110497237569061 0.502762430939227, 0.110497237569061 C 0.
            397790055248619, 0.430939226519337 0.397790055248619, 0.430939226519337 0.397790055248619, 0.
            430939226519337 C 0.535911602209945, 0.364640883977901 0.801104972375691, 0.469613259668508 0.
            801104972375691, 0.712707182320442 C 0.773480662983425, 1.011049723756910 0.375690607734807, 1.
            0939226519337 0.248618784530387, 0.850828729281768"
                android:strokeColor = "#0f0"
                android:strokeWidth = "0.08"/>
    </group></vector>
```

上面这段代码在 MyCode\MySample626\app\src\main\res\drawable\myvector.xml 文件中。在这段代码中,android:pathData 属性值表示绘制矢量图形数字 5 的数据。android:strokeWidth 表示矢量图的线条宽度。android:strokeColor 表示矢量图的线条颜色。此实例的完整项目在 MyCode\MySample626 文件夹中。

125 使用 ObjectAnimator 改变图像的色相值

此实例主要通过使用属性动画 ObjectAnimator 改变自定义控件 AnimImageView 的属性,实现以动画的形式动态改变 ImageView 控件的图像的 R、G、B 通道的色相值。当实例运行之后,单击"开始色相旋转"按钮,则将以动画的形式依次旋转图像的 R、G、B 通道的色相值,效果分别如图 125.1 的左图和右图所示。单击"暂停色相旋转"按钮,则将暂停 ImageView 控件的图像色相旋转动画。

图 125.1

主要代码如下:

```
public class MainActivity extends Activity {
    AnimImageView myImageView;
    ObjectAnimator myObjectAnimator;
    @Override
```

```
protected void onCreate(Bundle savedInstanceState) {
    super.onCreate(savedInstanceState);
    setContentView(R.layout.activity_main);
    myImageView = (AnimImageView) findViewById(R.id.myImageView);
    //自定义属性动画
    myObjectAnimator = ObjectAnimator.ofFloat(myImageView,"myDegree",0f,360f);
    //设置动画时长和重复次数,-1表示无限执行
    myObjectAnimator.setDuration(5000).setRepeatCount(-1);
}
//响应单击"开始色相旋转"按钮
public void onClickBtn1(View v) {
    if(myObjectAnimator.isPaused()){
        myObjectAnimator.resume();            //继续动画
    } else{
        myObjectAnimator.start();             //第一次执行动画
    } }
//响应单击"暂停色相旋转"按钮
public void onClickBtn2(View v) {
    if(myObjectAnimator.isStarted()){
        myObjectAnimator.pause();             //暂停动画
} } }
```

上面这段代码在 MyCode\MySample769\app\src\main\java\com\bin\luo\mysample\MainActivity.java 文件中。在这段代码中,myObjectAnimator = ObjectAnimator.ofFloat(myImageView,"myDegree",0f,360f)用于创建属性动画。ObjectAnimator.ofFloat()方法的语法声明如下:

```
public static ObjectAnimator ofFloat (Object target,
                        String propertyName, float... values)
```

其中,Object target 参数用于指定这个动画操作的控件。String propertyName 参数用于指定这个动画操作的属性。float... values 参数是可变长参数,即指示属性值改变范围,在此实例中,即是将 myImageView 控件的 myDegree 属性从 0f 改变到 360f。

myImageView = (AnimImageView)findViewById(R.id.myImageView)的 AnimImageView 是在 ImageView 控件的基础上创建的自定义控件,AnimImageView 类(控件)的主要代码如下:

```
public class AnimImageView extends ImageView {
    float myDegree = 0;
    ColorMatrix myMatrix;
    public float getMyDegree() { return myDegree; }
    public void setMyDegree(float degree) { myDegree = degree; }
    public AnimImageView(Context context, AttributeSet attrs) {
        super(context, attrs);
        myMatrix = new ColorMatrix();              //初始化 ColorMatrix 对象
    }
    @Override
    protected void onDraw(Canvas canvas) {
        super.onDraw(canvas);
        myMatrix.setRotate(0, myDegree);           //旋转 R 通道的色相值
        myMatrix.setRotate(1, myDegree);           //旋转 G 通道的色相值
        myMatrix.setRotate(2, myDegree);           //旋转 B 通道的色相值
        //在图像中应用 ColorMatrix 对象
        setColorFilter(new ColorMatrixColorFilter(myMatrix));
} }
```

上面这段代码在 MyCode\MySample769\app\src\main\java\com\bin\luo\mysample\AnimImageView.java 文件中。此实例的完整项目在 MyCode\MySample769 文件夹中。

126 使用 AnimatorSet 组合多个 ObjectAnimator

此实例主要通过使用 AnimatorSet 组合多个属性动画，实现图像同时产生旋转、缩放等多种动画效果。当实例运行之后，单击"开始播放动画"按钮，则图像（妖怪）将以自身为中心进行旋转、缩小直到消失，在图像缩小的过程中，图像的透明度也将逐渐变小，效果分别如图 126.1 的左图和右图所示。

图 126.1

主要代码如下：

```java
public class MainActivity extends Activity {
    ImageView myImageView;
    AnimatorSet myAnimatorSet;
    @Override
    protected void onCreate(Bundle savedInstanceState) {
        super.onCreate(savedInstanceState);
        setContentView(R.layout.activity_main);
        myImageView = (ImageView) findViewById(R.id.myImageView);
        //创建透明度动画
        ObjectAnimator myObjectAnimator1 =
                    ObjectAnimator.ofFloat(myImageView, "alpha", 1f, 0f);
        //创建旋转动画
        ObjectAnimator myObjectAnimator2 =
                    ObjectAnimator.ofFloat(myImageView,"rotation", 0f, 1800f);
        //创建水平缩放动画
        ObjectAnimator myObjectAnimator3 =
                    ObjectAnimator.ofFloat(myImageView, "scaleX", 1.0f, 0f);
        //创建垂直缩放动画
        ObjectAnimator myObjectAnimator4 =
```

```
                    ObjectAnimator.ofFloat(myImageView, "scaleY", 1.0f, 0f);
    //创建动画集合
    myAnimatorSet = new AnimatorSet();
    //在动画集合中添加四种动画
    myAnimatorSet.play(myObjectAnimator1).with(myObjectAnimator2).
                        with(myObjectAnimator3).with(myObjectAnimator4);
    //设置动画持续时间9秒
    myAnimatorSet.setDuration(9000);
    }
    //响应单击"开始播放动画"按钮
    public void onClickBtn1(View v) { myAnimatorSet.start(); }
    //响应单击"停止播放动画"按钮
    public void onClickBtn2(View v) { myAnimatorSet.cancel(); }
}
```

上面这段代码在 MyCode \ MySample161 \ app \ src \ main \ java \ com \ bin \ luo \ mysample \ MainActivity.java 文件中。在这段代码中，myAnimatorSet.play（myObjectAnimator1）.with（myObjectAnimator2）.with（myObjectAnimator3）.with（myObjectAnimator4）用于在 myAnimatorSet 动画集合加载 myObjectAnimator1、myObjectAnimator2、myObjectAnimator3 和 myObjectAnimator4 四种动画。AnimationSet 与 AnimatorSet 最大的不同在于，AnimationSet 使用的是 Animation 子类，AnimatorSet 使用的是 Animator 的子类。Animation 是针对视图外观的动画实现，动画被应用时外观改变，但视图的触发点不会发生变化，仍在原来定义的位置。Animator 是针对视图属性的动画实现，动画被应用时对象属性产生变化，最终导致视图外观变化。AnimationSet 最常用的操作是调用其 addAnimation（）方法将多个不同的动画组织起来，然后调用 view 对象的 startAnimation（）方法触发这些动画执行。AnimatorSet 最常用的操作是调用其 play（）、before（）、with（）、after（）等方法设置动画的执行顺序，然后调用 start（）方法触发动画。此实例的完整项目在 MyCode\MySample161 文件夹中。

127 使用 TypeEvaluator 实现颜色过渡动画

此实例主要通过使用 ObjectAnimator 创建颜色动画，并在 ofObject（）方法中自定义 TypeEvaluator，实现控件的背景从一种颜色自然过渡到另一种颜色。当实例运行之后，单击"开始播放动画"按钮，则背景颜色将在 5 秒内从灰色逐渐改变成绿色，效果分别如图 127.1 的左图和右图所示。单击"停止播放动画"按钮，则停止颜色动画的改变动作。

主要代码如下：

```
public class MainActivity extends Activity {
    ImageView myImageView;
    ObjectAnimator myObjectAnimator;
    LinearLayout myLayout;
    @Override
    protected void onCreate(Bundle savedInstanceState) {
        super.onCreate(savedInstanceState);
        setContentView(R.layout.activity_main);
        myImageView = (ImageView) findViewById(R.id.myImageView);
        myLayout = (LinearLayout) findViewById(R.id.myLayout);
        int startColor = 0x00ffffff;
        int endColor = 0xff008000;
```

```java
myObjectAnimator = ObjectAnimator.ofObject(myLayout,
    "backgroundColor", new TypeEvaluator() {            //创建背景颜色改变动画
    @Override
    public Object evaluate(float fraction, Object startValue, Object endValue){
      int startInt = (Integer) startValue;
      int startA = (startInt >> 24) & 0xff;
      int startR = (startInt >> 16) & 0xff;
      int startG = (startInt >> 8) & 0xff;
      int startB = startInt & 0xff;
      int endInt = (Integer) endValue;
      int endA = (endInt >> 24) & 0xff;
      int endR = (endInt >> 16) & 0xff;
      int endG = (endInt >> 8) & 0xff;
      int endB = endInt & 0xff;
      return (int) ((startA + (int) (fraction * (endA - startA))) << 24) |
             (int) ((startR + (int) (fraction * (endR - startR))) << 16) |
             (int) ((startG + (int) (fraction * (endG - startG))) << 8) |
             (int) ((startB + (int) (fraction * (endB - startB))));
    } }, startColor, endColor);
  myObjectAnimator.setDuration(5000);                   //设置动画持续时间 5 秒
}
//响应单击"开始播放动画"按钮
public void onClickBtn1(View v) { myObjectAnimator.start(); }
//响应单击"停止播放动画"按钮
public void onClickBtn2(View v) { myObjectAnimator.cancel(); }
}
```

图 127.1

上面这段代码在 MyCode \ MySample162 \ app \ src \ main \ java \ com \ bin \ luo \ mysample \ MainActivity.java 文件中。在这段代码中，myObjectAnimator = ObjectAnimator.ofObject(myLayout,"backgroundColor"，new TypeEvaluator() { }，startColor，endColor)用于创建一个颜色改变动画，myLayout 表示动画执行对象，backgroundColor 表示动画执行对象的属性名称，startColor 表示动画起始值，endColor 表示动画结束值。TypeEvaluator()用于自定义估值器配置颜

色从动画起始值以何种方式改变成动画结束值,在 Android 中,如果使用 ObjectAnimator 的 ofObject()方法创建属性动画,通常需要定义此行为。此实例的完整项目在 MyCode\MySample162 文件夹中。

128 通过 trimPathEnd 实现动态生成手指图形

此实例主要通过在 objectAnimator 动画中改变 trimPathEnd 属性值从 0 到 1(0%到 100%),实现动态生成手指矢量图形的特效。当实例运行之后,单击屏幕,则将从右端启动手指矢量图形的动态绘制过程,直到最后完成,效果分别如图 128.01 的左图和右图所示。

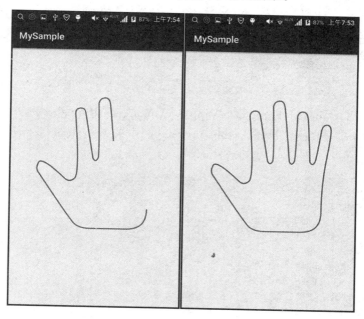

图 128.1

主要代码如下:

```java
public void onClickImageView(View v) {          //响应单击 ImageView 控件
    bChecked = !bChecked;
    final int[] myStateSet = {android.R.attr.state_checked * (bChecked ? 1 : -1)};
    myImageView.setImageState(myStateSet, true);
}
```

上面这段代码在 MyCode\MySample617\app\src\main\java\com\bin\luo\mysample\MainActivity.java 文件中。在这段代码中,myImageView.setImageState(myStateSet,true)用于根据当前 myStateSet 的值,确定在单击 myImageView 控件时是否执行或终止动画;因此 myImageView 的 src 属性不能直接为图像资源,而应是 XML 文件。主要代码如下:

```xml
<ImageView android:id="@+id/myImageView"
    android:layout_width="match_parent"
    android:layout_height="match_parent"
    android:layout_margin="50dp"
    android:layout_centerInParent="true"
    android:onClick="onClickImageView"
    android:src="@drawable/mystage"/>
```

上面这段代码在 MyCode\MySample617\app\src\main\res\layout\activity_main.xml 文件中。在这段代码中,android:src="@drawable/mystage"用于设置在 myImageView 控件上加载的 XML 动画 mystage。mystage 动画的主要内容如下:

```xml
<?xml version = "1.0" encoding = "utf-8"?>
<animated-selector xmlns:android = "http://schemas.android.com/apk/res/android">
  <item android:id = "@+id/myStart"
        android:state_checked = "true"
        android:drawable = "@drawable/myvector"/>
  <item android:id = "@+id/myFinish"
        android:drawable = "@drawable/myvector"/>
  <transition android:drawable = "@drawable/mytransition"
              android:fromId = "@id/myStart"
              android:toId = "@id/myFinish"/>
</animated-selector>
```

上面这段代码在 MyCode\MySample617\app\src\main\res\drawable\mystage.xml 文件中。在这段代码中,android:drawable="@drawable/myvector"中的 myvector 是矢量图形文件,myvector 文件的主要内容如下:

```xml
<vector xmlns:android = "http://schemas.android.com/apk/res/android"
        android:width = "240dp"
        android:height = "340dp"
        android:tint = "#f00"
        android:viewportWidth = "24"
        android:viewportHeight = "34">
  <path android:name = "myPathName"
        android:pathData = "M22,23 q0,4 -4,4 h-7 q-2,0 -3,-1 T1,16 q-0.6,-0.8 0,-2 t5,3 q1,1 2,0 T8,4 q0,-1 0.9,-1.1 t1.1,1 1.5,9 q0.25,0.5 0.5,0.5 t0.5,-0.5 0,-11 q0.2,-1 1.1,-1.1 t1.1,1.1 1,11 q0.25,0.5 0.5,0.5 t0.5,-0.5 0.5,-9 q0.2,-1 1,-1 t1,1 0.5,9 q0.25,0.5 0.5,0.5 t0.5,-0.5 1.2,-6.5 q0.3,-1 1,-1 t0.8,1 -0.8,6 T22,23"
        android:trimPathEnd = "0"
        android:strokeColor = "#000"
        android:strokeWidth = "0.2"/>
</vector>
```

上面这段代码在 MyCode\MySample617\app\src\main\res\drawable\myvector.xml 文件中。在这段代码中,android:pathData="M22,23 q0,4 −4,4 h−7 q−2,0 −3,−1 T1,16 q−0.6,−0.8 0,−2 t5,3 q1,1 2,0 T8,4 q0,−1 0.9,−1.1 t1.1,1 1.5,9 q0.25,0.5 0.5,0.5 t0.5,−0.5 0,−11 q0.2,−1 1.1,−1.1 t1.1,1.1 1,11 q0.25,0.5 0.5,0.5 t0.5,−0.5 0.5,−9 q0.2,−1 1,−1 t1,1 0.5,9 q0.25,0.5 0.5,0.5 t0.5,−0.5 1.2,−6.5 q0.3,−1 1,−1 t0.8,1 −0.8,6 T22,23"表示绘制手指矢量图的数据。android:strokeWidth="0.2"表示矢量图的线条宽度。android:strokeColor="#000"表示矢量图的线条颜色。android:trimPathEnd ="0"表示不显示该矢量图(因此启动应用时,屏幕显示空白),android:trimPathEnd ="1"表示完全显示该矢量图。android:name= "myPathName"表示该矢量图的自定义名称,在调用动画时需要此名称。

在 mystage.xml 文件中的代码 android:drawable="@drawable/mytransition"用于指定目标动画 mytransition,mytransition 动画的主要内容如下:

```xml
<animated-vector xmlns:android="http://schemas.android.com/apk/res/android"
                 android:drawable="@drawable/myvector">
    <target android:name="myPathName"
            android:animation="@animator/myanimated"/>
</animated-vector>
```

上面这段代码在 MyCode\MySample617\app\src\main\res\drawable\mytransition.xml 文件中。在这段代码中,android:name="myPathName" 表示自定义的路径(矢量图)名称,android:animation="@animator/myanimated" 中的 myanimated 文件用于指定 trimPathEnd 属性动画的具体参数,trimPathEnd 属性表示截掉从某个位置到终点的部分,保留剩下的部分。myanimated 文件的主要内容如下:

```xml
<?xml version="1.0" encoding="utf-8"?>
<set xmlns:android="http://schemas.android.com/apk/res/android">
<objectAnimator android:duration="5000"
                android:interpolator="@android:interpolator/linear"
                android:propertyName="trimPathEnd"
                android:valueFrom="0"
                android:valueTo="1"/>
</set>
```

上面这段代码在 MyCode\MySample617\app\src\main\res\animator\myanimated.xml 文件中。在这段代码中,android:propertyName 用于指定该矢量图将要发生改变的属性名,android:valueFrom 用于设置动画初始值,android:valueTo 用于设置动画结束值;android:valueFrom="0" 和 android:valueTo="1",表示矢量图从无到有;android:valueFrom="1" 和 android:valueTo="0" 表示矢量图从有到无。换句话说,如果 objectAnimator 的 android:propertyName="trimPathEnd",则 android:valueFrom="0" 和 android:valueTo="1" 表示路径(矢量图)从起点增长到终点,即初始截断部分是 100%,从起点开始逐渐缩小到终点,达到 0%;android:valueFrom="1" 和 android:valueTo="0" 表示路径(矢量图)从终点缩短到起点,即初始截断部分是 0%,从终点开始逐渐扩大到起点,达到 100%。此实例的完整项目在 MyCode\MySample617 文件夹中。

129 使用 ValueAnimator 动态改变扇形转角

此实例主要通过在自定义 View 中使用 ValueAnimator 动画控制扇形的转角变化,实现以动画的形式动态绘制不同角度的扇形。当实例运行之后,扇形的转角将沿着顺时针方向从小变大,再沿着逆时针方向从大变小,并且永不停歇,效果分别如图 129.1 的左图和右图所示。

主要代码如下:

```xml
<com.bin.luo.mysample.myImageView android:id="@+id/myView"
                                  android:layout_width="wrap_content"
                                  android:layout_height="wrap_content"
                                  android:scaleType="center"
                                  android:visibility="visible"/>
```

上面这段代码在 MyCode\MySample176\app\src\main\res\layout\activity_main.xml 文件中。在这段代码中,com.bin.luo.mysample.myImageView 即是用于以动画的形式改变扇形转角大小的自定义控件,myImageView 类(控件)的主要代码如下:

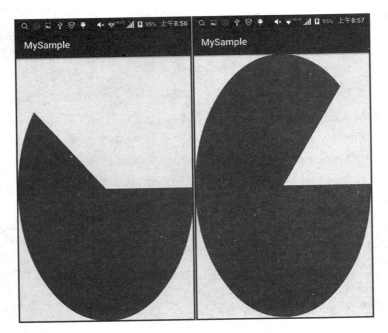

图 129.1

```java
public class myImageView extends View {
 private float myDegrees;
 private Paint myPaint;
 public myImageView(Context context, AttributeSet attrs) {
  super(context, attrs);
  myPaint = new Paint(Paint.ANTI_ALIAS_FLAG);
  myPaint.setColor(Color.BLUE);
  startAnimation();
 }
 @Override
 protected void onDraw(Canvas canvas) { myDrawText(canvas); }
 private void myDrawText(Canvas myCanvas) {
  int myHeight = getHeight();
  int myWidth = getWidth();
  myCanvas.drawArc(new RectF(0,0,myWidth,myHeight),
                   0f, myDegrees,true, myPaint);          //根据变化的角度绘制扇形
 }
 private void startAnimation() {
  float startDegrees = 0f;                                //开始角度
  float endDegrees = 315f;                                //结束角度
  ValueAnimator myValueAnimator =
                 ValueAnimator.ofFloat(startDegrees, endDegrees);
  //监听角度的改变
  myValueAnimator.addUpdateListener(
                    new ValueAnimator.AnimatorUpdateListener() {
   @Override
   public void onAnimationUpdate(ValueAnimator myAnimator) {
    myDegrees = (float)myAnimator.getAnimatedValue();     //获取当前角度大小
    invalidate();
   } });
  //设置动画重复模式为反向重复
  myValueAnimator.setRepeatMode(ValueAnimator.REVERSE);
```

```
    //设置动画重复次数为无限次
    myValueAnimator.setRepeatCount(ValueAnimator.INFINITE);
    myValueAnimator.setDuration(5000);              //设置动画持续时间为5秒
    myValueAnimator.start();                         //启动动画
}}
```

上面这段代码在 MyCode \ MySample176 \ app \ src \ main \ java \ com \ bin \ luo \ mysample \ myImageView.java 文件中。此实例的完整项目在 MyCode\MySample176 文件夹中。

130　使用 ValueAnimator 实现分段转圈动画

此实例主要通过在自定义 View 中使用 PathMeasure 的 getSegment()方法对圆形路径进行分段截取,实现显示分段圆弧转圈的动画效果。当实例运行之后,分段圆弧转圈效果分别如图 130.1 的左图和右图所示。

主要代码如下:

```
<com.bin.luo.mysample.PathView android:layout_width = "match_parent"
                               android:layout_height = "match_parent"/>
```

上面这段代码在 MyCode\MySample851\app\src\main\res\layout\activity_main.xml 文件中。在这段代码中,com.bin.luo.mysample.PathView 即是实现分段圆弧转圈动画的自定义控件,PathView 类的主要代码如下:

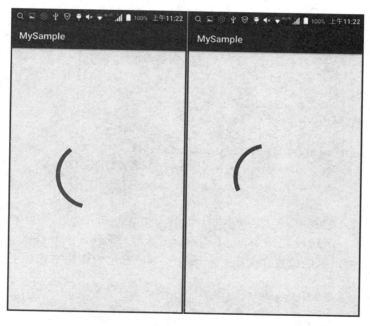

图　130.1

```
public class PathView extends View {
    private Path myPath;
    private Paint myPaint;
    private PathMeasure myPathMeasure;
```

```
private float myAnimatedValue;
private Path newPath;
private float myLength;
public PathView(Context context, AttributeSet attrs) {
  super(context, attrs);
  myPathMeasure = new PathMeasure();
  myPaint = new Paint(Paint.ANTI_ALIAS_FLAG);
  myPaint.setStyle(Paint.Style.STROKE);
  myPaint.setColor(Color.BLUE);
  myPaint.setStrokeWidth(25);
  myPath = new Path();
  myPath.addCircle(250, 300, 200, Path.Direction.CW);
  myPathMeasure.setPath(myPath, true);
  myLength = myPathMeasure.getLength();
  newPath = new Path();
  final ValueAnimator valueAnimator = ValueAnimator.ofFloat(0, 1);
  valueAnimator.addUpdateListener(new ValueAnimator.AnimatorUpdateListener(){
   @Override
   public void onAnimationUpdate(ValueAnimator valueAnimator) {
     myAnimatedValue = (float) valueAnimator.getAnimatedValue();
     invalidate();
   } });
  valueAnimator.setDuration(2000);
  valueAnimator.setRepeatCount(ValueAnimator.INFINITE);
  valueAnimator.start();
}
@Override
protected void onDraw(Canvas canvas) {
  super.onDraw(canvas);
  canvas.translate(250,500);                    //平移画布至(250,500)
  newPath.reset();
  newPath.lineTo(0,0);
  float myStop = myLength * myAnimatedValue;
  float myStart = (float) (myStop - ((0.5 -
                    Math.abs(myAnimatedValue - 0.5)) * myLength));
  myPathMeasure.getSegment(myStart, myStop, newPath, true);
  canvas.drawPath(newPath, myPaint);            //绘制各段圆弧(分割的路径)
} }
```

上面这段代码在 MyCode\MySample851\app\src\main\java\com\bin\luo\mysample\PathView.java 文件中。在这段代码中,myPathMeasure.getSegment(myStart,myStop,newPath,true)用于截取将要绘制圆弧的位置和大小。getSegment()方法的语法格式如下:

```
public boolean getSegment(float startD,
                float stopD, Path dst, boolean startWithMoveTo)
```

其中,参数 float startD 表示该(截取的圆弧)线段的起点距离原始 Path 起点的长度。参数 float stopD 表示该(截取的圆弧)线段的终点距离原始 Path 起点的长度,参数 Pathdst 表示截取的圆弧线段将添加到 dst 中,注意是添加,而不是替换。3 个参数的关系是:0≤startD<stopD≤Path。参数 boolean startWithMoveTo 表示起始点是否使用 moveTo 用于保证截取的 Path 第一个点位置不变。

getSegment()方法的返回值用于判断截取是否成功,如果返回值为 true 表示截取成功,结果存入

dst 中；如果返回值为 false 表示截取失败，不会改变 dst 中内容。此实例的完整项目在 MyCode\MySample851 文件夹中。

131　使用 ValueAnimator 在三维 Z 轴上平移图像

此实例主要通过使用 ValueAnimator 的 ofFloat()方法设置 z 轴平移参数的变化范围，再通过 ValueAnimator 的 getAnimatedValue()方法获取适时 z 轴平移参数值，实现图像在 z 轴上进行平移。当实例运行之后，单击"开始播放动画"按钮，则图像将在 z 轴上进行平移，由于 z 轴垂直于屏幕，因此平移图像与缩放图像的效果相当类似，效果分别如图 131.1 的左图和右图所示。

图　131.1

主要代码如下：

```
public void onClickButton(View v) {                           //响应单击"开始播放动画"按钮
    ValueAnimator myValueAnimator =
        ValueAnimator.ofFloat(0, 900);                        //0,900 表示图像在 z 轴的平移范围
    myValueAnimator.setDuration(3000);                        //设置动画持续时间 3 秒
    //设置重复次数为无限次
    myValueAnimator.setRepeatCount(ValueAnimator.INFINITE);
    myValueAnimator.setRepeatMode(ValueAnimator.REVERSE);     //设置重复模式为反向
    myValueAnimator.addUpdateListener(new ValueAnimator.AnimatorUpdateListener(){
        @Override
        public void onAnimationUpdate(ValueAnimator animation) {
            float myValue = (float) animation.getAnimatedValue();
            Bitmap myNewBmp = GetNewImage(myBitmap, myValue); //根据改变的平移值重绘图像
            myImageView.setImageBitmap(myNewBmp);             //显示新图像
        } });
    myValueAnimator.start();
}
public Bitmap GetNewImage(Bitmap myOldBmp, float myValue) {
```

```
BitmapDrawable myDrawable =
        (BitmapDrawable) getResources().getDrawable(R.mipmap.myimage);
Bitmap oldBmp = myDrawable.getBitmap();
myCamera.save();                                      //保存状态
Matrix myMatrix = new Matrix();
myCamera.translate(0, 0, myValue);
myCamera.getMatrix(myMatrix);
myCamera.restore();                                   //恢复状态
//设置图像处理的中心点
myMatrix.preTranslate(oldBmp.getWidth() >> 1, oldBmp.getHeight() >> 1);
Bitmap newBmp = Bitmap.createBitmap(oldBmp, 0, 0, oldBmp.getWidth(),
        oldBmp.getHeight(), myMatrix, true);          //通过矩阵生成新图像
return newBmp;
}
```

上面这段代码在 MyCode\MySample180\app\src\main\java\com\bin\luo\mysample\MainActivity.java 文件中。在这段代码中，myCamera.translate(0, 0, myValue)用于在 x、y、z 三个轴上平移图像，由于 x、y 轴的值均为 0，且动画只向 z 轴传值，因此只有 z 轴的平移运动。此实例的完整项目在 MyCode\MySample180 文件夹中。

132　使用 ValueAnimator 实现起飞转平飞动画

此实例主要通过在自定义 TypeEvaluator 的 evaluate()方法中监听运动路径的 y 坐标，并使用 ValueAnimator 操作该自定义 TypeEvaluator，实现在飞机图像开始（起飞）时以仰角 30 度姿势运动（飞行），在到达指定位置（$y=300$）时，以水平姿势运动（飞行）的动态变换姿势效果。当实例运行之后，单击"启动动画"按钮，则飞机图像在开始（起飞）时以仰角 30 度姿势运动（飞行），如图 132.1 的左图所示；在到达指定位置（$y=300$，细线为飞行轨迹）时，以水平姿势运动（飞行），如图 132.1 的右图所示。

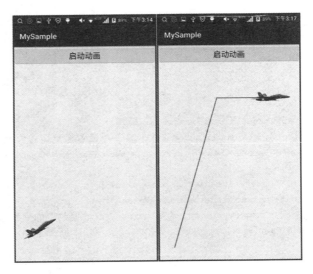

图　132.1

主要代码如下：

```java
//响应单击"启动动画"按钮
public void OnClickButton(View v) {
    myImageView.setVisibility(View.VISIBLE);
    //逆时针旋转30度，即飞机起飞
    myImageView.setRotation(-30);
    PointF myStartPoint = new PointF(0, 1400);
    PointF myMiddlePoint = new PointF(400, 300);
    PointF myEndPoint = new PointF(700, 300);
    TypeEvaluator<PointF> myPointEvaluator = new TypeEvaluator<PointF>() {
        @Override
        public PointF evaluate(float fraction, PointF startValue, PointF endValue) {
            float x = startValue.x + fraction * (endValue.x - startValue.x);
            float y = startValue.y + fraction * (endValue.y - startValue.y);
            //当y值达到300时，飞机（图像）保持水平
            if (y == 300)
                myImageView.setRotation(0);
            PointF myEvaluatePoint = new PointF(x, y);
            return myEvaluatePoint;
        }
    };
    //运动（飞行）路径为myStartPoint, myMiddlePoint, myEndPoint连成的折线
    ValueAnimator myValueAnimator = ValueAnimator.ofObject(myPointEvaluator,
                        myStartPoint, myMiddlePoint, myEndPoint);
    myValueAnimator.setDuration(5000);
    myValueAnimator.setInterpolator(new DecelerateInterpolator());
    myValueAnimator.addUpdateListener(new ValueAnimator.AnimatorUpdateListener(){
        @Override
        public void onAnimationUpdate(ValueAnimator animation) {
            //根据变化的路径值设置ImageView的坐标
            PointF myCurrentPoint = (PointF) animation.getAnimatedValue();
            myImageView.setX(myCurrentPoint.x);
            myImageView.setY(myCurrentPoint.y);
        }
    });
    myValueAnimator.start();                //启动动画
}
```

上面这段代码在 MyCode\MySample853\app\src\main\java\com\bin\luo\mysample\MainActivity.java 文件中。此实例的完整项目在 MyCode\MySample853 文件夹中。

133 自定义 TypeEvaluator 以 GIF 动画显示图像

此实例主要通过创建自定义估值器 MyTypeEvaluator 配置图像数组索引的规则，并使用该自定义估值器创建 ValueAnimator 动画，从而实现以 GIF 动画风格显示图像。当实例运行之后，单击"启动动画"按钮，则将以 GIF 动画的风格显示图像数组中的每幅图像，效果分别如图 133.1 的左图和右图所示。

主要代码如下：

```java
public class MainActivity extends Activity {
    ImageView myImageView;
    int[] myImages = {R.mipmap.mygif_layer1, R.mipmap.mygif_layer2,
```

```
            R.mipmap.mygif_layer3,R.mipmap.mygif_layer4,R.mipmap.mygif_layer5,
            R.mipmap.mygif_layer6,R.mipmap.mygif_layer7,R.mipmap.mygif_layer8};
@Override
protected void onCreate(Bundle savedInstanceState) {
  super.onCreate(savedInstanceState);
  setContentView(R.layout.activity_main);
  myImageView = (ImageView)findViewById(R.id.myImageView);
}
public void onClickButton(View v) {                          //响应单击"启动动画"按钮
    //设置 myValueAnimator 动画显示的图像数组开始索引是 0,结束索引是 7
    ValueAnimator myValueAnimator =
            ValueAnimator.ofObject(new MyTypeEvaluator(),0,7);
  myValueAnimator.addUpdateListener(new ValueAnimator.AnimatorUpdateListener(){
    @Override
    public void onAnimationUpdate(ValueAnimator animation) {
      int myIndex = (int)animation.getAnimatedValue();
      //根据时间线确定图像数组索引
      myImageView.setImageResource(myImages[myIndex]);
    }});
  myValueAnimator.setDuration(500);          //动画持续时间是 500ms
  myValueAnimator.setRepeatCount(-1);        //无限重复
  myValueAnimator.start();                   //启动动画
}
//自定义 TypeEvaluator 制定数组索引规则
public class MyTypeEvaluator implements TypeEvaluator<Integer> {
  @Override
  public Integer evaluate(float fraction, Integer startValue, Integer endValue) {
    int myCurrentIndex = (int)(startValue + fraction*(endValue - startValue));
    return myCurrentIndex;
}}}
```

图 133.1

上面这段代码在 MyCode\MySample860\app\src\main\java\com\bin\luo\mysample\MainActivity.java 文件中。在这段代码中，myValueAnimator = ValueAnimator.ofObject（new MyTypeEvaluator(),0,7）表示根据自定义类 MyTypeEvaluator 的规则按照图像数组索引顺序在 myIndex =（int）animation.getAnimatedValue()中获得当前显示图像的索引，然后将该图像显示出来。此实例的完整项目在 MyCode\MySample860 文件夹中。

134　使用 Animation 实现图像围绕自身中心旋转

此实例主要通过使用补间动画实现图像围绕自身中心旋转。当实例运行之后，单击"开始播放动画"按钮，则图像（车轮）将围绕自身中心进行旋转，如图 134.1 的左图和右图所示。单击"停止播放动画"按钮，则停止旋转。

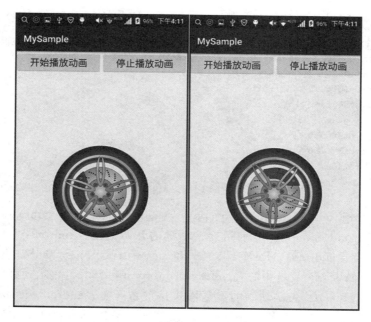

图　134.1

主要代码如下：

```
public class MainActivity extends Activity {
 Animation myAnimation;
 ImageView myImageView;
 @Override
 protected void onCreate(Bundle savedInstanceState) {
  super.onCreate(savedInstanceState);
  setContentView(R.layout.activity_main);
  //获取动画资源
  myAnimation = AnimationUtils.loadAnimation(this, R.anim.myanimation);
  myAnimation.setFillAfter(true);             //设置动画停留在最后的状态
  myImageView = (ImageView)findViewById(R.id.myImageView);
  Button myBtnPlay = (Button)findViewById(R.id.myBtnPlay);
  myBtnPlay.setOnClickListener(new View.OnClickListener(){
   public void onClick(View arg0){            //响应单击"开始播放动画"按钮
    myImageView.startAnimation(myAnimation);  //在控件上实施动画
  } });
```

```
        Button myBtnStop = (Button)findViewById(R.id.myBtnStop);
        myBtnStop.setOnClickListener(new View.OnClickListener(){
          public void onClick(View arg0){              //响应单击"停止播放动画"按钮
            myImageView.clearAnimation();              //在控件上清除动画
          } });
      } }
```

上面这段代码在 MyCode\MySample150\app\src\main\java\com\bin\luo\mysample\MainActivity.java 文件中。在这段代码中,myAnimation = AnimationUtils.loadAnimation(this,R.anim.myanimation)用于加载在 res\anim 目录中的动画资源 myanimation。myImageView.startAnimation(myAnimation)用于在 ImageView 控件上执行 myAnimation 动画,myImage.clearAnimation()用于在 ImageView 控件上清除动画效果。myanimation 动画的主要内容如下:

```xml
<?xml version = "1.0" encoding = "utf-8"?>
<set xmlns:android = "http://schemas.android.com/apk/res/android"
     android:interpolator = "@android:anim/accelerate_interpolator">
<rotate android:repeatMode = "restart"
        android:repeatCount = "infinite"
        android:fromDegrees = "0"
        android:toDegrees = "1800"
        android:pivotX = "50%"
        android:pivotY = "50%"
        android:duration = "3000"/>
</set>
```

上面这段代码在 MyCode\MySample150\app\src\main\res\anim\myanimation.xml 文件中。在这段代码中,android:fromDegrees="0"表示旋转动画的开始角度。android:toDegrees="1800"表示旋转动画的结束角度。android:pivotX="50%"和 android:pivotY="50%"表示旋转中心的水平坐标和垂直坐标(以图像中心为旋转中心)。android:duration="3000"表示动画的持续时间是 3 秒。android:repeatMode="restart"表示动画的重复模式是重新开始。android:repeatCount="infinite"表示动画的重复次数是无限次。此实例的完整项目在 MyCode\MySample150 文件夹中。

135 自定义 Animation 实现旋转切换扑克牌正反面

此实例主要通过在自定义类 Rotate3DAnimation 中使用 Camera 和 Matrix,并使用自定义类 Rotate3DAnimationListener 动态切换指定(扑克牌的正反面)图像,实现扑克牌正反面的旋转切换。当实例运行之后,单击"显示正面"按钮,则扑克牌反面图像将围绕 y 轴旋转 180°,切换到扑克牌正面,如图 135.1 的左图所示。单击"显示反面"按钮,则扑克牌正面图像将围绕 y 轴旋转 180°,切换到扑克牌反面,如图 135.1 的右图所示。

主要代码如下:

```
    public void onClickBtn1(View v) {                    //响应单击"显示正面"按钮
      myRotate3DAnimation = new Rotate3DAnimation(180, 90);   //开始前半段旋转动画
      myRotate3DAnimation.setAnimationListener(new Rotate3DAnimationListener() {
        @Override
        public void onAnimationEnd(Animation animation) {
          //切换显示(扑克牌正面)图像
```

```java
        myImageView.setImageDrawable(getDrawable(R.mipmap.mybmpfront));
        //重新初始化对象,以重置旋转参数
        myRotate3DAnimation = new Rotate3DAnimation(90, 0);
        myImageView.startAnimation(myRotate3DAnimation);          //执行后半段动画
    } });
    myImageView.startAnimation(myRotate3DAnimation);              //启动动画
}
public void onClickBtn2(View v) {                                 //响应单击"显示反面"按钮
    myRotate3DAnimation = new Rotate3DAnimation(0, 90);
    myRotate3DAnimation.setAnimationListener(new Rotate3DAnimationListener() {
        @Override
        public void onAnimationEnd(Animation animation) {
            //切换显示(扑克牌反面)图像
            myImageView.setImageDrawable(getDrawable(R.mipmap.mybmpback));
            //重新初始化对象,以重置旋转参数
            myRotate3DAnimation = new Rotate3DAnimation(90, 180);
            myImageView.startAnimation(myRotate3DAnimation);      //启动后半段动画
    } });
    myImageView.startAnimation(myRotate3DAnimation);              //启动动画
}
public class Rotate3DAnimation extends Animation {
    int myCenterX, myCenterY;                                     //旋转中心位置坐标
    Camera myCamera = new Camera();
    int myEndDegrees, myStartDegrees;                             //旋转前后的角度
    public Rotate3DAnimation(int startDegrees, int endDegrees) {
        myStartDegrees = startDegrees;
        myEndDegrees = endDegrees;
    }
    @Override
    public void initialize(int width, int height, int parentWidth, int parentHeight){
        super.initialize(width, height, parentWidth, parentHeight);
        myCenterX = width / 2;                                    //设置旋转中心位置坐标值
        myCenterY = height / 2;
        setFillAfter(true);                                       //设置在动画完成后是否保留状态
        setDuration(2000);                                        //设置动画时长2000ms
        //设置先加速后减速的时间插值器
        setInterpolator(new AccelerateDecelerateInterpolator());
    }
    @Override
    protected void applyTransformation(float interpolatedTime, Transformation t) {
        Matrix myMatrix = t.getMatrix();                          //获取矩阵对象
        myCamera.save();
        myCamera.rotateY(myStartDegrees);                         //设置旋转前角度
        myCamera.rotateY((myEndDegrees - myStartDegrees) * interpolatedTime);
        myCamera.getMatrix(myMatrix);                             //获取变换矩阵
        myMatrix.preTranslate(-myCenterX, -myCenterY);
        myMatrix.postTranslate(myCenterX, myCenterY);
        myCamera.restore();
} }
public class Rotate3DAnimationListener implements Animation.AnimationListener {
    @Override
    public void onAnimationStart(Animation animation){ }
    @Override
    public void onAnimationEnd(Animation animation){ }
    @Override
    public void onAnimationRepeat(Animation animation){ }
}
```

上面这段代码在 MyCode \ MySample871 \ app \ src \ main \ java \ com \ bin \ luo \ mysample \ MainActivity.java 文件中。此实例的完整项目在 MyCode\MySample871 文件夹中。

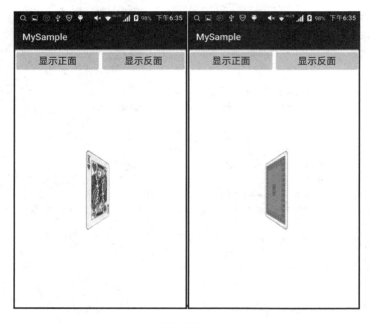

图　135.1

136　使用 AnimationSet 实现组合多个不同的动画

此实例主要通过使用 AnimationSet 的 playTogether() 方法叠加多个动画，实现多个动画同时执行的效果。当实例运行之后，单击"开始播放动画"按钮，则图像将在 3 秒内同时执行缩放和旋转的动作，效果分别如图 136.1 的左图和右图所示。

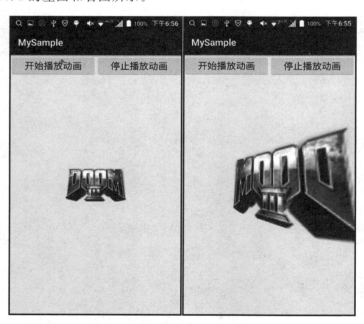

图　136.1

主要代码如下:

```java
public class MainActivity extends Activity {
    ImageView myImageView;
    AnimatorSet myAnimatorSet;
    @Override
    protected void onCreate(Bundle savedInstanceState) {
        super.onCreate(savedInstanceState);
        setContentView(R.layout.activity_main);
        myImageView = (ImageView) findViewById(R.id.myImageView);
        ObjectAnimator myObjectAnimator1 =
                ObjectAnimator.ofFloat(myImageView, "scaleX", 1.0f, 3.0f, 1.0f);
        ObjectAnimator myObjectAnimator2 =
                ObjectAnimator.ofFloat(myImageView, "scaleY", 1.0f, 3.0f, 1.0f);
        ObjectAnimator myObjectAnimator3 =
                ObjectAnimator.ofFloat(myImageView, "rotationY", 0.0f, 720.0f);
        myAnimatorSet = new AnimatorSet();
        myAnimatorSet.playTogether(myObjectAnimator1, myObjectAnimator2,
                                    myObjectAnimator3);
        myAnimatorSet.setDuration(3000);
        myAnimatorSet.start();
    }
    //响应单击"开始播放动画"按钮
    public void onClickButton1(View v){ myAnimatorSet.start(); }
    //响应单击"停止播放动画"按钮
    public void onClickButton2(View v){ myAnimatorSet.cancel(); }
}
```

上面这段代码在 MyCode\MySample171\app\src\main\java\com\bin\luo\mysample\MainActivity.java 文件中。在这段代码中,myObjectAnimator1 = ObjectAnimator.ofFloat(myImageView, "scaleX", 1.0f, 3.0f, 1.0f)表示在指定的时间内,myImageView 控件在水平方向放大 3 倍,然后还原。myObjectAnimator2 = ObjectAnimator.ofFloat (myImageView, "scaleY", 1.0f, 3.0f, 1.0f)表示在指定的时间内,myImageView 控件在垂直方向放大 3 倍,然后还原。myObjectAnimator3 = ObjectAnimator.ofFloat (myImageView, "rotationY", 0.0f, 720.0f)表示在指定的时间内,myImageView 控件围绕 y 轴旋转 720 度。myAnimatorSet.playTogether(myObjectAnimator1, myObjectAnimator2,myObjectAnimator3)表示将 myAnimator1、myAnimator2、myAnimator3 三个动画加载到 myAnimatorSet 动画集合中,每个动画的所有动作(旋转和缩放)在指定的时间内同时执行。此实例的完整项目在 MyCode\MySample171 文件夹中。

137 使用 Animation 实现按照顺序显示网格 Item

此实例主要通过使用 GridLayoutAnimationController 加载 XML 文件中的透明度改变动画,实现按照指定的顺序和方向在 GridView 的每个单元格(Item)中实现透明度从无到有显示动画。当实例运行之后,将按照从右到左、从下向上的顺序逐个显示在 GridView 的每个单元格中的图像,效果分别如图 137.1 的左图和右图所示。

主要代码如下：

```java
public class MainActivity extends Activity {
    @Override
    public void onCreate(Bundle savedInstanceState) {
        super.onCreate(savedInstanceState);
        setContentView(R.layout.activity_main);
        GridView myGridView = (GridView) findViewById(R.id.myGridView);
        ArrayList<HashMap<String,Object>> myArray = new ArrayList<>();
        HashMap<String,Object> myItem = new HashMap<>();
        //为节省篇幅，此处省略了加载 Item 代码，详细内容请看源文件
        //......
        SimpleAdapter myAdapter = new SimpleAdapter(this, myArray,
                    R.layout.myitem, new String[]{"myImage","myName"},
                    new int[]{R.id.myImage,R.id.myName});
        myGridView.setAdapter(myAdapter);
        Animation myAnimation = AnimationUtils.loadAnimation(this, R.anim.myanim);
        GridLayoutAnimationController myGridLayoutAnimationController =
                            new GridLayoutAnimationController(myAnimation);
        //设置动画方向为从右向左、从下向上
        myGridLayoutAnimationController.setDirection(
                GridLayoutAnimationController.DIRECTION_BOTTOM_TO_TOP|
                GridLayoutAnimationController.DIRECTION_RIGHT_TO_LEFT);
        myGridLayoutAnimationController.setDirectionPriority(
                GridLayoutAnimationController.PRIORITY_ROW);        //设置行优先级
        //在 myGridView 控件上设置动画
        myGridView.setLayoutAnimation(myGridLayoutAnimationController);
    }
}
```

图　137.1

上面这段代码在 MyCode\MySample862\app\src\main\java\com\bin\luo\mysample\ MainActivity. java 文件中。在这段代码中，Animation myAnimation ＝ AnimationUtils. loadAnimation(this, R. anim.

myanim)用于根据 myanim.xml 文件创建透明度动画,GridView 控件的(单元格)每个 Item 的透明度改变动画则由 myAnimation 实现,myanim.xml 文件的主要内容如下:

```
<?xml version = "1.0" encoding = "utf-8"?>
<set xmlns:android = "http://schemas.android.com/apk/res/android">
  <alpha android:duration = "1500"
         android:fromAlpha = "0.0"
         android:interpolator = "@android:anim/accelerate_interpolator"
         android:toAlpha = "1.0" />
</set>
```

上面这段代码在 MyCode\MySample862\app\src\main\res\anim\myanim.xml 文件中。在这段代码中,android:fromAlpha="0.0" 表示在动画开始时完全透明,android:toAlpha="1.0" 表示在动画结束时完全不透明,android:duration="1500" 表示动画持续时间是 1500ms。此实例的完整项目在 MyCode\MySample862 文件夹中。

138 使用 windowAnimations 实现缩放对话框窗口

此实例主要通过在以 DialogFragment 类为基类的自定义类 MyDialogFragment 中重写 onStart() 方法,实现在弹出或关闭提示框时显示缩放窗口动画。当实例运行之后,直接按下后退键,则将由小到大弹出一个提示框,单击该提示框中的"返回"按钮,则提示框将由大到小,直到消失,效果分别如图 138.1 的左图和右图所示。

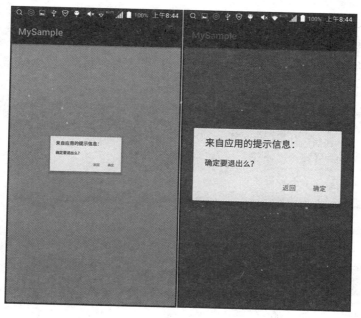

图 138.1

主要代码如下:

```
public class MainActivity extends Activity {
  @Override
  protected void onCreate(Bundle savedInstanceState) {
```

```
    super.onCreate(savedInstanceState);
    setContentView(R.layout.activity_main);
  }
  @Override
  public boolean onKeyDown(int keyCode,KeyEvent event){
    if(event.getKeyCode() == KeyEvent.KEYCODE_BACK){          //按下后退键显示对话框
      MyDialogFragment myDialogFragment = new MyDialogFragment();
      myDialogFragment.show(getFragmentManager(),"MyDialog");
      return true;
    }
    return super.onKeyDown(keyCode,event);
} }
```

上面这段代码在 MyCode \ MySample743 \ app \ src \ main \ java \ com \ bin \ luo \ mysample \ MainActivity.java 文件中。在这段代码中，myDialogFragment = new MyDialogFragment( ) 中的 MyDialogFragment 是自定义类，用于实现缩放窗口，该类的主要代码如下：

```
public class MyDialogFragment extends DialogFragment {
  @Override
  public void onStart() {
    super.onStart();
    Window myWindow = getDialog().getWindow();
    WindowManager.LayoutParams myAttrs = myWindow.getAttributes();
    myAttrs.windowAnimations = R.style.MyDialogAnim;
    myWindow.setAttributes(myAttrs);
  }
  @Override
  public Dialog onCreateDialog(Bundle savedInstanceState) {
    AlertDialog.Builder myBuilder = new AlertDialog.Builder(getActivity());
    myBuilder.setTitle("来自应用的提示信息:");
    myBuilder.setMessage("确定要退出么?");
    myBuilder.setPositiveButton("确定",new DialogInterface.OnClickListener(){
      @Override
      public void onClick(DialogInterface dialog, int which) {
        getActivity().finish();
        dialog.dismiss();
      } });
    myBuilder.setNegativeButton("返回", new DialogInterface.OnClickListener() {
      @Override
      public void onClick(DialogInterface dialog, int which){dialog.dismiss(); }
    });
    return myBuilder.create();
} }
```

上面这段代码在 MyCode \ MySample743 \ app \ src \ main \ java \ com \ bin \ luo \ mysample \ MyDialogFragment.java 文件中。在这段代码中，myAttrs.windowAnimations = R.style.MyDialogAnim 表示提示框在显示或关闭时使用 R.style.MyDialogAnim 指定的动画。MyDialogAnim 在 Styles 中指定，主要代码如下：

```
<resources>
<style name = "AppTheme" parent = "android:Theme.Material.Light.DarkActionBar"/>
```

```
<style name="MyDialogAnim">
  <item name="@android:windowEnterAnimation">@anim/myenter</item>
  <item name="@android:windowExitAnimation">@anim/myexit</item>
</style>
</resources>
```

上面这段代码在 MyCode\MySample743\app\src\main\res\values\styles.xml 文件中。在这段代码中，<item name="@android:windowEnterAnimation">@anim/myenter</item>表示进场动画使用 myenter 动画。<item name="@android:windowExitAnimation">@anim/myexit</item>表示退场动画使用 myexit 动画。myenter 动画文件的主要内容如下：

```
<?xml version="1.0" encoding="utf-8"?>
<set xmlns:android="http://schemas.android.com/apk/res/android">
 <scale android:duration="5000"
        android:fromXScale="0.0"
        android:fromYScale="0.0"
        android:pivotX="50%"
        android:pivotY="50%"
        android:toXScale="1.0"
        android:toYScale="1.0"/>
</set>
```

上面这段代码在 MyCode\MySample743\app\src\main\res\anim\myenter.xml 文件中。myexit 动画文件的主要内容如下：

```
<?xml version="1.0" encoding="utf-8"?>
<set xmlns:android="http://schemas.android.com/apk/res/android">
 <scale android:duration="5000"
        android:fromXScale="1.0"
        android:fromYScale="1.0"
        android:pivotX="50%"
        android:pivotY="50%"
        android:toXScale="0.0"
        android:toYScale="0.0"/>
</set>
```

上面这段代码在 MyCode\MySample743\app\src\main\res\anim\myexit.xml 文件中。此实例的完整项目在 MyCode\MySample743 文件夹中。

139 使用 AnimationDrawable 播放多幅图像

此实例主要通过使用 AnimationDrawable 存储多幅图像，并设置每幅图像之间的间隔时长，实现类似于 GIF 格式的动图效果。当实例运行之后，单击"开始播放"按钮，则开始逐帧播放多幅图像；单击"停止播放"按钮，则停止播放图像，效果分别如图 139.1 的左图和右图所示。

主要代码如下：

```
public class MainActivity extends Activity {
  AnimationDrawable myAnimationDrawable;
  @Override
```

```java
protected void onCreate(Bundle savedInstanceState) {
    super.onCreate(savedInstanceState);
    setContentView(R.layout.activity_main);
    ImageView myImageView = (ImageView) findViewById(R.id.myImageView);
    myAnimationDrawable = new AnimationDrawable();
    Drawable myDrawable1 = getResources().getDrawable(R.mipmap.mybmp001);
    Drawable myDrawable2 = getResources().getDrawable(R.mipmap.mybmp002);
    Drawable myDrawable3 = getResources().getDrawable(R.mipmap.mybmp003);
    Drawable myDrawable4 = getResources().getDrawable(R.mipmap.mybmp004);
    Drawable myDrawable5 = getResources().getDrawable(R.mipmap.mybmp005);
    Drawable myDrawable6 = getResources().getDrawable(R.mipmap.mybmp006);
    Drawable myDrawable7 = getResources().getDrawable(R.mipmap.mybmp007);
    Drawable myDrawable8 = getResources().getDrawable(R.mipmap.mybmp008);
    //将图像添加至AnimationDrawable对象中,并设定显示时长
    myAnimationDrawable.addFrame(myDrawable1, 100);
    myAnimationDrawable.addFrame(myDrawable2, 100);
    myAnimationDrawable.addFrame(myDrawable3, 1000);
    myAnimationDrawable.addFrame(myDrawable4, 100);
    myAnimationDrawable.addFrame(myDrawable5, 100);
    myAnimationDrawable.addFrame(myDrawable6, 100);
    myAnimationDrawable.addFrame(myDrawable7, 100);
    myAnimationDrawable.addFrame(myDrawable8, 100);
    myAnimationDrawable.setOneShot(false);              //一直播放该动画
    //在ImageView控件上应用该AnimationDrawable对象
    myImageView.setImageDrawable(myAnimationDrawable);
}
//响应单击"开始播放"按钮
public void onClickButton1(View v) {myAnimationDrawable.start(); }
//响应单击"停止播放"按钮
public void onClickButton2(View v) {myAnimationDrawable.stop();}
}
```

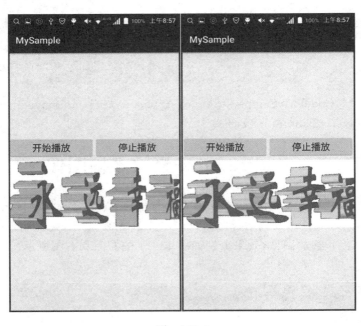

图 139.1

上面这段代码在 MyCode\MySample771\app\src\main\java\com\bin\luo\mysample\MainActivity.java 文件中。在这段代码中，myAnimationDrawable.addFrame(myDrawable1，100)表示将 myDrawable1（一幅图像）添加到 myAnimationDrawable 中，并且停留 100ms。myAnimationDrawable.setOneShot(false)表示一直播放该动画，如果参数值为 true，则在播放一遍后停止。此实例的完整项目在 MyCode\MySample771 文件夹中。

140　使用 AnimationDrawable 创建爆炸动画

此实例主要通过使用 AnimationDrawable 管理动画，并在 FrameLayout 的 OnTouch 事件响应方法中控制动画的显示位置，实现单击屏幕任意位置即可实现播放爆炸动画。当实例运行之后，单击屏幕任意位置即可播放爆炸动画，效果分别如图 140.1 的左图和右图所示。

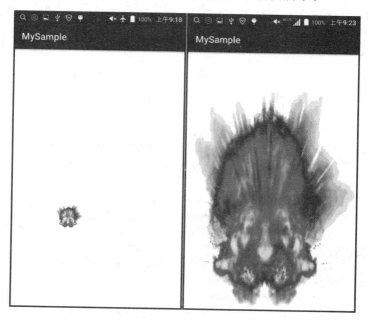

图　140.1

主要代码如下：

```java
public class MainActivity extends Activity {
    private MyView myView;
    private AnimationDrawable myAnimationDrawable;
    @Override
    protected void onCreate(Bundle savedInstanceState) {
        super.onCreate(savedInstanceState);
        FrameLayout myFrameLayout = new FrameLayout(this);
        myFrameLayout.setBackgroundColor(Color.WHITE);
        setContentView(myFrameLayout);
        myView = new MyView(this);
        myView.setBackgroundResource(R.drawable.mybmps);          //设置 myView 显示爆炸动画
        myView.setVisibility(View.INVISIBLE);                      //设置 myView 默认隐藏
        myAnimationDrawable = (AnimationDrawable)myView.getBackground();
        myFrameLayout.addView(myView);
        myFrameLayout.setOnTouchListener(new View.OnTouchListener(){
```

```java
    @Override
    public boolean onTouch(View source,MotionEvent event){
     if(event.getAction() == MotionEvent.ACTION_DOWN){
      myAnimationDrawable.stop();                          //停止动画播放
      float x = event.getX();
      float y = event.getY();
      myView.setLocation((int)y - 90 , (int)x - 90);       //控制动画的显示位置
      myView.setVisibility(View.VISIBLE);
      myAnimationDrawable.start();                         //开始动画播放
     }
     return false;
}});}
class MyView extends ImageView{                            //自定义 View,用于承载爆炸动画
 public MyView(Context context) { super(context); }
 public void setLocation(int top, int left){               //控制 MyView 的显示位置
  this.setFrame(left, top, left + 180, top + 180);
 }
 @Override
 protected void onDraw(Canvas canvas){ super.onDraw(canvas); }
} }
```

上面这段代码在 MyCode\ MySample147 \ app \ src \ main \ java \ com \ bin \ luo \ mysample \ MainActivity.java 文件中。在这段代码中,myView.setBackgroundResource(R.drawable.mybmps) 用于将动画资源 mybmps 设置为 myView 的背景资源。mybmps 是负责配置动画的 XML 文件,其主要代码如下:

```xml
<?xml version = "1.0" encoding = "utf - 8"?>
< animation - list xmlns:android = "http://schemas.android.com/apk/res/android"
              android:oneshot = "true">
 < item android:drawable = "@drawable/mybmp001" android:duration = "60" />
 < item android:drawable = "@drawable/mybmp002" android:duration = "60" />
 < item android:drawable = "@drawable/mybmp003" android:duration = "60" />
 < item android:drawable = "@drawable/mybmp004" android:duration = "60" />
 < item android:drawable = "@drawable/mybmp005" android:duration = "60" />
 < item android:drawable = "@drawable/mybmp006" android:duration = "60" />
 < item android:drawable = "@drawable/mybmp007" android:duration = "60" />
 < item android:drawable = "@drawable/mybmp008" android:duration = "60" />
 < item android:drawable = "@drawable/mybmp009" android:duration = "60" />
 < item android:drawable = "@drawable/mybmp010" android:duration = "60" />
 < item android:drawable = "@drawable/mybmp011" android:duration = "60" />
 < item android:drawable = "@drawable/mybmp012" android:duration = "60" />
 < item android:drawable = "@drawable/mybmp013" android:duration = "60" />
 < item android:drawable = "@drawable/mybmp014" android:duration = "60" />
</animation - list >
```

上面这段代码在 MyCode\MySample147\app\src\main\res\drawable\mybmps.xml 文件中。在这段代码中,< animation-list >元素为根节点,< item >节点定义了每一帧的参数,表示一个 drawable 资源的帧(动画图像)和帧间隔。设置 android:oneshot 属性为 true,表示此动画只执行一次,并停留在最后一帧。设置 android:oneshot 为 false,则表示动画循环播放。此实例的完整项目在 MyCode\MySample147 文件夹中。

141 使用 RotateAnimation 实现围绕自身中心旋转

此实例主要通过使用 RotateAnimation 创建旋转动画，实现图像以自身为中心进行旋转。当实例运行之后，单击"开始播放动画"按钮，则图像将以顺时针方向进行旋转，且永不停歇，效果分别如图 141.1 的左图和右图所示。

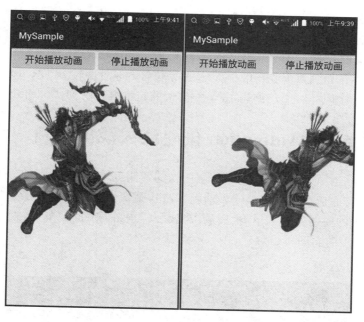

图 141.1

主要代码如下：

```java
public class MainActivity extends Activity {
 ImageView myImageView;
 RotateAnimation myRotateAnimation;
 @Override
 protected void onCreate(Bundle savedInstanceState) {
  super.onCreate(savedInstanceState);
  setContentView(R.layout.activity_main);
  myImageView = (ImageView) findViewById(R.id.myImageView);
  myImageView.setVisibility(View.GONE);
  myRotateAnimation = new RotateAnimation(0, 360, 540, 960);    //创建旋转动画
  myRotateAnimation.setDuration(3000);                           //设置动画持续时间3秒
  myRotateAnimation.setRepeatCount(Animation.INFINITE);          //设置动画无限次重复
  myImageView.setAnimation(myRotateAnimation);                   //在 ImageView 图像控件上添加动画
 }
 //响应单击"开始播放动画"按钮
 public void onClickButton1(View v) {
  myImageView.setVisibility(View.VISIBLE);
  myRotateAnimation.startNow();
 }
 //响应单击"停止播放动画"按钮
 public void onClickButton2(View v) {myRotateAnimation.cancel(); }
}
```

上面这段代码在 MyCode\MySample158\app\src\main\java\com\bin\luo\mysample\MainActivity.java 文件中。在这段代码中，myRotateAnimation = new RotateAnimation(0，360，540，960)用于创建一个旋转动画，RotateAnimation()方法的语法声明如下：

```
RotateAnimation(float fromDegrees,
                float toDegrees, float pivotX, float pivotY)
```

其中，float fromDegrees 表示开始角度。float toDegrees 表示结束角度。float pivotX 和 float pivotY 表示旋转中心点的坐标，pivotX 为距离左侧的偏移量，pivotY 为距离顶部的偏移量，即为相对于 View 左上角(0,0)的坐标。如果 View 的 width 是 100px，height 是 100px，则 RotateAnimation(0,10,100,100)将以右下角顶点为旋转中心点，从原始位置顺时针旋转 10 度；RotateAnimation(0,10,0,0)将以 View 的左上角为旋转中心点，旋转 10°。此实例的完整项目在 MyCode\MySample158 文件夹中。

142　使用 AlphaAnimation 创建淡入淡出动画

此实例主要通过使用 AlphaAnimation 控制图像的透明度，实现以动画的形式淡入淡出地显示或隐藏图像。当实例运行之后，单击"播放淡入动画"按钮，则图像（妖怪）将以淡入的动画风格在 5 秒内显示出来，如图 142.1 的左图所示。单击"播放淡出动画"按钮，则图像（妖怪）将以淡出的动画风格在 5 秒内消失，如图 142.1 的右图所示。

主要代码如下：

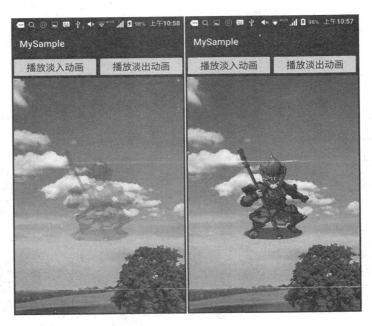

图　142.1

```
public class MainActivity extends Activity {
    AlphaAnimation myAnimationIn,myAnimationOut;
    ImageView myImageView;
    @Override
    protected void onCreate(Bundle savedInstanceState) {
        super.onCreate(savedInstanceState);
```

```
setContentView(R.layout.activity_main);
//配置淡入动画
myAnimationIn = new AlphaAnimation(0.0f, 1.0f);
myAnimationIn.setDuration(5000);
myAnimationIn.setFillAfter(true);
//配置淡出动画
myAnimationOut = new AlphaAnimation(1.0f, 0.0f);
myAnimationOut.setDuration(5000);
myAnimationOut.setFillAfter(true);
myImageView = (ImageView)findViewById(R.id.myImageView);
}
public void onClickButton1(View v) {           //响应单击"播放淡入动画"按钮
    myImageView.clearAnimation();              //在 myImageView 上清除动画
    //在 myImageView 上实施淡入动画
    myImageView.startAnimation(MainActivity.this.myAnimationIn);
}
public void onClickButton2(View v) {           //响应单击"播放淡出动画"按钮
    myImageView.clearAnimation();              //在 myImageView 上清除动画
    //在 myImageView 上实施淡出动画
    myImageView.startAnimation(MainActivity.this.myAnimationOut);
} }
```

上面这段代码在 MyCode \ MySample151 \ app \ src \ main \ java \ com \ bin \ luo \ mysample \ MainActivity.java 文件中。在这段代码中，myAnimationIn = new AlphaAnimation(0.0f，1.0f)用于构建一个透明度动画，参数 0.0f 为动画开始参数，表示完全透明；参数 1.0f 为动画结束参数，表示完全不透明。myAnimationIn.setDuration(5000)表示动画持续时间为 5 秒。myAnimationIn.setFillAfter(true)表示在动画结束时保持最后的状态。此实例的完整项目在 MyCode\MySample151 文件夹中。

143 使用 ScaleAnimation 创建缩放图像动画

此实例主要通过使用 ScaleAnimation 创建缩放动画，从而使图像以自身为中心在水平方向和垂直方向上进行缩放。当实例运行之后，单击"开始播放动画"按钮，则图像将以自身为中心进行放大，效果分别如图 143.1 的左图和右图所示。

主要代码如下：

```
//响应单击"开始播放动画"按钮
public void onClickButton1(View v) {
    //清除 myImageView 控件的动画
    myImageView.clearAnimation();
    myScaleAnimation = new ScaleAnimation(0.0f, 1.4f, 0.0f, 1.4f,
            Animation.RELATIVE_TO_SELF, 0.5f, Animation.RELATIVE_TO_SELF, 0.5f);
    //设置动画持续时间 3 秒
    myScaleAnimation.setDuration(3000);
    //在 myImageView 控件上添加动画
    myImageView.setAnimation(myScaleAnimation);
    myScaleAnimation.start();
}
//响应单击"停止播放动画"按钮
public void onClickButton2(View v) { myScaleAnimation.cancel(); }
```

图 143.1

上面这段代码在 MyCode\MySample159\app\src\main\java\com\bin\luo\mysample\MainActivity.java 文件中。在这段代码中，myScaleAnimation = new ScaleAnimation(0.0f, 1.4f, 0.0f, 1.4f, Animation.RELATIVE_TO_SELF, 0.5f, Animation.RELATIVE_TO_SELF, 0.5f)用于创建一个缩放动画，ScaleAnimation()构造函数的语法声明如下：

```
public ScaleAnimation(float fromX, float toX, float fromY, float toY,
    int pivotXType, float pivotXValue, int pivotYType, float pivotYValue)
```

其中，参数 float fromX 表示 x 坐标上的起始尺寸。参数 float toX 表示 x 坐标上的结束尺寸。参数 float fromY 表示 y 坐标上的起始尺寸。参数 float toY 表示 y 坐标上的结束尺寸。参数 int pivotXType 表示 x 轴的伸缩模式，可以取值为 ABSOLUTE、RELATIVE_TO_SELF、RELATIVE_TO_PARENT。参数 float pivotXValue 表示 x 坐标的伸缩值。参数 int pivotYType 表示 y 轴的伸缩模式，可以取值为 ABSOLUTE、RELATIVE_TO_SELF、RELATIVE_TO_PARENT。参数 float pivotYValue 表示 y 坐标的伸缩值。此实例的完整项目在 MyCode\MySample159 文件夹中。

144 在 ViewPager 中实现上下滑动的转场动画

此实例主要通过使用 ViewPager.PageTransformer 接口创建自定义转场动画类 VerticalPageTransformer，从而在 ViewPager 中实现以上滑下滑的转场动画切换页面。当实例运行之后，如果手指在屏幕上从左向右滑动，则上一张图像将从上向下滑入；如果手指在屏幕上从右向左滑动，则下一张图像将从下向上滑入，效果分别如图 144.1 的左图和右图所示。

主要代码如下：

```
private class VerticalPageTransformer implements ViewPager.PageTransformer {
    @Override
    public void transformPage(View view, float position) {
        if (position <= 1) {
            view.setTranslationX(view.getWidth() * -position);
            float myPosition = position * view.getHeight();
            view.setTranslationY(myPosition);                    //在垂直方向上平移图像
        } } }
```

上面这段代码在 MyCode \ MySample521 \ app \ src \ main \ java \ com \ bin \ luo \ mysample \ MainActivity.java 文件中。需要说明的是，在 Android Studio 中，使用此实例的相关控件需要在 gradle 中引入 compile 'com.android.support:design:23.3.0'依赖项。此实例的完整项目在 MyCode \MySample521 文件夹中。

图　144.1

145　通过下拉手指实现两个 Activity 的相互切换

此实例主要通过在 onTouchEvent()事件响应方法中计算手指在按下和抬起之间移动的垂直距离，判断手指是否执行了下拉动作，并据此从一个 Activity 跳转到另一个 Activity。当实例运行之后，MainActivity 如图 145.1 的左图所示；如果手指在 MainActivity 向下滑动，则将跳转到 NextActivity，如图 145.1 的右图所示。

图　145.1

主要代码如下：

```java
public class MainActivity extends Activity {
    //手指按下的点为(x1, y1),手指离开的点为(x2, y2)
    float x1 = 0, x2 = 0, y1 = 0, y2 = 0;
    @Override
    protected void onCreate(Bundle savedInstanceState) {
        super.onCreate(savedInstanceState);
        setContentView(R.layout.activity_main);
    }
    @Override
    public boolean onTouchEvent(MotionEvent event) {
        if (event.getAction() == MotionEvent.ACTION_DOWN) {           //当手指按下的时候
            x1 = event.getX();
            y1 = event.getY();
        }
        if (event.getAction() == MotionEvent.ACTION_UP) {             //当手指离开的时候
            x2 = event.getX();
            y2 = event.getY();
            if (y1 - y2 > 50) {
                //Toast.makeText(MainActivity.this, "向上滑", Toast.LENGTH_SHORT).show();
            } else if (y2 - y1 > 50) {
                //Toast.makeText(MainActivity.this, "向下滑", Toast.LENGTH_SHORT).show();
                //在下滑手指时实现由 MainActivity 切换到 NextActivity
                Intent myIntent = new Intent(MainActivity.this, NextActivity.class);
                startActivity(myIntent);
            } else if (x1 - x2 > 50) {
                Toast.makeText(MainActivity.this, "向左滑", Toast.LENGTH_SHORT).show();
            } else if (x2 - x1 > 50) {
                Toast.makeText(MainActivity.this, "向右滑", Toast.LENGTH_SHORT).show();
            }}
        return true;
    }}
```

上面这段代码在 MyCode\MySample425\app\src\main\java\com\bin\luo\mysample\MainActivity.java 文件中。在这段代码中，onTouchEvent()方法用于响应手指在屏幕上的滑动，y2－y1＞50 表示手指在垂直方向滑动的距离大于 50 像素。myIntent＝new Intent(MainActivity.this, NextActivity.class)表示在执行 startActivity(intent)方法后，将从 MainActivity 跳转到 NextActivity。此外需要注意的是，当在应用中新增 NextActivity 之后，需要在 AndroidManifest.xml 文件中添加＜activity android：name＝".NextActivity"/＞代码进行注册。此实例的完整项目在 MyCode\MySample425 文件夹中。

146　在应用启动时使用进场动画启动 Activity

此实例主要通过在 onCreate()方法中调用 overridePendingTransition()方法，实现在启动应用时显示进场动画。当实例运行之后，单击桌面上的当前应用图标，则应用将从小到大膨胀，如图 146.1 的左图所示；直到铺满整个屏幕，如图 146.1 的右图所示。

图 146.1

主要代码如下：

```java
public class MainActivity extends Activity {
 @Override
 protected void onCreate(Bundle savedInstanceState) {
  super.onCreate(savedInstanceState);
  overridePendingTransition(R.anim.myenter,R.anim.myexit);
  setContentView(R.layout.activity_main);
 }
}
```

上面这段代码在 MyCode \ MySample509 \ app \ src \ main \ java \ com \ bin \ luo \ mysample \ MainActivity.java 文件中。在这段代码中，overridePendingTransition（R.anim.myenter，R.anim.myexit）表示在启动应用时执行转场动画 myenter 和 myexit，myenter 动画（进场动画）的作用是使当前应用窗口由小变大，myexit 动画（退场动画）的作用是使手机桌面窗口由大变小。myenter 动画文件的主要内容如下：

```xml
<?xml version = "1.0" encoding = "utf-8"?>
<set xmlns:android = "http://schemas.android.com/apk/res/android">
 <scale android:duration = "5000"
        android:fromXScale = "0.0"
        android:fromYScale = "0.0"
        android:pivotX = "50%"
        android:pivotY = "50%"
        android:toXScale = "1.0"
        android:toYScale = "1.0"/>
</set>
```

上面这段代码在 MyCode\MySample509\app\src\main\res\anim\myenter.xml 文件中。myexit 动画文件的主要内容如下：

```xml
<?xml version = "1.0" encoding = "utf-8"?>
<set xmlns:android = "http://schemas.android.com/apk/res/android">
  <scale android:duration = "5000"
        android:fromXScale = "1.0"
        android:fromYScale = "1.0"
        android:pivotX = "50%"
        android:pivotY = "50%"
        android:toXScale = "0.0"
        android:toYScale = "0.0"/>
</set>
```

上面这段代码在 MyCode\MySample509\app\src\main\res\anim\myexit.xml 文件中。此实例的完整项目在 MyCode\MySample509 文件夹中。

147　以左入右出的动画效果切换两个 Activity

此实例主要通过使用 overridePendingTransition()方法，实现以从左向右滑入的动画效果切换两个 Activity。当实例运行之后，图 147.1 左图所示的图像（MainActivity 的背景图像）将向右滑出，图 147.1 右图所示的图像（SecondActivity 的背景图像）将从左滑入。

图　147.1

主要代码如下：

```java
public class MainActivity extends Activity {
  @Override
  protected void onCreate(Bundle savedInstanceState) {
    super.onCreate(savedInstanceState);
    setContentView(R.layout.activity_main);
    new Handler().postDelayed(new Runnable(){
      @Override
```

```
public void run() {
    Intent myIntent = new Intent(MainActivity.this,SecondActivity.class);
    MainActivity.this.startActivity(myIntent);
    MainActivity.this.finish();
    overridePendingTransition(android.R.anim.slide_in_left,
        android.R.anim.slide_out_right); } },1000);      //实现由左向右滑入的动画效果
} }
```

上面这段代码在 MyCode\MySample297\app\src\main\java\com\bin\luo\mysample\ MainActivity. java 文件中。在这段代码中，overridePendingTransition()方法用于在参数中加载滑入和滑出两个动画，该方法通常在 startActivity()或者 finish()执行之后调用。overridePendingTransition()方法的语法声明如下：

```
overridePendingTransition(int enterAnim, int exitAnim)
```

其中，参数 int enterAnim 表示跳转目标 Activity 的进场动画。参数 int exitAnim 表示当前 Activity 的离场动画。

此实例的完整项目在 MyCode\MySample297 文件夹中。

148 以收缩扩张的动画效果切换两个 Activity

此实例主要通过使用 overridePendingTransition()方法，实现以收缩扩张的动画效果切换两个 Activity。当实例运行之后，在图 148.1 左图所示的图像中，飞机所代表的 MainActivity 将从大到小逐渐收缩，同时轮船所代表的 SecondActivity 将从小到大逐渐扩张，直到完全铺满屏幕，如图 148.1 的右图所示。

图 148.1

主要代码如下：

```java
public class MainActivity extends Activity {
  @Override
  protected void onCreate(Bundle savedInstanceState) {
    super.onCreate(savedInstanceState);
    setContentView(R.layout.activity_main);
    new Handler().postDelayed(new Runnable() {
      @Override
      public void run() {
        Intent myIntent = new Intent(MainActivity.this, SecondActivity.class);
        MainActivity.this.startActivity(myIntent);
        MainActivity.this.finish();
        overridePendingTransition(R.anim.myzoomin, R.anim.myzoomout);
      } }, 500);              //以收缩和扩张动画切换两个 Activity
} }
```

上面这段代码在 MyCode\MySample299\app\src\main\java\com\bin\luo\mysample\MainActivity.java 文件中。在这段代码中，overridePendingTransition()方法用于在参数中加载收缩和扩张两个动画，扩张动画 myzoomin 的主要内容如下：

```xml
<?xml version = "1.0" encoding = "utf-8"?>
<set xmlns:android = "http://schemas.android.com/apk/res/android" >
<scale android:duration = "5000"
       android:fromXScale = "0.1"
       android:fromYScale = "0.1"
       android:pivotX = "50%p"
       android:pivotY = "50%p"
       android:toXScale = "1.0"
       android:toYScale = "1.0" />
</set>
```

上面这段代码在 MyCode\MySample299\app\src\main\res\anim\myzoomin.xml 文件中。收缩动画 myzoomout 的主要内容如下：

```xml
<?xml version = "1.0" encoding = "utf-8"?>
<set xmlns:android = "http://schemas.android.com/apk/res/android">
 <scale android:duration = "5000"
        android:fromXScale = "1.0"
        android:fromYScale = "1.0"
        android:pivotX = "50%p"
        android:pivotY = "50%p"
        android:toXScale = "0.1"
        android:toYScale = "0.1" />
</set>
```

上面这段代码在 MyCode\MySample299\app\src\main\res\anim\myzoomout.xml 文件中。此实例的完整项目在 MyCode\MySample299 文件夹中。

149 使用转场动画 Explode 切换两个 Activity

此实例主要演示了在 startActivity()方法中设置转场动画,实现以 Explode 动画方式切换两个 Activity。当实例运行之后,单击"以 Explode 动画进入第二个 Activity"按钮,如图 149.1 的左图所示,则第二个 Activity 将以收缩的效果出现。在第二个 Activity 中单击"以 Explode 动画返回到第一个 Activity"按钮,如图 149.1 的右图所示,则第一个 Activity 将以扩散的效果出现。

图 149.1

主要代码如下:

```
//响应单击"以 Explode 动画进入第二个 Activity"按钮
public void onClickButton1(View v) {
    Intent myIntent = new Intent(MainActivity.this,SecondActivity.class);
    startActivity(myIntent,
            ActivityOptions.makeSceneTransitionAnimation(this).toBundle());
}
```

上面这段代码在 MyCode\MySample503\app\src\main\java\com\bin\luo\mysample\ MainActivity.java 文件中。在这段代码中,startActivity(myIntent, ActivityOptions. makeSceneTransition-Animation(this). toBundle()) 表示从 MainActivity 跳转到 SecondActivity 时使用转场动画。SecondActivity 的主要代码如下:

```
public class SecondActivity extends Activity {
    @Override
    protected void onCreate(Bundle savedInstanceState) {
        super.onCreate(savedInstanceState);
        setContentView(R.layout.activity_second);
        getWindow().setEnterTransition(new Explode().setDuration(2000));   //进场动画
        getWindow().setExitTransition(new Explode().setDuration(2000));    //退场动画
```

```
}
//响应单击"以 Explode 动画返回到第一个 Activity"按钮
public void onClickButton2(View v) {
    this.onBackPressed();
}
}
```

上面这段代码在 MyCode\MySample503\app\src\main\java\com\bin\luo\mysample\SecondActivity.java 文件中。在这段代码中，getWindow(). setEnterTransition(new Explode(). setDuration(2000)) 表示在进入 SecondActivity 时，使用转场动画 Explode，持续时间是 2 秒。getWindow(). setExitTransition(new Explode(). setDuration(2000)) 表示在退出 SecondActivity 时，使用转场动画 Explode，持续时间是 2 秒。此外，当在项目中新增了 SecondActivity，则需要在 AndroidManifest.xml 文件中注册该 Activity，如< activity android:name=". SecondActivity"/>。此实例的完整项目在 MyCode\MySample503 文件夹中。

150　使用转场动画 Slide 切换两个 Activity

此实例主要通过在转场动画 Slide 的 setSlideEdge() 方法中设置滑动标志，实现以定制的方向切换两个 Activity。当实例运行之后，单击"以 Slide 动画进入到第二个 Activity"按钮，则第二个 Activity 将从屏幕右边滑向左边至全屏。在第二个 Activity 中单击"以 Slide 动画进入到第一个 Activity"按钮，则第二个 Activity 将从屏幕左边滑向右边至消失，从而显示第一个 Activity，效果分别如图 150.1 的左图和右图所示。

图　150.1

主要代码如下：

```
//响应单击"以 Slide 动画进入到第二个 Activity"按钮
public void onClickButton1(View v) {
```

```java
Intent myIntent = new Intent(MainActivity.this,SecondActivity.class);
startActivity(myIntent,
        ActivityOptions.makeSceneTransitionAnimation(this).toBundle());
}
```

上面这段代码在 MyCode\MySample535\app\src\main\java\com\bin\luo\mysample\MainActivity.java 文件中。在这段代码中，startActivity（myIntent，ActivityOptions.makeSceneTransitionAnimation(this).toBundle()）表示从 MainActivity 跳转到 SecondActivity 时使用转场动画。SecondActivity 的主要代码如下：

```java
public class SecondActivity extends Activity {
@Override
protected void onCreate(Bundle savedInstanceState) {
    super.onCreate(savedInstanceState);
    setContentView(R.layout.activity_second);
    Slide mySlide = new Slide();
    mySlide.setDuration(2000);
    mySlide.setSlideEdge(Gravity.RIGHT);
    getWindow().setEnterTransition(mySlide);      //进场动画
    getWindow().setExitTransition(mySlide);       //退场动画
}
//响应单击"以Slide动画进入到第一个Activity"按钮
public void onClickButton2(View v) { this.onBackPressed(); }
}
```

上面这段代码在 MyCode\MySample535\app\src\main\java\com\bin\luo\mysample\SecondActivity.java 文件中。在这段代码中，mySlide.setSlideEdge(Gravity.RIGHT)表示转场动画的滑动方向是从右向左。Gravity 的成员较多，setSlideEdge()方法仅支持下列成员：Gravity.LEFT、Gravity.TOP、Gravity.RIGHT、Gravity.BOTTOM、Gravity.START、Gravity.END。此外，当在项目中新增了 SecondActivity，则需要在 AndroidManifest.xml 文件中注册该 Activity，如< activity android:name=".SecondActivity"/>。此实例的完整项目在 MyCode\MySample535 文件夹中。

151 以指定位置的转场动画切换两个 Activity

此实例主要通过使用 ActivityOptionsCompat 的 makeScaleUpAnimation()方法，实现以指定位置和大小的放大动画来切换两个 Activity。当实例运行之后，在 MainActivity 中单击任意一幅图像（如左下角的那幅图像），则将以该图像的位置和大小放大该图像（SecondActivity）至全屏，效果分别如图 151.1 的左图和右图所示。

主要代码如下：

```java
//单击某图像之后在SecondActivity中显示对应的大图
public void onClickImageView(View v) {
    Intent myIntent = new Intent(this, SecondActivity.class);
    myIntent.putExtra("MyImageID", v.getId());
    ActivityOptionsCompat myActivityOptionsCompat =
                ActivityOptionsCompat.makeScaleUpAnimation(v, 0, 0,
                        v.getMeasuredWidth(), v.getMeasuredHeight());
    ActivityCompat.startActivity(MainActivity.this, myIntent,
                        myActivityOptionsCompat.toBundle());
}
```

图 151.1

上面这段代码在 MyCode\MySample536\app\src\main\java\com\bin\luo\mysample\MainActivity.java 文件中。在这段代码中，myActivityOptionsCompat = ActivityOptionsCompat.makeScaleUpAnimation(v，0，0，v.getMeasuredWidth()，v.getMeasuredHeight())用于在两个 Activity 切换时，指定执行放大动画的位置和大小，makeScaleUpAnimation()方法的语法声明如下：

```
makeScaleUpAnimation(View source, int startX,
                     int startY, int startWidth, int startHeight)
```

其中，参数 View source 是一个 view 对象，用于确定新 Activity 启动的初始坐标。参数 int startX 表示新 Activity 出现的初始 x 坐标，这个坐标相对于 source 的左上角 x 坐标。参数 int startY 表示新 Activity 出现的初始 y 坐标，这个坐标相对于 source 的左上角 y 坐标。参数 int startWidth 表示新 Activity 初始的宽度。参数 startHeight 表示新 Activity 初始的高度。

myIntent = new Intent(this，SecondActivity.class)表示新建一个 Intent，该 Intent 实现从 MainActivity 切换到 SecondActivity。SecondActivity 的主要代码如下：

```
public class SecondActivity extends Activity {
  ImageView myImageView;
  @Override
  protected void onCreate(Bundle savedInstanceState) {
    super.onCreate(savedInstanceState);
    getWindow().setFlags(WindowManager.LayoutParams.FLAG_FULLSCREEN,
        WindowManager.LayoutParams.FLAG_FULLSCREEN);                //设置全屏
    setContentView(R.layout.activity_second);
    int myImageID = getIntent().getExtras().getInt("MyImageID");
    myImageView = (ImageView)findViewById(R.id.myImageView);
    switch(myImageID){
      case R.id.myView1 :
        myImageView.setImageResource(R.mipmap.myimage1);
```

```
            break;
        case R.id.myView2:
            myImageView.setImageResource(R.mipmap.myimage2);
            break;
        case R.id.myView3:
            myImageView.setImageResource(R.mipmap.myimage3);
            break;
        case R.id.myView4:
            myImageView.setImageResource(R.mipmap.myimage4);
            break;
    } }
    //单击大图像之后返回 MainActivity
    public void onClickBigImage(View v) { this.onBackPressed(); }
}
```

上面这段代码在 MyCode \ MySample536 \ app \ src \ main \ java \ com \ bin \ luo \ mysample \ SecondActivity.java 文件中。需要说明的是,使用此实例的相关类需要在 gradle 中引入 compile 'com.android.support:design:25.0.1'依赖项。此外,当在项目中新增了 SecondActivity,则需要在 AndroidManifest.xml 文件中注册该 Activity,如< activity android:name= ".SecondActivity"/>。此实例的完整项目在 MyCode\MySample536 文件夹中。

152　在切换 Activity 时叠加缩放动画和转场动画

此实例主要演示了在 startActivity()方法中设置转场动画,并为 ImageView 控件增加缩放动画,实现在从 MainActivity 跳转到 SecondActivity 时同时显示转场动画和缩放动画。当实例运行之后,MainActivity 将显示 4 幅电影海报图像,如图 152.1 的左图所示;单击其中任意一幅电影海报图像,如右上角的那幅图像,则 SecondActivity 将从屏幕顶部滑向底部,同时该电影海报图像将从小变大,直到填满整个屏幕,如图 152.1 的右图所示;单击后退按钮,则 SecondActivity 将从屏幕底部滑向顶部,直到消失,从而显示第一个 Activity。

图　152.1

主要代码如下：

```java
//单击某图像之后在 SecondActivity 中显示对应的大图
public void onClickmyBtn1(View v) {
    ActivityOptions myActivityOptions =
            ActivityOptions.makeSceneTransitionAnimation(this, v, "");
    Intent myIntent = new Intent(this, SecondActivity.class);
    myIntent.putExtra("MyImageID", v.getId());
    startActivity(myIntent, myActivityOptions.toBundle());
}
```

上面这段代码在 MyCode\MySample506\app\src\main\java\com\bin\luo\mysample\MainActivity.java 文件中。在这段代码中，startActivity(myIntent, myActivityOptions.toBundle())表示从 MainActivity 跳转到 SecondActivity 时使用转场动画。SecondActivity 的主要代码如下：

```java
public class SecondActivity extends Activity {
    ImageView myImageView;
    @Override
    protected void onCreate(Bundle savedInstanceState) {
        super.onCreate(savedInstanceState);
        getWindow().setFlags(WindowManager.LayoutParams.FLAG_FULLSCREEN,
                WindowManager.LayoutParams.FLAG_FULLSCREEN);          //设置全屏
        setContentView(R.layout.activity_second);
        getWindow().setEnterTransition(new Explode().setDuration(2000));   //进场动画
        getWindow().setExitTransition(new Explode().setDuration(2000));    //退场动画
        int myImageID = getIntent().getExtras().getInt("MyImageID");
        myImageView = (ImageView)findViewById(R.id.myImageView);
        switch(myImageID){
            case R.id.myView1 :
                myImageView.setImageResource(R.mipmap.myimage1);
                break;
            case R.id.myView2:
                myImageView.setImageResource(R.mipmap.myimage2);
                break;
            case R.id.myView3:
                myImageView.setImageResource(R.mipmap.myimage3);
                break;
            case R.id.myView4:
                myImageView.setImageResource(R.mipmap.myimage4);
                break;
        }
        ScaleAnimation myScaleAnimation = new ScaleAnimation(0.0f, 1.0f,
                0.0f, 1.0f,Animation.RELATIVE_TO_SELF, 0.5f,
                Animation.RELATIVE_TO_SELF, 0.5f);          //创建缩放动画
        myScaleAnimation.setDuration(2000);                 //设置动画持续时间2秒
        myImageView.setAnimation(myScaleAnimation);         //在 ImageView 上添加缩放动画
        myScaleAnimation.start();
    }}
```

上面这段代码在 MyCode\MySample506\app\src\main\java\com\bin\luo\mysample\SecondActivity.java 文件中。在这段代码中，getWindow().setEnterTransition(new Explode().setDuration(2000))表示在进入 SecondActivity 时，使用转场动画 Explode，持续时间是2秒；该转场

动画将与 ImageView 控件的缩放动画 myScaleAnimation 同时执行。getWindow().setExitTransition(new Explode().setDuration(2000))表示在退出 SecondActivity 时,使用转场动画 Explode,持续时间是 2 秒。此外,当在项目中新增了 SecondActivity,则需要在 AndroidManifest.xml 文件中注册该 Activity,如< activity android:name = ".SecondActivity"/>。此实例的完整项目在 MyCode\ MySample506 文件夹中。

153　在切换 Activity 的转场动画中共享多对元素

此实例主要通过在 ActivityOptions 的 makeSceneTransitionAnimation()方法参数中新建 Pair 共享元素键值对,从而实现在两个不同 Activity 的多对元素间产生转场动画效果。当实例运行之后,由于 MainActivity 的左右两个 ImageView 控件(两幅电影海报)和 SecondActivity 的上下两个 ImageView 控件(两幅电影海报对应的大图像)被设置为两对共享元素,因此在单击 MainActivity 的 "进入 SecondActivity"按钮之后,在 SecondActivity 中将出现在 MainActivity 左上角的图像向右拉伸,在 MainActivity 右上角的图像向左下角拉伸的效果,效果分别如图 153.1 的左图和右图所示。

图　153.1

主要代码如下:

```
public void onClickmyBtn1(View v) {              //响应单击"进入 SecondActivity"按钮
    Intent myIntent = new Intent(this, SecondActivity.class);
    myIntent.putExtra("MyImage1", R.mipmap.myimage1);
    myIntent.putExtra("MyImage2", R.mipmap.myimage2);
    ActivityOptions options = ActivityOptions.makeSceneTransitionAnimation
        (MainActivity.this,new Pair<View,String>(myImageView1,"myShare1"),
            new Pair<View,String>(myImageView2,"myShare2"));
    startActivity(myIntent, options.toBundle());
}
```

上面这段代码在 MyCode\MySample537\app\src\main\java\com\bin\luo\mysample\

MainActivity.java 文件中。在这段代码中，Pair < View，String >（myImageView1，"myShare1"）与 SecondActivity 中的 ViewCompat.setTransitionName（myImageViewA，"myShare1"）形成一对共享元素；Pair < View，String >（myImageView2，"myShare2"）与 SecondActivity 中的 ViewCompat.setTransitionName（myImageViewB，"myShare2"）形成一对共享元素。SecondActivity 的主要代码如下：

```java
public class SecondActivity extends Activity {
    ImageView myImageViewA, myImageViewB;
    @Override
    protected void onCreate(Bundle savedInstanceState) {
        super.onCreate(savedInstanceState);
        getWindow().setFlags(WindowManager.LayoutParams.FLAG_FULLSCREEN,
                WindowManager.LayoutParams.FLAG_FULLSCREEN);
        setContentView(R.layout.activity_second);
        myImageViewA = (ImageView)findViewById(R.id.myImageViewA);
        myImageViewB = (ImageView)findViewById(R.id.myImageViewB);
        ViewCompat.setTransitionName(myImageViewA, "myShare1");
        ViewCompat.setTransitionName(myImageViewB, "myShare2");
        int myID1 = getIntent().getExtras().getInt("MyImage1");
        int myID2 = getIntent().getExtras().getInt("MyImage2");
        myImageViewA.setImageResource(myID1);
        myImageViewB.setImageResource(myID2);
    }
    //单击大图像之后返回 MainActivity
    public void onClickBigImage(View v) { this.onBackPressed(); }
}
```

上面这段代码在 MyCode\MySample537\app\src\main\java\com\bin\luo\mysample\SecondActivity.java 文件中。需要说明的是，使用此实例的相关类需要在 gradle 中引入 compile 'com.android.support:design:25.0.1'依赖项。此外，当在项目中新增了 SecondActivity，则需要在 AndroidManifest.xml 文件中注册该 Activity，如< activity android:name= ".SecondActivity"/>。此实例的完整项目在 MyCode\MySample537 文件夹中。

154　使用 FragmentTransaction 自定义转场动画

此实例主要通过使用 FragmentTransaction 的 setCustomAnimations（）方法，实现在切换 Fragment 时显示自定义的转场动画。当实例运行之后，将显示默认的 Fragment"异形魔怪"；单击第二个 Fragment 的标签"头脑特工队"，则 Fragment"异形魔怪"将向右滑出，Fragment"头脑特工队"将从左滑入，如图 154.1 的左图所示；单击第三个 Fragment 的标签"赛车总动员"，则 Fragment"头脑特工队"将向右滑出，Fragment"赛车总动员"将从左滑入，如图 154.1 的右图所示。

主要代码如下：

```java
public class MainActivity extends FragmentActivity
                            implements View.OnClickListener{
    private RadioButton myRadioButton1;
    private RadioButton myRadioButton2;
    private RadioButton myRadioButton3;
    private RadioGroup myRadioGroup;
```

```java
    private android.support.v4.app.Fragment myFragment1, myFragment2, myFragment3;
    private android.support.v4.app.FragmentManager myFragmentManager;
    private android.support.v4.app.FragmentTransaction myFragmentTransaction;
    @Override
    protected void onCreate(Bundle savedInstanceState) {
        super.onCreate(savedInstanceState);
        requestWindowFeature(Window.FEATURE_NO_TITLE);
        setContentView(R.layout.activity_main);
        myFragmentManager = getSupportFragmentManager();
        myFragmentTransaction = myFragmentManager.beginTransaction();
        myRadioButton1 = (RadioButton) findViewById(R.id.myRadioButton1);
        myRadioButton2 = (RadioButton) findViewById(R.id.myRadioButton2);
        myRadioButton3 = (RadioButton) findViewById(R.id.myRadioButton3);
        myRadioGroup = (RadioGroup) findViewById(R.id.rg);
        //为三个按钮(标签)添加监听
        myRadioButton1.setOnClickListener(this);
        myRadioButton2.setOnClickListener(this);
        myRadioButton3.setOnClickListener(this);
        //启动默认选中第一个
        myRadioGroup.check(R.id.myRadioButton1);
        myFragment1 = new Fragment1();
        myFragmentTransaction.replace(R.id.myContent, myFragment1);
        myFragmentTransaction.commit();
    }
    @Override
    public void onClick(View v) {
        myFragmentManager = getSupportFragmentManager();
        myFragmentTransaction = myFragmentManager.beginTransaction();
        myFragmentTransaction.setCustomAnimations(android.R.anim.slide_in_left,
                android.R.anim.slide_out_right);         //添加转场动画
        switch (v.getId()) {
          case R.id.myRadioButton1:
            myFragment1 = new Fragment1();
            myFragmentTransaction.replace(R.id.myContent, myFragment1);
            break;
          case R.id.myRadioButton2:
            myFragment2 = new Fragment2();
            myFragmentTransaction.replace(R.id.myContent, myFragment2);
            break;
          case R.id.myRadioButton3:
            myFragment3 = new Fragment3();
            myFragmentTransaction.replace(R.id.myContent, myFragment3);
            break;
          default:
            break;
        }
        myFragmentTransaction.commit();
    }}
```

上面这段代码在 MyCode\MySample513\app\src\main\java\com\bin\luo\mysample\MainActivity.java 文件中。在这段代码中，myFragmentTransaction.setCustomAnimations(android.R.anim.slide_in_left,android.R.anim.slide_out_right)用于自定义三个 Fragment 的转场

图 154.1

动画，android.R.anim.slide_in_left 表示进场动画，android.R.anim.slide_out_right 表示退场动画。需要说明的是，在 Android Studio 中使用此实例的相关类需要在 gradle 中引入 compile 'com.android.support:support-v4:24.2.0' 依赖项。此实例的完整项目在 MyCode\MySample513 文件夹中。

155 使用 TransitionManager 实现上下滑动动画

此实例主要通过指定 TransitionManager 的 beginDelayedTransition()方法参数为容器（控件的父节点），然后在修改布局参数后调用 setLayoutParams()方法，从而实现在修改布局参数后的子节点（控件）产生位移效果的过渡动画。当实例运行之后，小鸡图像（ImageView 控件）位于屏幕顶端，单击屏幕，则小鸡图像将沿着箭头方向从屏幕顶部滑向底部，如图 155.1 的左图所示。再次单击屏幕，则小鸡图像沿着箭头方向从屏幕底部滑向顶部，如图 155.1 的右图所示。

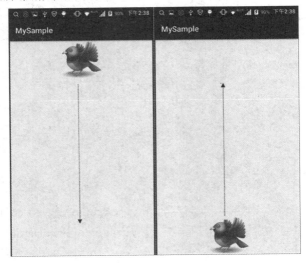

图 155.1

主要代码如下：

```java
public class MainActivity extends Activity {
  RelativeLayout myLayout;
  boolean bChanged = false;
  @Override
  protected void onCreate(Bundle savedInstanceState) {
    super.onCreate(savedInstanceState);
    setContentView(R.layout.activity_main);
    myLayout = (RelativeLayout)findViewById(R.id.myLayout);
  }
  //单击屏幕实现 ALIGN_PARENT_TOP 和 ALIGN_PARENT_BOTTOM 过渡动画切换
  public void onClickScreen(View v) {
    ViewGroup myRootView = (ViewGroup)findViewById(R.id.myRootView);
    TransitionManager.beginDelayedTransition(myRootView);
    RelativeLayout.LayoutParams lp = new RelativeLayout.LayoutParams(
            myLayout.getMeasuredWidth(),myLayout.getMeasuredHeight());
    if (bChanged) { lp.addRule(RelativeLayout.ALIGN_PARENT_TOP); }
    else { lp.addRule(RelativeLayout.ALIGN_PARENT_BOTTOM); }
    bChanged = ! bChanged;
    myLayout.setLayoutParams(lp);
} }
```

上面这段代码在 MyCode\MySample592\app\src\main\java\com\bin\luo\mysample\MainActivity.java 文件中。在这段代码中，TransitionManager.beginDelayedTransition(myRootView)表示在父容器 myRootView 中进行过渡动画。myRootView =（ViewGroup）findViewById(R.id.myRootView)用于根据 activity_main 布局中的 RelativeLayout 创建动画容器。activity_main 布局的主要内容如下：

```xml
<?xml version = "1.0" encoding = "utf-8"?>
<RelativeLayout xmlns:android = "http://schemas.android.com/apk/res/android"
                xmlns:tools = "http://schemas.android.com/tools"
                android:id = "@+id/activity_main"
                android:layout_width = "match_parent"
                android:layout_height = "match_parent"
                android:onClick = "onClickScreen"
                tools:context = "com.bin.luo.mysample.MainActivity">
<!-- 这个 RelativeLayout 用来做动画的父布局 -->
<RelativeLayout android:id = "@+id/myRootView"
                android:gravity = "center_horizontal"
                android:layout_width = "match_parent"
                android:layout_height = "match_parent">
  <include layout = "@layout/myscene1"></include>
</RelativeLayout></RelativeLayout>
```

上面这段代码在 MyCode\MySample592\app\src\main\res\layout\activity_main.xml 文件中。在这段代码中，< include layout = "@layout/myscene1"></include >表示容器中的子布局是 myscene1，myscene1 布局的主要内容如下：

```xml
<?xml version = "1.0" encoding = "utf-8"?>
<RelativeLayout xmlns:android = "http://schemas.android.com/apk/res/android"
                android:id = "@+id/myLayout"
                android:layout_width = "wrap_content"
                android:layout_height = "wrap_content">
```

```
< ImageView android:layout_width = "100dp"
            android:layout_height = "100dp"
            android:src = "@mipmap/myimage1"/>
</RelativeLayout>
```

上面这段代码在 MyCode\MySample592\app\src\main\res\layout\myscene1.xml 文件中。需要说明的是，使用此实例的相关类需要在 gradle 中引入 compile 'com.android.support:design:25.0.1'依赖项。此实例的完整项目在 MyCode\MySample592 文件夹中。

156　使用 TransitionManager 实现围绕 Y 轴旋转

此实例主要通过指定 TransitionManager 的 beginDelayedTransition()方法参数分别为容器（控件的父节点）和 ChangeTransform 动画，并修改控件的 RotationY 属性，从而实现控件围绕 y 轴旋转的动画效果。当实例运行之后，ImageView 控件（电影海报图像）位于屏幕中心，单击屏幕，则 ImageView 控件（电影海报图像）将围绕 y 轴正向旋转 45°，如图 156.1 的右图所示。再次单击屏幕，则 ImageView 控件（电影海报图像）将围绕 y 轴反向旋转 45°，如图 156.1 的左图所示。

图　156.1

主要代码如下：

```
//单击屏幕实现图像围绕 y 轴旋转的动画切换
public void onClickScreen(View v) {
  ViewGroup myRootView = (ViewGroup)findViewById(R.id.myRootView);
  TransitionManager.beginDelayedTransition(myRootView,
                  new ChangeTransform().setDuration(2000));
  if (bChanged) { myImageView.setRotationY(0f);}
  else { myImageView.setRotationY(45f); }
  bChanged = !bChanged;
}
```

上面这段代码在 MyCode\MySample600\app\src\main\java\com\bin\luo\mysample\MainActivity.java 文件中。在这段代码中，TransitionManager.beginDelayedTransition(myRootView,new ChangeTransform().setDuration(2000))表示在父容器 myRootView 中执行 ChangeTransform 过渡动画，ChangeTransform 过渡动画主要用于改变 ImageView 控件的缩放比例和旋转角度。myRootView =（ViewGroup）findViewById(R.id.myRootView)用于根据 activity_main 布局中的 RelativeLayout 创建动画容器。activity_main 布局的主要内容如下：

```xml
<?xml version = "1.0" encoding = "utf-8"?>
<RelativeLayout xmlns:android = "http://schemas.android.com/apk/res/android"
                xmlns:tools = "http://schemas.android.com/tools"
                android:id = "@+id/activity_main"
                android:layout_width = "match_parent"
                android:layout_height = "match_parent"
                android:onClick = "onClickScreen"
                tools:context = "com.bin.luo.mysample.MainActivity">
<!-- 这个 RelativeLayout 用来做动画的父布局 -->
<RelativeLayout android:id = "@+id/myRootView"
                android:gravity = "center_vertical|center_horizontal"
                android:layout_width = "match_parent"
                android:layout_height = "match_parent">
  <ImageView android:layout_width = "wrap_content"
             android:layout_height = "wrap_content"
             android:padding = "50dp"
             android:src = "@mipmap/myimage1"
             android:id = "@+id/myImageView"/>
</RelativeLayout></RelativeLayout>
```

上面这段代码在 MyCode\MySample600\app\src\main\res\layout\activity_main.xml 文件中。需要说明的是，使用此实例的相关类需要在 gradle 中引入 compile 'com.android.support:design:25.0.1'依赖项。此实例的完整项目在 MyCode\MySample600 文件夹中。

157 使用 TransitionManager 实现 Fade 动画效果

此实例主要通过指定 TransitionManager 的 beginDelayedTransition()方法的参数分别为容器（控件的父节点）和 Fade 动画，然后通过自定义方法改变控件的 Visibility 属性，从而实现控件在显示或隐藏时产生淡入淡出的过渡动画效果。当实例运行之后，两个 ImageView 控件（电影海报图像）位于屏幕中心，如图 157.1 的左图所示；单击屏幕，则当前的两个 ImageView 控件（电影海报图像）淡出，显示另外两个 ImageView 控件（电影海报图像），如图 157.1 的右图所示。再次单击屏幕，则以前隐藏的两个 ImageView 控件（电影海报图像）淡入，如图 157.1 的左图所示。

主要代码如下：

```java
public class MainActivity extends Activity {
 ImageView myImageView1,myImageView2,myImageView3,myImageView4;
 @Override
 protected void onCreate(Bundle savedInstanceState) {
  super.onCreate(savedInstanceState);
  setContentView(R.layout.activity_main);
  myImageView1 = (ImageView)findViewById(R.id.myImageView1);
```

```
    myImageView2 = (ImageView)findViewById(R.id.myImageView2);
    myImageView3 = (ImageView)findViewById(R.id.myImageView3);
    myImageView4 = (ImageView)findViewById(R.id.myImageView4);
}
//单击屏幕实现淡入淡出动画切换
public void onClickScreen(View v) {
    ViewGroup myRootView = (ViewGroup)findViewById(R.id.myRootView);
    TransitionManager.beginDelayedTransition( myRootView,
                                new Fade().setDuration(5000));
    toggleVisibility(myImageView3,myImageView4);
}
private static void toggleVisibility(View... views){
    for (View v:views) {
        boolean isVisible = v.getVisibility() == View.VISIBLE;
        v.setVisibility(isVisible?View.INVISIBLE:View.VISIBLE);
} } }
```

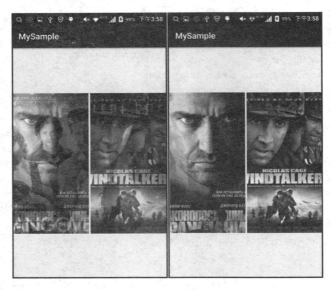

图 157.1

上面这段代码在 MyCode\MySample594\app\src\main\java\com\bin\luo\mysample\MainActivity.java 文件中。在这段代码中，TransitionManager.beginDelayedTransition（myRootView，new Fade().setDuration（5000））表示在父容器 myRootView 中执行淡出淡入过渡动画。ViewGroup myRootView＝(ViewGroup)findViewById（R.id.myRootView）用于根据 activity_main 布局中的 RelativeLayout 创建动画容器。activity_main 布局的主要内容如下：

```xml
<?xml version = "1.0" encoding = "utf-8"?>
<RelativeLayout xmlns:android = "http://schemas.android.com/apk/res/android"
                xmlns:tools = "http://schemas.android.com/tools"
                android:id = "@+id/activity_main"
                android:layout_width = "match_parent"
                android:layout_height = "match_parent"
                android:onClick = "onClickScreen"
                tools:context = "com.bin.luo.mysample.MainActivity">
<!-- 这个 RelativeLayout 用来做动画的父布局 -->
```

```xml
<RelativeLayout android:id = "@+id/myRootView"
                android:gravity = "center_vertical"
                android:layout_width = "match_parent"
                android:layout_height = "match_parent">
    <include layout = "@layout/myscene1"></include>
</RelativeLayout></RelativeLayout>
```

上面这段代码在 MyCode\MySample594\app\src\main\res\layout\activity_main.xml 文件中。在这段代码中，<include layout = "@layout/myscene1"></include> 表示容器中的子布局是 myscene1，myscene1 布局的主要内容如下：

```xml
<?xml version = "1.0" encoding = "utf-8"?>
<RelativeLayout xmlns:android = "http://schemas.android.com/apk/res/android"
                android:id = "@+id/myLayout"
                android:layout_width = "wrap_content"
                android:layout_height = "wrap_content">
<LinearLayout android:layout_width = "match_parent"
              android:layout_height = "wrap_content"
              android:orientation = "horizontal">
    <ImageView android:id = "@+id/myImageView1"
               android:layout_width = "match_parent"
               android:layout_height = "match_parent"
               android:layout_margin = "2dp"
               android:layout_weight = "1"
               android:scaleType = "centerCrop"
               android:src = "@mipmap/myimage1"/>
    <ImageView android:id = "@+id/myImageView2"
               android:layout_width = "match_parent"
               android:layout_height = "match_parent"
               android:layout_margin = "2dp"
               android:layout_weight = "1"
               android:scaleType = "centerCrop"
               android:src = "@mipmap/myimage2"/>
</LinearLayout>
<LinearLayout android:layout_width = "match_parent"
              android:layout_height = "wrap_content"
              android:orientation = "horizontal">
    <ImageView android:id = "@+id/myImageView3"
               android:layout_width = "match_parent"
               android:layout_height = "match_parent"
               android:layout_margin = "2dp"
               android:layout_weight = "1"
               android:scaleType = "centerCrop"
               android:src = "@mipmap/myimage3"/>
    <ImageView android:id = "@+id/myImageView4"
               android:layout_width = "match_parent"
               android:layout_height = "match_parent"
               android:layout_margin = "2dp"
               android:layout_weight = "1"
               android:scaleType = "centerCrop"
               android:src = "@mipmap/myimage4"/>
</LinearLayout>
</RelativeLayout>
```

上面这段代码在 MyCode\MySample594\app\src\main\res\layout\myscene1.xml 文件中。需

要说明的是，使用此实例的相关类需要在 gradle 中引入 compile 'com.android.support:design:25.0.1'依赖项。此实例的完整项目在 MyCode\MySample594 文件夹中。

158　使用 TransitionManager 组合多个不同动画

此实例主要通过在 TransitionManager 的 beginDelayedTransition()方法中组合 Explode 和 ChangeBounds 动画，实现放大缩小图像和四散飞出的特殊效果。当实例运行之后，四幅图像并列居中，如图 158.1 的左图所示。单击任意一幅图像，如右上角的图像，则该图像放大 2.5 倍，其他图像沿着各自的方向飞出屏幕，如图 158.1 的右图所示。单击放大的图像，则该图像恢复原状，其他图像沿着各自的方向飞入屏幕。

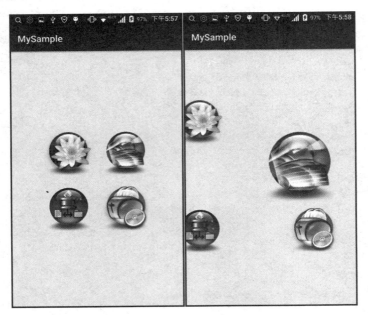

图　158.1

主要代码如下：

```java
public class MainActivity extends Activity implements View.OnClickListener{
    private ImageView myImageView1,myImageView2, myImageView3, myImageView4;
    private boolean bLarge;
    private ViewGroup myRootView;
    private int mySize;
    @Override
    protected void onCreate(Bundle savedInstanceState) {
        super.onCreate(savedInstanceState);
        setContentView(R.layout.activity_main);
        myRootView = (ViewGroup) findViewById(R.id.myRootView);
        myImageView1 = (ImageView) findViewById(R.id.myImageView1);
        myImageView2 = (ImageView) findViewById(R.id.myImageView2);
        myImageView3 = (ImageView) findViewById(R.id.myImageView3);
        myImageView4 = (ImageView) findViewById(R.id.myImageView4);
        mySize = myImageView1.getLayoutParams().width;
        myImageView1.setOnClickListener(this);
```

```java
    myImageView2.setOnClickListener(this);
    myImageView3.setOnClickListener(this);
    myImageView4.setOnClickListener(this);
}
//响应单击当前的 ImageView 控件
public void onClick(View view) {
    Explode myExplode = new Explode();
    ChangeBounds myChangeBounds = new ChangeBounds();
    TransitionSet myTransitions = new TransitionSet();
    myTransitions.addTransition(myExplode).addTransition(myChangeBounds);
    myTransitions.setDuration(500);
    TransitionManager.beginDelayedTransition(myRootView, myTransitions);
    changeSize(view);
    changeVisibility(myImageView1, myImageView2, myImageView3, myImageView4);
    view.setVisibility(View.VISIBLE);
}
//放大当前单击的 ImageView 控件(导致 ChangeBounds 动画发生)
private void changeSize(View view) {
    bLarge = !bLarge;
    ViewGroup.LayoutParams layoutParams = view.getLayoutParams();
    if(bLarge){
        layoutParams.width = (int)(2.5 * mySize);
        layoutParams.height = (int)(2.5 * mySize);
    }else {
        layoutParams.width = mySize;
        layoutParams.height = mySize;
    }
    view.setLayoutParams(layoutParams);
}
//隐藏(或显示)非当前单击的 ImageView 控件(导致 Explode 动画发生)
private void changeVisibility(View ...views){
    for (View view:views){view.setVisibility(view.getVisibility() ==
                View.VISIBLE?View.INVISIBLE:View.VISIBLE); }
} }
```

上面这段代码在 MyCode \ MySample628 \ app \ src \ main \ java \ com \ bin \ luo \ mysample \ MainActivity.java 文件中。在这段代码中，TransitionManager.beginDelayedTransition(myRootView，myTransitions)中的 myTransitions 动画在控件尺寸发生改变或 Visibility 属性发生变化时触发。myRootView =（ViewGroup）findViewById(R.id.myRootView)用于根据 activity_main 布局中的 RelativeLayout 创建动画容器。activity_main 布局的主要内容如下：

```xml
<RelativeLayout android:id = "@ + id/myRootView"
            android:layout_width = "match_parent"
            android:layout_height = "match_parent"
            android:layout_centerInParent = "true"
            android:gravity = "center_horizontal|center_vertical">
    <ImageView android:id = "@ + id/myImageView1"
            android:layout_width = "100dp"
            android:layout_height = "100dp"
            android:layout_margin = "10dp"
            android:src = "@mipmap/myimage1"/>
    <ImageView android:id = "@ + id/myImageView2"
```

```
                android:layout_width = "100dp"
                android:layout_height = "100dp"
                android:layout_margin = "10dp"
                android:layout_toRightOf = "@id/myImageView1"
                android:src = "@mipmap/myimage2"/>
    < ImageView android:id = "@ + id/myImageView3"
                android:layout_width = "100dp"
                android:layout_height = "100dp"
                android:layout_below = "@id/myImageView1"
                android:layout_margin = "10dp"
                android:src = "@mipmap/myimage3"/>
    < ImageView android:id = "@ + id/myImageView4"
                android:layout_width = "100dp"
                android:layout_height = "100dp"
                android:layout_below = "@id/myImageView1"
                android:layout_margin = "10dp"
                android:layout_toRightOf = "@id/myImageView3"
                android:src = "@mipmap/myimage4"/>
</RelativeLayout >
```

上面这段代码在 MyCode\MySample628\app\src\main\res\layout\activity_main.xml 文件中。此实例的完整项目在 MyCode\MySample628 文件夹中。

159 使用 TransitionManager 实现单布局过渡动画

此实例主要通过使用 setLayoutParams()方法修改控件的 LayoutParams 布局参数，实现使 TransitionManager 的 beginDelayedTransition()方法中的 ChangeBounds 过渡动画参数（图像）产生动画效果。当实例运行之后，小鸟图像（ImageView 控件）位于屏幕左上角，单击屏幕，则小鸟图像将沿着弧线路径（细实线，实际不可见）从屏幕左上角滑向右下角，如图 159.1 的左图所示。再次单击屏幕，则小鸟图像将沿弧线路径从屏幕右下角滑向屏幕左上角，如图 159.1 的右图所示。

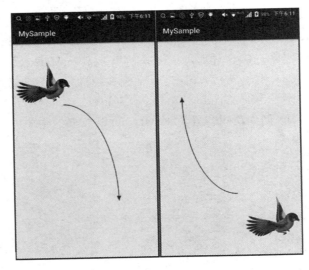

图 159.1

主要代码如下:

```
//单击屏幕实现左上角和右下角的弧线路径动画切换
public void onClickScreen(View v) {
    ChangeBounds myChangeBounds = new ChangeBounds();
    myChangeBounds.setPathMotion(new ArcMotion());          //设置弧线路径
    myChangeBounds.setDuration(5000);                        //设置动画时间 5 秒
    ViewGroup myRootView = (ViewGroup) findViewById(R.id.myRootView);
    TransitionManager.beginDelayedTransition(myRootView, myChangeBounds);
    FrameLayout.LayoutParams myLayoutParams =
            (FrameLayout.LayoutParams) myImageView.getLayoutParams();
    myLayoutParams.gravity = bChanged ? (Gravity.LEFT | Gravity.TOP) :
            (Gravity.BOTTOM | Gravity.RIGHT);
    myImageView.setLayoutParams(myLayoutParams);
    bChanged = !bChanged;
}
```

上面这段代码在 MyCode\MySample610\app\src\main\java\com\bin\luo\mysample\MainActivity.java 文件中。在这段代码中,TransitionManager.beginDelayedTransition(myRootView,myChangeBounds)表示在父容器 myRootView 中执行 myChangeBounds 动画。在 Android 中,ChangeBounds 动画通常在 LayoutParams 属性发生改变的情况下执行。myLayoutParams.gravity = bChanged ? (Gravity.LEFT | Gravity.TOP) :(Gravity.BOTTOM | Gravity.RIGHT)和 myImageView.setLayoutParams(myLayoutParams)表示如果当前 ImageView 控件在屏幕左上角,则通过设置 LayoutParams 属性将其置于右下角;如果当前 ImageView 控件在屏幕右下角,则通过设置 LayoutParams 属性将其置于左上角;一旦控件的 LayoutParams 属性发生改变即触发 ChangeBounds 动画。myRootView =(ViewGroup)findViewById(R.id. myRootView)用于根据 activity_main 布局中的 FrameLayout 创建动画容器。activity_main 布局的主要内容如下:

```
<FrameLayout android:id = "@ + id/myRootView"
            android:layout_width = "match_parent"
            android:layout_height = "match_parent">
    <ImageView android:layout_width = "150dp"
            android:layout_height = "200dp"
            android:padding = "2dp"
            android:layout_alignParentLeft = "true"
            android:layout_alignParentTop = "true"
            android:src = "@mipmap/myimage1"
            android:id = "@ + id/myImageView"/>
</FrameLayout>
```

上面这段代码在 MyCode\MySample610\app\src\main\res\layout\activity_main.xml 文件中。此实例的完整项目在 MyCode\MySample610 文件夹中。

160 使用 TransitionManager 实现平移过渡动画

此实例主要通过使用 TransitionManager 的 go()方法,使两个布局(场景)实现以 Slide 动画风格滑出滑入。当实例运行之后,将全屏显示重庆图像(布局),单击重庆图像,则该图像将向下滑出,同时上海图像(布局)则向上滑入至全屏显示,效果分别如图 160.1 的左图和右图所示。

图 160.1

主要代码如下：

```
//单击重庆图像(布局)之后过渡到上海图像(布局)
public void onClickImageView1(View v) {
    ViewGroup myMainLayout = (ViewGroup)findViewById(R.id.activity_main);
    Scene mySecondScene = Scene.getSceneForLayout(myMainLayout,
                                        R.layout.layout_second, this);
    Slide mySlide = new Slide();
    mySlide.setDuration(5000);
    mySlide.setSlideEdge(Gravity.BOTTOM);
    Transition myTransition = mySlide;
    TransitionManager.go(mySecondScene, myTransition);
}
//单击上海图像(布局)之后过渡到重庆图像(布局)
public void onClickImageView2(View v) {
    ViewGroup mySecondLayout = (ViewGroup)findViewById(R.id.layout_second);
    Scene myMainScene = Scene.getSceneForLayout(mySecondLayout,
                                        R.layout.activity_main, this);
    Slide mySlide = new Slide();
    mySlide.setDuration(5000);
    mySlide.setSlideEdge(Gravity.BOTTOM);
    Transition myTransition = mySlide;
    TransitionManager.go(myMainScene, myTransition);
}
```

上面这段代码在 MyCode\MySample544\app\src\main\java\com\bin\luo\mysample\MainActivity.java 文件中。在这段代码中，mySlide.setSlideEdge(Gravity.BOTTOM)表示当前布局（图像）从底部滑出、其他图像从底部滑入；如果是 mySlide.setSlideEdge(Gravity.TOP)，则表示当前布局（图像）从顶部滑出、其他图像从顶部滑入；如果是 mySlide.setSlideEdge(Gravity.LEFT)，则表示当前布局（图像）从左边滑出、其他图像从左边滑入；如果是 mySlide.setSlideEdge(Gravity.RIGHT)，则表示当前布局（图像）从右边滑出、其他图像从右边滑入。TransitionManager.go （mySecondScene，myTransition）表示从 activity_main 布局过渡到 layout_second 布局。关于这两个布局文件的主要代码请参考源代码中的 MyCode\MySample544\app\src\main\res\layout\activity_main.xml 文件和 MyCode\MySample544\app\src\main\res\layout\layout_second.xml 文件。需要说明的是，使用此实例的相关类需要在 gradle 中引入 compile 'com.android.support:design:25.0.1' 依赖项。此实例的完整项目在 MyCode\MySample544 文件夹中。

161　使用 TransitionManager 实现缩放部分图像

此实例主要通过使用 setClipBounds()方法指定场景过渡动画的（裁剪）范围，实现在两个布局过渡的过程中仅放大或缩小控件（ImageView）在范围内的部分（图像）。当实例运行之后，ImageView 控件（图像）位于屏幕的正中（即 myscene1），单击屏幕，则 ImageView 控件（图像的右下角部分）将由小变大，直到铺满屏幕的右下角，如图 161.1 的右图所示（即 myscene2）。在 myscene2 中单击屏幕，则 ImageView 控件（图像的右下角部分）将由大变小，如图 161.1 的左图所示（即 myscene1）。

图　161.1

主要代码如下：

```
//单击 myscene1 过渡到 myscene2(图像的右下角由小变大)
public void onClickScene1(View v) {
    ViewGroup myRootView = (ViewGroup)findViewById(R.id.myRootView);
    //设置过渡动画的范围在右下角
    myRootView.setClipBounds(new Rect(myRootView.getWidth()/2,
        myRootView.getHeight()/2,myRootView.getWidth(),myRootView.getHeight()));
    Scene myScene2 = Scene.getSceneForLayout(myRootView, R.layout.myscene2, this);
    TransitionManager.go(myScene2, new ChangeBounds().setDuration(5000));
}
//单击 myscene2 过渡到 myscene1(图像的右下角由大变小)
public void onClickScene2(View v) {
    ViewGroup myRootView = (ViewGroup)findViewById(R.id.myRootView);
```

```
    //设置过渡动画的范围在右下角
    myRootView.setClipBounds(new Rect(myRootView.getWidth()/2,
        myRootView.getHeight()/2,myRootView.getWidth(),myRootView.getHeight()));
    Scene myScene1 = Scene.getSceneForLayout(myRootView, R.layout.myscene1, this);
    TransitionManager.go(myScene1, new ChangeBounds().setDuration(5000));
}
```

上面这段代码在 MyCode\MySample554\app\src\main\java\com\bin\luo\mysample\MainActivity.java 文件中。在这段代码中，TransitionManager.go(myScene2，new ChangeBounds().setDuration(5000))表示以 ChangeBounds 动画风格在 5 秒内从 myScene1 过渡到 myScene2，即图像的右下角由小变大。TransitionManager.go(myScene1，new ChangeBounds().setDuration(5000))表示以 ChangeBounds 动画风格在 5 秒内从 myScene2 过渡到 myScene1，即图像的右下角由大变小。ViewGroup myRootView =（ViewGroup）findViewById（R.id.myRootView）用于根据 activity_main 布局中的 RelativeLayout 创建动画容器。myRootView.setClipBounds（new Rect（myRootView.getWidth（）/2，myRootView.getHeight（）/2，myRootView.getWidth（），myRootView.getHeight（）））表示 TransitionManager.go()方法在执行过渡动画时，仅在裁剪的 Rect 中执行（即图像的右下角）。关于 activity_main、myscene1、myscene2 三个布局的详细内容请参考源代码中的 MyCode\MySample554\app\src\main\res\layout\activity_main.xml 文件、MyCode\MySample554\app\src\main\res\layout\myscene1.xml 文件和 MyCode\MySample554\ app\src\main\res\layout\myscene2.xml 文件。此外，使用此实例的相关类需要在 gradle 中引入 compile 'com.android.support:design:25.0.1'依赖项。此实例的完整项目在 MyCode\MySample554 文件夹中。

162 使用 TransitionManager 实现矢量路径动画

此实例主要通过在 TransitionManager 的 go()方法中加载在 XML 文件中定制的过渡动画，实现在两个布局（myscene1 和 myscene2）切换时，ImageView 控件沿着指定的折线路径平移。当实例运行之后，蜘蛛图像（ImageView 控件）位于屏幕顶端，如图 162.1 的左图所示（即 myscene1）。单击屏幕，则蜘蛛图像将沿着细实线（实际不可见）所示的折线路径平移到屏幕底部，如图 162.1 的右图所示（即 myscene2）。在 myscene2 中单击屏幕，则蜘蛛图像将沿着细实线所示的折线路径（与前次的平移方向相反）平移到屏幕顶部，如图 162.1 的左图所示（即 myscene1）。

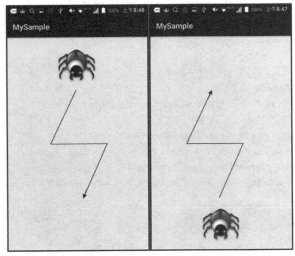

图 162.1

主要代码如下：

```
//单击 myscene1 过渡到 myscene2(图像沿折线向下平移)
public void onClickScene1(View v) {
  ViewGroup myRootView = (ViewGroup)findViewById(R.id.myRootView);
  Scene myScene2 = Scene.getSceneForLayout(myRootView, R.layout.myscene2, this);
  TransitionManager.go(myScene2, TransitionInflater.from(MainActivity.this).
inflateTransition(R.transition.mytransition));
}
//单击 myscene2 过渡到 myscene1(图像沿折线向上平移)
public void onClickScene2(View v) {
  ViewGroup myRootView = (ViewGroup)findViewById(R.id.myRootView);
  Scene myScene1 = Scene.getSceneForLayout(myRootView, R.layout.myscene1, this);
  TransitionManager.go(myScene1, TransitionInflater.from(MainActivity.this).
inflateTransition(R.transition.mytransition));
}
```

上面这段代码在 MyCode \ MySample586 \ app \ src \ main \ java \ com \ bin \ luo \ mysample \ MainActivity.java 文件中。在这段代码中，TransitionManager.go（myScene2，TransitionInflater.from(MainActivity.this).inflateTransition(R.transition.mytransition)）表示使用在 mytransition 过渡动画中定制的 ChangeBounds 动画在 5 秒内沿着折线路径从 myScene1 过渡到 myScene2。mytransition 过渡动画的主要内容如下：

```
<?xml version = "1.0" encoding = "utf-8"?>
<transitionSet xmlns:android = "http://schemas.android.com/apk/res/android"
               android:duration = "5000"
               android:interpolator = "@android:interpolator/decelerate_cubic">
  <changeBounds>
    <patternPathMotion
       android:patternPathData = "M200 0 L50 250 L300 250 L200 500"/>
  </changeBounds>
</transitionSet>
```

上面这段代码在 MyCode\MySample586\app\src\main\res\transition\mytransition.XML 文件中。在这段代码中，android:patternPathData = "M200 0 L50 250 L300 250 L200 500"代表图 162.1 所示的折线数据，该 XML 文件实际上相当于下面这段代码：

```
Path myPath = new Path();
myPath.moveTo(200,0);
myPath.lineTo(50,250);
myPath.lineTo(300,250);
myPath.lineTo(200,500);
PatternPathMotion myMotion = new PatternPathMotion();
myMotion.setPatternPath(myPath);
ChangeBounds myChangeBounds = new ChangeBounds();
myChangeBounds.setPathMotion(myMotion);
myChangeBounds.setDuration(5000);
```

需要说明的是，transition 过渡动画通常需要一个容器(也叫 RootView)，用于在不同的情形下加载不同的布局。在 MainActivity.java 文件中，myRootView＝(ViewGroup) findViewById(R.id.myRootView)即是用于根据 activity_main 布局中的 RelativeLayout 创建动画容器。activity_main 布

局的主要内容如下：

```xml
<?xml version="1.0" encoding="utf-8"?>
<RelativeLayout xmlns:android="http://schemas.android.com/apk/res/android"
                xmlns:tools="http://schemas.android.com/tools"
                android:id="@+id/activity_main"
                android:layout_width="match_parent"
                android:layout_height="match_parent"
                tools:context="com.bin.luo.mysample.MainActivity">
<!-- 这个RelativeLayout用来做动画的父布局 -->
<RelativeLayout android:id="@+id/myRootView"
                android:layout_width="match_parent"
                android:layout_height="wrap_content"
                android:layout_centerInParent="true">
<include layout="@layout/myscene1"></include>
</RelativeLayout></RelativeLayout>
```

上面这段代码在 MyCode\MySample586\app\src\main\res\layout\activity_main.xml 文件中。在这段代码中，<include layout="@layout/myscene1"></include> 表示容器中的子布局是 myscene1，myscene1 布局的主要内容如下：

```xml
<?xml version="1.0" encoding="utf-8"?>
<RelativeLayout xmlns:android="http://schemas.android.com/apk/res/android"
                android:onClick="onClickScene1"
                android:padding="20dp"
                android:gravity="center_horizontal"
                android:layout_width="match_parent"
                android:layout_height="match_parent">
<ImageView android:layout_alignParentTop="true"
           android:layout_width="100dp"
           android:layout_height="100dp"
           android:src="@mipmap/myimage1"
           android:id="@+id/myImageView1"/>
</RelativeLayout>
```

上面这段代码在 MyCode\MySample586\app\src\main\res\layout\myscene1.xml 文件中。在使用 TransitionManager.go() 方法实现过渡动画时，通常需要两个布局（场景），即此实例的 myscene1 和 myscene2，当 myscene1 的某个控件的 ID 值与 myscene2 的某个控件的 ID 值相同，且位置或大小不同时，在从 myscene1 过渡到 myscene2 时，这对控件就会执行平移或缩放动画。myscene2 布局的详细内容如下：

```xml
<RelativeLayout xmlns:android="http://schemas.android.com/apk/res/android"
                android:padding="20dp"
                android:gravity="center_horizontal"
                android:onClick="onClickScene2"
                android:layout_width="match_parent"
                android:layout_height="match_parent">
<ImageView android:layout_width="100dp"
           android:layout_height="100dp"
           android:src="@mipmap/myimage1"
           android:layout_alignParentBottom="true"
           android:id="@+id/myImageView1"/>
</RelativeLayout>
```

上面这段代码在 MyCode\MySample586\app\src\main\res\layout\myscene2.xml 文件中。需要说明的是,使用此实例的相关类需要在 gradle 中引入 compile 'com.android.support:design:25.0.1'依赖项。此实例的完整项目在 MyCode\MySample586 文件夹中。

163 使用 TransitionManager 同时实现多种动画

此实例主要通过在 XML 文件中将 explode 动画和 fade 动画添加到 transitionSet 过渡动画集合，并将其传递到 TransitionManager 的 beginDelayedTransition()方法中,实现控件在执行动画时产生 explode 动画和 fade 动画的叠加效果。当实例运行之后,怪兽图像(ImageView 控件)位于屏幕正中，单击屏幕,则怪兽图像将以随机路径滑出屏幕,在滑出屏幕的过程中,其透明度将渐渐减弱,直到完全消失;再次单击屏幕,则怪兽图像将以随机路径滑入屏幕,在滑入屏幕的过程中,其透明度将渐渐增强,直到完全显示,效果分别如图 163.1 的左图和右图所示。

图 163.1

主要代码如下：

```java
//单击屏幕,图像将组合 explode 和 fade 两种动画实现显示或消失
public void onClickScreen(View v) {
    ViewGroup myRootView = (ViewGroup)findViewById(R.id.myRootView);
    TransitionManager.beginDelayedTransition(myRootView, TransitionInflater.from(MainActivity.this).
inflateTransition(R.transition.mytransition));
    toggleVisibility(myImageView);
}
private static void toggleVisibility(View... views){
    for (View v:views) {
        boolean isVisible = v.getVisibility() == View.VISIBLE;
        v.setVisibility(isVisible?View.INVISIBLE:View.VISIBLE);
    }
}
```

上面这段代码在 MyCode\MySample607\app\src\main\java\com\bin\luo\mysample\

MainActivity.java 文件中。在这段代码中,TransitionManager.beginDelayedTransition(myRootView, TransitionInflater.from(MainActivity.this).inflateTransition(R.transition.mytransition))表示在父容器 myRootView 中执行 mytransition 中的 explode 和 fade 叠加动画。mytransition 过渡动画的主要内容如下:

```xml
<?xml version = "1.0" encoding = "utf-8"?>
<transitionSet xmlns:android = "http://schemas.android.com/apk/res/android"
               android:duration = "5000"
               android:interpolator = "@android:interpolator/decelerate_cubic">
    <explode/><fade/>
</transitionSet>
```

上面这段代码在 MyCode\MySample607\app\src\main\res\transition\mytransition.xml 文件中。在这段代码中,<explode/>即表示 explode 动画,可以在标签内设置其属性;同理,也可以在<fade/>标签内设置 fade 动画的属性。在 Android 中,Fade、Explode、Slide 动画作用于 View 的 Visibility 属性改变的时候。transition 过渡动画通常需要一个容器(也叫 RootView),用于在不同的情形下加载不同的布局。在 MainActivity.java 文件中,myRootView = (ViewGroup)findViewById(R.id.myRootView)用于根据 activity_main 布局中的 RelativeLayout 创建动画容器。activity_main 布局的主要内容如下:

```xml
<?xml version = "1.0" encoding = "utf-8"?>
<RelativeLayout xmlns:android = "http://schemas.android.com/apk/res/android"
                xmlns:tools = "http://schemas.android.com/tools"
                android:id = "@+id/activity_main"
                android:layout_width = "match_parent"
                android:layout_height = "match_parent"
                android:onClick = "onClickScreen"
                tools:context = "com.bin.luo.mysample.MainActivity">
<!-- 这个 RelativeLayout 用来做动画的父布局 -->
<RelativeLayout android:id = "@+id/myRootView"
                android:gravity = "center_horizontal|center_vertical"
                android:layout_width = "match_parent"
                android:layout_height = "match_parent">
<ImageView android:layout_width = "150dp"
           android:layout_height = "200dp"
           android:scaleType = "centerCrop"
           android:src = "@mipmap/myimage1"
           android:id = "@+id/myImageView"/>
</RelativeLayout></RelativeLayout>
```

上面这段代码在 MyCode\MySample607\app\src\main\res\layout\activity_main.xml 文件中。需要说明的是,使用此实例的相关类需要在 gradle 中引入 compile 'com.android.support:design:25.0.1'依赖项。此实例的完整项目在 MyCode\MySample607 文件夹中。

164 使用 TransitionManager 实现 XML 定制动画

此实例主要通过使用过渡动画(fade)的 targetId 属性指定控件,实现在两个布局(myscene1 和 myscene2)切换时,仅有指定的控件执行过渡动画。实例运行之后,当从 myscene1 切换到 myscene2 时,仅有 myImageView3 控件(即下面的电影海报图像)执行淡入动画,如图 164.1 的左图所示(即 myscene2)。当从 myscene2 切换到 myscene1 时,也仅有 myImageView3 控件(即下面的电影海报图像)执行淡出动画,如图 164.1 的右图所示(即 myscene1)。

图 164.1

主要代码如下:

```
//单击 myscene1 过渡到 myscene2(仅下面的图像(myImageView3)执行过渡动画)
public void onClickScene1(View v) {
  ViewGroup myRootView = (ViewGroup)findViewById(R.id.myRootView);
  Scene myScene2 = Scene.getSceneForLayout(myRootView, R.layout.myscene2, this);
  TransitionManager.go(myScene2, TransitionInflater.from(MainActivity.this).
inflateTransition(R.transition.mytransition));
}
//单击 myscene2 过渡到 myscene1(仅下面的图像(myImageView3)执行过渡动画)
public void onClickScene2(View v) {
  ViewGroup myRootView = (ViewGroup)findViewById(R.id.myRootView);
  Scene myScene1 = Scene.getSceneForLayout(myRootView, R.layout.myscene1, this);
  TransitionManager.go(myScene1, TransitionInflater.from(MainActivity.this).inflateTransition(R.
transition.mytransition));
}
```

上面这段代码在 MyCode \ MySample590 \ app \ src \ main \ java \ com \ bin \ luo \ mysample \ MainActivity.java 文件中。在这段代码中,TransitionManager.go(myScene2, TransitionInflater.from(MainActivity.this).inflateTransition(R.transition.mytransition))表示在从 myScene1 切换到 myScene2 时,执行在 mytransition 过渡动画中定制的 myImageView3 控件淡入动画。mytransition 过渡动画的主要内容如下:

```
<?xml version = "1.0" encoding = "utf - 8"?>
<transitionSet xmlns:android = "http://schemas.android.com/apk/res/android"
               android:duration = "5000"
               android:interpolator = "@android:interpolator/decelerate_cubic">
<!-- 仅 myImageView3 执行淡入过渡动画 -->
<fade android:fadingMode = "fade_in">
  <targets><target android:targetId = "@id/myImageView3"/></targets>
```

```
</fade>
<!-- 仅 myImageView3 执行淡出过渡动画 -->
<fade android:fadingMode = "fade_out">
 <targets><target android:targetId = "@id/myImageView3"/></targets>
</fade>
</transitionSet>
```

上面这段代码在 MyCode\MySample590\app\src\main\res\transition\mytransition.xml 文件中。在这段代码中，<target android:targetId="@id/myImageView3"/>表示在目标布局 myScene2 中，仅有 myImageView3 控件执行淡入（或淡出）动画，其他控件则按照默认的方式进行切换。过渡动画通常需要一个容器（也叫 RootView），用于在不同的情形下加载不同的布局。关于 activity_main、myscene1、myscene2 三个布局的详细内容请在源代码中查看 MyCode\MySample590\app\src\main\res\layout\activity_main.xml 文件、MyCode\MySample590\app\src\main\res\layout\myscene1.xml 文件和 MyCode\MySample590\app\src\main\res\layout\myscene2.xml 文件。需要说明的是，使用此实例的相关类需要在 gradle 中引入 compile 'com.android.support:design:25.0.1' 依赖项。此实例的完整项目在 MyCode\MySample590 文件夹中。

165　使用 TransitionManager 指定控件执行动画

此实例主要通过使用过渡动画类的 addTarget()方法指定需要执行过渡动画的控件，实现两个布局在使用过渡动画切换时，仅有指定控件执行过渡动画，其他控件则不执行过渡动画。当实例运行之后，单击 activity_main 布局的 ImageView 控件（电影海报图像），如图 165.1 的左图所示，则将以 Fade 动画风格切换到 activity_second 布局；在此实例中，由于 activity_second 布局上面的 ImageView 控件（《风语者》电影海报图像）被指定执行过渡动画，因此在从 activity_main 布局切换到 activity_second 布局时，仅 activity_second 布局上面的 ImageView 控件（《风语者》电影海报图像）有 Fade 过渡动画效果，activity_second 布局下面的 ImageView 控件（电影海报图像）则没有 Fade 过渡动画效果，如图 165.1 的右图所示。

图　165.1

主要代码如下:

```
public void onClickmyImageEnter(View v) {          //响应单击图像进入 activity_second 布局
    ViewGroup myMainLayout = (ViewGroup)findViewById(R.id.activity_main);
    Scene mySecondScene = Scene.getSceneForLayout(myMainLayout,
                                    R.layout.activity_second, this);
    Fade myFade = new Fade();
    myFade.setDuration(5000);
    myFade.addTarget(R.id.myImageView2);           //仅在 myImageView2 控件上执行过渡动画
    TransitionManager.go(mySecondScene, myFade);
}
public void onClickmyImageExit(View v) {           //响应单击图像进入 activity_main 布局
    ViewGroup mySecondLayout = (ViewGroup)findViewById(R.id.activity_second);
    Scene myMainScene = Scene.getSceneForLayout(mySecondLayout,
                                    R.layout.activity_main, this);
    Fade myFade = new Fade();
    myFade.setDuration(5000);
    TransitionManager.go(myMainScene, myFade);
}
```

上面这段代码在 MyCode\MySample579\app\src\main\java\com\bin\luo\mysample\MainActivity.java 文件中。在这段代码中，myFade.addTarget（R.id.myImageView2）表示 myImageView2 控件必须执行 myFade 过渡动画。TransitionManager.go(mySecondScene，myFade) 表示在从 activity_main 布局切换到 activity_second 布局时，执行 myFade 规定的过渡动画。关于 activity_main、activity_second 布局的详细内容请参考源代码中的 MyCode\MySample579\app\src\main\res\layout\activity_main.xml 文件和 MyCode\MySample579\app\src\main\res\layout\layout_second.xml 文件。此实例的完整项目在 MyCode\MySample579 文件夹中。

166 使用 TransitionManager 实现列表项滑入动画

此实例主要通过在 TransitionManager 的 beginDelayedTransition() 方法中设置 RecyclerView 控件为过渡动画 Slide 的容器，实现在显示 RecyclerView 时出现各个 Item(列表项)从右端滑入的动画效果，以及在隐藏 RecyclerView 时出现各个 Item(列表项)向右端滑出的动画效果。当实例运行之后，单击"执行 Slide 动画"按钮，则 RecyclerView 的各个 Item 将向右端滑出，直到完全消失；再次单击"执行 Slide 动画"按钮，则 RecyclerView 的各个 Item 将从右端滑入，直到完全显示，效果分别如图 166.1 的左图和右图所示。

主要代码如下:

```
public void onClickmyBtn1(View v) {                //响应单击"执行 Slide 动画"按钮
    TransitionSet myTransitionSets = new TransitionSet();
    Slide mySlide = new Slide();
    mySlide.setSlideEdge(Gravity.RIGHT);
    myTransitionSets.addTransition(mySlide);
    myTransitionSets.setDuration(2000);
    TransitionManager.beginDelayedTransition(myRecyclerView, myTransitionSets);
    if (bChanged) { myRecyclerView.setAdapter(myAdapter); }
    else { myRecyclerView.setAdapter(null); }
    bChanged = !bChanged;
}
```

上面这段代码在 MyCode\MySample609\app\src\main\java\com\bin\luo\mysample\MainActivity.java 文件中。在这段代码中，TransitionManager.beginDelayedTransition(myRecyclerView，myTransitionSets)表示在 myRecyclerView 中执行 myTransitionSets 中的 Slide 动画。Slide 动画通常在显示或隐藏控件的时候就执行，无须其他代码。在执行 Slide 动画时通常需要一个容器(一般为布局管理器)，但是 RecyclerView 控件天然就是一个容器，它里面可以容纳多个 Item，因此也无须其他代码，从 activity_main 布局中就可以看出实现此动画没有其他的容器。关于 activity_main 布局的详细内容请参考源代码中的 MyCode\MySample609\app\src\main\res\layout\activity_main.xml 文件。需要说明的是，在 Android Studio 中，使用此实例的相关类和控件需要在 gradle 中引入 compile 'com.android.support:recyclerview-v7:25.2.0'依赖项。此实例的完整项目在 MyCode\MySample609 文件夹中。

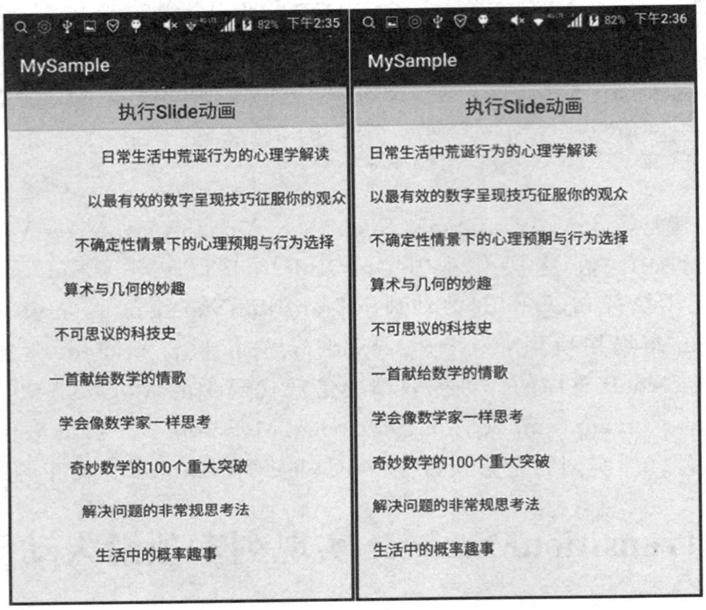

图 166.1

167　使用 TransitionManager 实现弧线路径动画

此实例主要通过在 TransitionManager 的 go()方法中加载在 XML 文件中定制的弧线路径动画，实现在两个布局(myscene1 和 myscene2)切换时，ImageView 控件沿着指定的弧线路径平移。当实例运行之后，蜘蛛图像(ImageView 控件)位于屏幕左端，如图 167.1 的左图所示(即 myscene1)。单击屏幕，则蜘蛛图像将沿着细实线所示的弧线路径平移到屏幕右端，如图 167.1 的右图所示(即 myscene2)。在 myscene2 中单击屏幕，则蜘蛛图像将沿着细实线所示的弧线路径平移到屏幕左端，如图 167.1 的左图所示(即 myscene1)。

主要代码如下：

```java
//单击myscene1过渡到myscene2(图像沿着上弧线向右平移)
public void onClickScene1(View v) {
    ViewGroup myRootView = (ViewGroup)findViewById(R.id.myRootView);
    Scene myScene2 = Scene.getSceneForLayout(myRootView, R.layout.myscene2, this);
```

```
    TransitionManager.go(myScene2, TransitionInflater.from(MainActivity.this).
    inflateTransition(R.transition.mytransition));
}
//单击myscene2过渡到myscene1(图像沿着下弧线向左平移)
public void onClickScene2(View v) {
    ViewGroup myRootView = (ViewGroup)findViewById(R.id.myRootView);
    Scene myScene1 = Scene.getSceneForLayout(myRootView, R.layout.myscene1, this);
    TransitionManager.go(myScene1, TransitionInflater.from(MainActivity.this).
    inflateTransition(R.transition.mytransition));
}
```

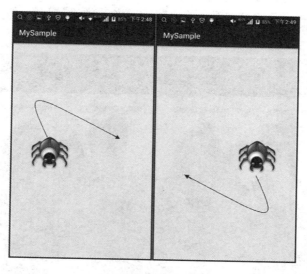

图 167.1

上面这段代码在 MyCode\MySample588\app\src\main\java\com\bin\luo\mysample\MainActivity.java 文件中。在这段代码中，TransitionManager.go(myScene2，TransitionInflater.from(MainActivity.this).inflateTransition(R.transition.mytransition))表示使用在mytransition过渡动画中定制的ChangeBounds动画在5秒内沿着弧线路径从myScene1过渡到myScene2。mytransition过渡动画的主要内容如下：

```xml
<?xml version = "1.0" encoding = "utf-8"?>
<transitionSet xmlns:android = "http://schemas.android.com/apk/res/android"
           android:duration = "5000"
           android:interpolator = "@android:interpolator/decelerate_cubic">
 <changeBounds>
  <patternPathMotion android:patternPathData = "M8,10 a4,6 0 1,1 6 6"/>
 </changeBounds></transitionSet>
```

上面这段代码在 MyCode\MySample588\app\src\main\res\transition\mytransition.XML 文件中。在这段代码中，android:patternPathData="M8,10 a4,6 0 1,1 6 6"代表图167.1所示的弧线数据。<changeBounds>标签表示使用ChangeBounds动画。transition过渡动画通常需要一个容器(也叫RootView)，用于在不同的情形下加载不同的布局。关于此实例的activity_main、myscene1、myscene2 三个布局的详细内容请参考源代码中的 MyCode\MySample588\app\src\main\res\layout\activity_main.xml文件、MyCode\MySample588\app\src\main\res\layout\myscene1.xml文件和

MyCode\MySample588\ app\src\main\res\ layout\myscene2.xml 文件。需要说明的是，使用此实例的相关类需要在 gradle 中引入 compile 'com.android.support:design:25.0.1'依赖项。此实例的完整项目在 MyCode\MySample588 文件夹中。

168　使用 TransitionManager 实现裁剪区域动画

此实例主要通过在 TransitionManager 的 go()方法中设置 ChangeClipBounds 动画，实现在两个布局(myscene1 和 myscene2)过渡时，裁剪的矩形区域将实现平移动画。当实例运行之后，单击屏幕，则在绿色背景左上角的裁剪矩形将平移到右下角，如图 168.1 的右图所示（即 myscene2）。在 myscene2 中单击屏幕，则在绿色背景右下角的裁剪矩形将平移到左上角，如图 168.1 的左图所示（即 myscene1）。

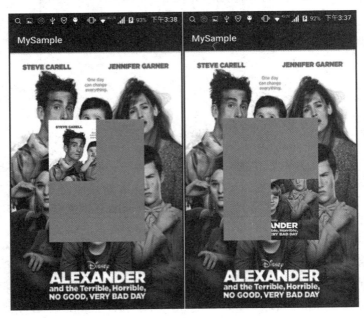

图　168.1

主要代码如下：

```java
public class MainActivity extends Activity {
    ImageView myImageView1, myImageView2;
    ViewGroup myRootView;
    Scene myScene1, myScene2;
    View myInflate1, myInflate2;
    @Override
    protected void onCreate(Bundle savedInstanceState) {
        super.onCreate(savedInstanceState);
        setContentView(R.layout.activity_main);
        myRootView = (ViewGroup)findViewById(R.id.myRootView);
        myInflate1 = LayoutInflater.from(this).inflate(R.layout.myscene1, null);
        myScene1 = new Scene(myRootView, myInflate1);
        myImageView1 = (ImageView) myInflate1.findViewById(R.id.myImageView1);
        myImageView1.setClipBounds(new Rect(0,0,300,400));
        TransitionManager.go(myScene1);
```

```
    }
    //单击 myscene1 过渡到 myscene2(裁剪矩形从左上角平移到右下角)
    public void onClickScene1(View v) {
        myInflate2 = LayoutInflater.from(this).inflate(R.layout.myscene2, null);
        myScene2 = new Scene(myRootView,myInflate2);
        myImageView2 = (ImageView) myInflate2.findViewById(R.id.myImageView2);
        myImageView2.setClipBounds(new Rect(300,400,600,800));
        TransitionManager.go(myScene2, new ChangeClipBounds().setDuration(5000));
    }
    //单击 myscene2 过渡到 myscene1(裁剪矩形从右下角平移到左上角)
    public void onClickScene2(View v) {
        TransitionManager.go(myScene1, new ChangeClipBounds().setDuration(5000));
    } }
```

上面这段代码在 MyCode\MySample583\app\src\main\java\com\bin\luo\mysample\MainActivity.java 文件中。在这段代码中，TransitionManager.go(myScene2, new ChangeClipBounds().setDuration(5000))表示以 ChangeClipBounds 动画风格(即裁剪矩形平移)在 5 秒内从 myScene1 过渡到 myScene2。ChangeClipBounds 与 ChangeBounds 的过渡动画功能比较类似，不同的是，ChangeBounds 针对的是 view 控件，而 ChangeClipBounds 针对的是 view 控件的剪切区域，即使用 setClipBound(Rect rect)方法设置的 rect，如果没有设置 rect，则没有动画效果。关于此实例的 activity_main、myscene1、myscene2 三个布局的详细内容请参考源代码中的 MyCode\MySample583\app\src\main\res\layout\ activity_main.xml 文件、MyCode\MySample583\app\src\main\res\layout\myscene1.xml 文件和 MyCode\MySample583\app\src\main\res\layout\myscene2.xml 文件。需要说明的是，使用此实例的相关类需要在 gradle 中引入 compile 'com.android.support:design:25.0.1'依赖项。此实例的完整项目在 MyCode\MySample583 文件夹中。

169 通过设置和获取控件的 Tag 确定动画过渡行为

此实例主要通过使用 setTag()和 getTag()方法为控件添加标记和获取控件的标记，使控件实现根据当前的标记决定如何执行过渡动画。当实例运行之后，单击 ImageView 控件(电影海报图像)，由于 ImageView 控件的当前标记是 myimage1，因此执行从 myimage1 透明过渡到 myimage2 的动画；再次单击 ImageView 控件，由于此时 ImageView 控件的当前标记已经被设置为 myimage2，因此执行从 myimage2 透明过渡到 myimage1 的动画，效果分别如图 169.1 的左图和右图所示。

主要代码如下：

```
public void onClickImage(View v) {                    //单击图像之后透明过渡到另一幅图像,或者相反
    //如果当前 ImageView 控件加载图像是 myimage1,则从 myimage1 过渡到 myimage2
    if ((Integer) myImageView.getTag() == R.mipmap.myimage1) {
        myTransitionDrawable = new TransitionDrawable(new Drawable[]
                {getDrawable(R.mipmap.myimage1), getDrawable(R.mipmap.myimage2)});
        myImageView.setTag(R.mipmap.myimage2);       //设置标记 myimage2
    //如果当前 ImageView 控件加载图像是 myimage2,则从 myimage2 过渡到 myimage1
    } else if ((Integer) myImageView.getTag() == R.mipmap.myimage2) {
        myTransitionDrawable = new TransitionDrawable(new Drawable[]
                {getDrawable(R.mipmap.myimage2), getDrawable(R.mipmap.myimage1)});
        myImageView.setTag(R.mipmap.myimage1);       //设置标记 myimage1
    }
```

```
    myImageView.setImageDrawable(myTransitionDrawable);
    myTransitionDrawable.startTransition(2000);    //设置动画时间为 2 秒
}
```

图　169.1

上面这段代码在 MyCode\MySample642\app\src\main\java\com\bin\luo\mysample\MainActivity.java 文件中。在这段代码中，myImageView.setTag(R.mipmap.myimage1)用于在 ImageView 控件上设置标记，myImageView.getTag()用于获取在 ImageView 控件上设置的标记，一般情况下，getTag()方法的返回值应该与 setTag()方法的参数值的数据类型一致。此实例的完整项目在 MyCode\MySample642 文件夹中。

170　在 TransitionSet 中指定多个动画的执行顺序

此实例主要通过设置 TransitionSet 的 setOrdering()方法的参数，实现在两个布局过渡的过程中多个动画或是同时执行，或是顺序执行。当实例运行之后，Revolting Rhymes 电影海报图像位于屏幕的正中，如图 170.1 的左图所示（即 myscene1）；单击屏幕，则将淡出 Revolting Rhymes 电影海报图像，然后淡入 Trolls Holiday 电影海报图像（即 myscene2），如图 170.1 的右图所示。在 myscene2 中单击屏幕，则将淡出 Trolls Holiday 电影海报图像，然后淡入 Revolting Rhymes 电影海报图像，如图 170.1 的左图所示。

主要代码如下：

```
//单击 myscene1 过渡到 myscene2(先淡出 myscene1,后淡入 myscene2)
public void onClickScene1(View v) {
    ViewGroup myRootView = (ViewGroup)findViewById(R.id.myRootView);
    Scene myScene2 = Scene.getSceneForLayout(myRootView, R.layout.myscene2, this);
    Fade myFadeIn = new Fade(Fade.IN);
    Fade myFadeOut = new Fade(Fade.OUT);
```

```
        TransitionSet myTransitionSet = new TransitionSet();
        myTransitionSet.addTransition(myFadeOut).addTransition(myFadeIn);
        myTransitionSet.setOrdering(TransitionSet.ORDERING_SEQUENTIAL);
        myTransitionSet.setDuration(5000);
        TransitionManager.go(myScene2, myTransitionSet);
    }
    //单击myscene2过渡到myscene1(先淡出myscene2,后淡入myscene1)
    public void onClickScene2(View v) {
        ViewGroup myRootView = (ViewGroup)findViewById(R.id.myRootView);
        Scene myScene1 = Scene.getSceneForLayout(myRootView, R.layout.myscene1, this);
        Fade myFadeIn = new Fade(Fade.IN);
        Fade myFadeOut = new Fade(Fade.OUT);
        TransitionSet myTransitionSet = new TransitionSet();
        myTransitionSet.addTransition(myFadeOut).addTransition(myFadeIn);
        myTransitionSet.setOrdering(TransitionSet.ORDERING_SEQUENTIAL);
        myTransitionSet.setDuration(5000);
        TransitionManager.go(myScene1, myTransitionSet);
    }
```

图 170.1

上面这段代码在MyCode\MySample555\app\src\main\java\com\bin\luo\mysample\MainActivity.java文件中。在这段代码中，myTransitionSet.addTransition(myFadeOut).addTransition(myFadeIn)用于在myTransitionSet过渡动画集合中增加myFadeOut和myFadeIn动画。myTransitionSet.setOrdering(TransitionSet.ORDERING_SEQUENTIAL)表示此实例的两个淡出淡入动画将按照添加顺序执行，如果是myTransition.setOrdering(TransitionSet.ORDERING_TOGETHER)，则两个动画将同时执行。关于activity_main、myscene1、myscene2三个布局的详细内容请参考源代码中的MyCode\MySample555\app\src\main\res\layout\activity_main.xml文件、MyCode\MySample555\app\src\main\res\layout\myscene1.xml文件和MyCode\MySample555\app\src\main\res\layout\myscene2.xml文件。此外，使用此实例的相关类需要在gradle中引入compile 'com.android.support:design:25.0.1'依赖项。此实例的完整项目在MyCode\MySample555文件夹中。

171 使用 TransitionDrawable 透明切换两幅图像

此实例主要通过使用 TransitionDrawable 创建图像过渡动画,实现在指定的时间内从一幅图像切换到另一幅图像。当实例运行之后,单击屏幕,则在 5 秒内从一幅电影海报图像渐渐切换到(改变透明度)另一幅电影海报图像,效果分别如图 171.1 的左图和右图所示。

图　171.1

主要代码如下:

```
public void onClickImage(View v) {          //单击图像之后透明过渡到另一幅图像
  TransitionDrawable myTransitionDrawable =
    (TransitionDrawable)getResources().getDrawable(R.drawable.mytransition);
  ImageView myImageView = (ImageView)findViewById(R.id.myImageView);
  myImageView.setImageDrawable(myTransitionDrawable);
  myTransitionDrawable.startTransition(5000);
}
```

上面这段代码在 MyCode\MySample541\app\src\main\java\com\bin\luo\mysample\MainActivity.java 文件中。在这段代码中,myTransitionDrawable =（TransitionDrawable）getResources().getDrawable(R.drawable.mytransition)中的 mytransition 用于设置淡入淡出过渡动画的源图像和目标图像。mytransition 过渡动画的主要内容如下:

```
<transition xmlns:android = "http://schemas.android.com/apk/res/android">
  <item android:drawable = "@drawable/myimage1"/>
  <item android:drawable = "@drawable/myimage2"/>
</transition>
```

上面这段代码在 MyCode\MySample541\app\src\main\res\drawable\mytransition.xml 文件中。<item android:drawable = "@drawable/myimage1"/>表示切换的源图像是 myimage1,<item

android:drawable="@drawable/myimage2"/>表示切换的目标图像是 myimage2。此实例的完整项目在 MyCode\MySample541 文件夹中。

172 使用 AnimatedVectorDrawable 实现转圈动画

此实例主要通过使用 AnimatedVectorDrawable 加载在 SVG 中实现的转圈动画,并作为 ImageView 控件的背景,实现高仿 Chrome 进度条的转圈加载效果。当实例运行之后,屏幕中心的天蓝色进度条将沿着顺时针方向进行旋转,并伴有视差,类似于 Chrome 浏览器的标签页加载进度条,效果分别如图 172.1 的左图和右图所示。

图 172.1

主要代码如下:

```
public class MainActivity extends Activity {
 @Override
 protected void onCreate(Bundle savedInstanceState) {
  super.onCreate(savedInstanceState);
  setContentView(R.layout.activity_main);
  ImageView myImageView = (ImageView) findViewById(R.id.myImageView);
  //动态加载 SVG 动画
  AnimatedVectorDrawable myAnimatedVectorDrawable =
      (AnimatedVectorDrawable) getDrawable(R.drawable.mysvganim);
  //设置 ImageView 控件的背景为 SVG 动画
  myImageView.setBackground(myAnimatedVectorDrawable);
  myAnimatedVectorDrawable.start();       //启动动画
 } }
```

上面这段代码在 MyCode\MySample870\app\src\main\java\com\bin\luo\mysample\ MainActivity.java 文件中。在这段代码中,myAnimatedVectorDrawable =(AnimatedVectorDrawable) getDrawable(R. drawable. mysvganim)用于根据矢量动画 mysvganim 创建 AnimatedVectorDrawable 对象,以实现转圈动

画。矢量动画 mysvganim 是一个 XML 格式的文件，它的主要作用是将动画 XML 文件和矢量图形 XML 文件组合在一起，矢量动画 mysvganim 的主要内容如下：

```xml
<animated-vector xmlns:android="http://schemas.android.com/apk/res/android"
                 android:drawable="@drawable/myprogressbar">
<!-- 该动画通过更改 SVG 图形的路径对象的 trimPathOffset 值，以实现匀速旋转特效 -->
<target android:name="myprogressbar"
        android:animation="@animator/myrotateanim"/>
<!-- 该动画通过更改 SVG 图形的路径对象的 trimPathEnd 值，以实现旋转视差特效 -->
<target android:name="myprogressbar"
        android:animation="@animator/myrotateaccelerateanim"/>
</animated-vector>
```

上面这段代码在 MyCode\MySample870\app\src\main\res\drawable\mysvganim.xml 文件中。在这段代码中，android:drawable="@drawable/myprogressbar" 的 myprogressbar 是矢量图形文件。myprogressbar.xml 文件的主要内容如下：

```xml
<vector xmlns:android="http://schemas.android.com/apk/res/android"
        android:width="200dp"
        android:height="200dp"
        android:viewportHeight="200.0"
        android:viewportWidth="200.0">
<path android:name="myprogressbar"
      android:fillColor="#fff"
      android:pathData="M100,100m-40,0a40,40 0,1 1,80 0a40,40 0,1 1,-80 0"
      android:strokeColor="#00BFFF"
      android:strokeWidth="5"
      android:trimPathStart="0.5"/>
</vector>
```

上面这段代码在 MyCode\MySample870\app\src\main\res\drawable\myprogressbar.xml 文件中。在 mysvganim.xml 文件中，android:animation="@animator/myrotateanim" 的 myrotateanim 是动画文件，该动画通过更改 SVG 图形的路径对象的 trimPathOffset 值，以实现匀速旋转特效。动画文件 myrotateanim.xml 的主要内容如下：

```xml
<objectAnimator xmlns:android="http://schemas.android.com/apk/res/android"
                android:duration="2500"
                android:interpolator="@android:interpolator/linear"
                android:propertyName="trimPathOffset"
                android:repeatCount="-1"
                android:valueFrom="0"
                android:valueTo="1"
                android:valueType="floatType"/>
```

上面这段代码在 MyCode\MySample870\app\src\main\res\animator\myrotateanim.xml 文件中。在 mysvganim.xml 文件中，android:animation="@animator/myrotateaccelerateanim" 的 myrotateaccelerateanim 是动画文件，该动画文件通过更改 SVG 图形的路径对象的 trimPathEnd 值，以实现旋转视差特效。myrotateaccelerateanim.xml 动画文件的主要内容如下：

```xml
<objectAnimator xmlns:android="http://schemas.android.com/apk/res/android"
    android:duration="2500"
    android:interpolator="@android:anim/accelerate_decelerate_interpolator"
    android:propertyName="trimPathEnd"
    android:repeatCount="-1"
    android:valueFrom="0"
    android:valueTo="1"
    android:valueType="floatType"/>
```

上面这段代码在 MyCode\MySample870\app\src\main\res\animator\ myrotateaccelerateanim. xml 文件中。此实例的完整项目在 MyCode\MySample870 文件夹中。

173 创建 AnimatedVectorDrawableCompat 动画

此实例主要通过设置动画集合(set)的 android:ordering 属性为 together，实现同时执行该集合中的所有动画。当实例运行之后，单击"执行动画"按钮，则矢量图形"√"符号将同时执行 rotation、scaleX、scaleY 三种动画，效果分别如图 173.1 的左图和右图所示。

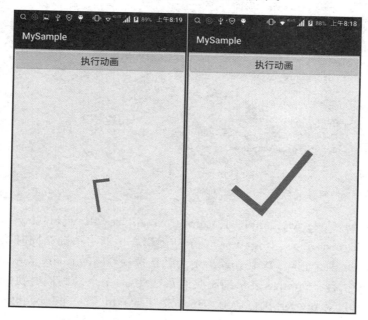

图 173.1

主要代码如下：

```java
public class MainActivity extends Activity {
    ImageView myImageView;
    @Override
    protected void onCreate(Bundle savedInstanceState) {
        super.onCreate(savedInstanceState);
        setContentView(R.layout.activity_main);
        myImageView = (ImageView)findViewById(R.id.myImageView);
        AnimatedVectorDrawableCompat myAnimatedVectorDrawableCompat =
                AnimatedVectorDrawableCompat.create(this, R.drawable.myanimated);
        myImageView.setImageDrawable(myAnimatedVectorDrawableCompat);
```

```
}
    public void onClickmyBtn1(View v) {                    //响应单击"执行动画"按钮
        ((Animatable) myImageView.getDrawable()).start();
    }
}
```

上面这段代码在MyCode\MySample613\app\src\main\java\com\bin\luo\mysample\ MainActivity.java文件中。在这段代码中，myAnimatedVectorDrawableCompat = AnimatedVectorDrawableCompat.create(this,R.drawable.myanimated)用于加载动画文件myanimated.xml中的矢量动画。myanimated.xml动画文件的主要内容如下：

```xml
<animated-vector xmlns:android="http://schemas.android.com/apk/res/android"
                 xmlns:aapt="http://schemas.android.com/aapt"
                 android:drawable="@drawable/myvector">
<target android:name="mytick">
  <aapt:attr name="android:animation">
    <set android:ordering="together">
      <objectAnimator android:duration="500"
              android:interpolator="@android:interpolator/linear"
              android:propertyName="rotation"
              android:valueFrom="180"
              android:valueTo="0"/>
      <objectAnimator android:duration="500"
              android:interpolator="@android:interpolator/linear"
              android:propertyName="scaleX"
              android:valueFrom="0.2"
              android:valueTo="1"/>
      <objectAnimator android:duration="500"
              android:interpolator="@android:interpolator/linear"
              android:propertyName="scaleY"
              android:valueFrom="0.2"
              android:valueTo="1"/>
</set></aapt:attr></target></animated-vector>
```

上面这段代码在MyCode\MySample613\app\src\main\res\drawable\myanimated.xml文件中。在这段代码中，android:ordering="together"表示动画集合中的多个动画同时执行，ordering属性有sequentially和together两个选项，其中together为默认选项，sequentially表示动画集合(set)中的多个动画，按照先后顺序执行，together表示动画集合(set)中的多个动画同时执行。<target android:name="mytick">表示仅对mytick执行动画。android:drawable="@drawable/myvector"用于指定矢量图形文件myvector.xml。矢量图形文件myvector.xml的主要代码如下：

```xml
<?xml version="1.0" encoding="utf-8"?>
<vector xmlns:android="http://schemas.android.com/apk/res/android"
        android:width="240dp"
        android:height="240dp"
        android:viewportHeight="24.0"
        android:viewportWidth="24.0">
    <group android:name="mytick"
        android:pivotX="12"
        android:pivotY="12">
      <path android:strokeColor="#FFFF0000"
            android:strokeWidth="1.5"
            android:pathData="M 4,11 L 10,16 L 20 3">
</path></group></vector>
```

上面这段代码在 MyCode\MySample613\app\src\main\res\drawable\myvector.xml 文件中。需要说明的是，使用此实例的相关类需要在 gradle 中引入 compile 'com.android.support：appcompat-v7:24.0.0'依赖项。此实例的完整项目在 MyCode\MySample613 文件夹中。

174　使用 ViewPropertyAnimator 创建多个动画

此实例主要通过直接在 ViewPropertyAnimator 上创建多个属性动画，实现在图像控件上同时产生多种动画效果。当实例运行之后，单击"开始播放动画"按钮，则图像（妖怪）控件将从左上角沿着对角线方向平移到右下角，效果分别如图 174.1 的左图和右图所示。

图　174.1

主要代码如下：

```
//响应单击"开始播放动画"按钮
public void onClickButton1(View v) {
    myViewPropertyAnimator = myImageView.animate();
    myViewPropertyAnimator.translationX(600)
            .translationY(1100).setDuration(5000).start();
}
//响应单击"停止播放动画"按钮
public void onClickButton2(View v) {
    myViewPropertyAnimator.cancel();
}
```

上面这段代码在 MyCode\MySample166\app\src\main\java\com\bin\luo\mysample\MainActivity.java 文件中。在这段代码中，myViewPropertyAnimator = myImageView.animate()用于返回 ViewPropertyAnimator 对象。myViewPropertyAnimator.translationX(600).translationY(1100).setDuration(5000).start()表示在 5 秒内 myImageView 控件在水平方向上移动 600，同时在垂直方向上移动 1100，合成之后近似于屏幕对角线方向。此实例的完整项目在 MyCode\MySample166 文件夹中。

175 自定义 selector 实现以动画形式改变阴影大小

此实例主要通过设置 CardView 控件的 stateListAnimator 属性为自定义 selector，并在自定义 selector 中使用 objectAnimator 改变 translationZ 属性值，实现以动画的形式改变控件的阴影大小。当实例运行之后，单击 CardView 控件，则在单击该控件时增大该控件的 translationZ 属性值，从而产生上浮的阴影效果，如图 175.1 的左图所示；在离开该控件时减少（恢复）至正常状态，则阴影消失，如图 175.1 的右图所示。

图 175.1

主要代码如下：

```xml
<android.support.v7.widget.CardView
    xmlns:card_view="http://schemas.android.com/apk/res-auto"
    android:layout_width="200dp"
    android:layout_height="wrap_content"
    android:layout_centerInParent="true"
    android:clickable="true"
    android:stateListAnimator="@animator/myselector"
    card_view:cardCornerRadius="20dp"
    card_view:cardPreventCornerOverlap="true">
    <ImageView android:layout_width="250dp"
        android:layout_height="350dp"
        android:background="@mipmap/myimage1"/>
</android.support.v7.widget.CardView>
```

上面这段代码在 MyCode\MySample575\app\src\main\res\layout\activity_main.xml 文件中。在这段代码中，android:stateListAnimator="@animator/myselector" 用于设置动态改变阴影大小的自定义选择器 selector。自定义选择器 selector 的主要内容如下：

```xml
<?xml version = "1.0" encoding = "utf-8"?>
<selector xmlns:android = "http://schemas.android.com/apk/res/android">
<item android:state_pressed = "true">
  <set>
    <objectAnimator android:duration = "@android:integer/config_shortAnimTime"
                android:propertyName = "translationZ"
                android:valueTo = "20dp"
                android:valueType = "floatType" /></set></item>
  <item>
  <set>
    <objectAnimator android:duration = "@android:integer/config_shortAnimTime"
                android:propertyName = "translationZ"
                android:valueTo = "0dp"
                android:valueType = "floatType" /></set></item>
</selector>
```

上面这段代码在 MyCode\MySample575\app\src\main\res\animator\myselector.xml 文件中。在这段代码中，<item android:state_pressed = "true">表示在控件被单击时执行此标签中的动画。android:propertyName = "translationZ"表示改变 translationZ 属性值，android:valueTo = "20dp"表示将 translationZ 属性值改变为 20dp。需要说明的是，如果当前项目的 res 目录下不存在 animator 子目录，则应该先在 res 目录下创建 animator 子目录，然后创建 myselector.xml 文件。此外，在 Android Studio 中，使用此实例的相关控件需要在 gradle 中引入 compile 'com.android.support：cardview-v7：25.3.1'依赖项。此实例的完整项目在 MyCode\MySample575 文件夹中。

176 使用 ripple 标签创建中心波纹扩散动画

此实例主要通过在 XML 文件中使用 ripple 标签创建指定颜色和形状的中心波纹扩散动画并设置为控件的背景，实现在单击控件时产生波纹扩散效果。当实例运行之后，单击 ImageView 控件（地球图像），则在该控件上产生蓝色的中心波纹扩散的动画，效果分别如图 176.1 的左图和右图所示。

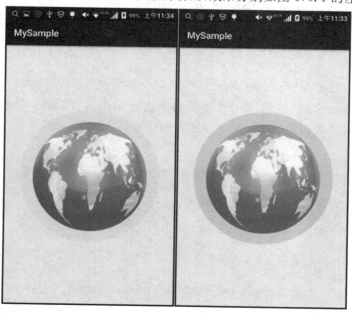

图 176.1

主要代码如下：

```
//响应单击图像产生波纹扩散的效果
public void onClickImageView(View myView) {
    myView.setBackground(getDrawable(R.drawable.myripple));
}
```

上面这段代码在 MyCode\MySample565\app\src\main\java\com\bin\luo\mysample\MainActivity.java 文件中。在这段代码中，myView.setBackground（getDrawable（R.drawable.myripple））表示设置动画资源文件 myripple 作为控件 myView 的背景。动画资源文件 myripple 的主要内容如下：

```
<?xml version="1.0" encoding="utf-8"?>
<ripple xmlns:android="http://schemas.android.com/apk/res/android"
        android:color="?android:attr/colorControlHighlight">
<item>
  <inset android:insetBottom="16dp"
         android:insetLeft="14dp"
         android:insetRight="14dp"
         android:insetTop="16dp">
    <shape android:shape="rectangle">
      <corners android:radius="150dp"/>
      <solid android:color="#D1EEEE"/>
      <padding android:bottom="14dp"
               android:left="18dp"
               android:right="18dp"
               android:top="14dp"/>
    </shape></inset></item></ripple>
```

上面这段代码在 MyCode\MySample565\app\src\main\res\drawable\myripple.xml 文件中。此实例的完整项目在 MyCode\MySample565 文件夹中。

177 使用 GLSurfaceView 实现 3D 地球的自转

此实例主要通过自定义 GLSurfaceView 控件的渲染器 Renderer，实现 3D 地球的自转动画。当实例运行之后，单击"开始自转"按钮，则 3D 地球将围绕 y 轴顺时针旋转；单击"停止自转"按钮，则 3D 地球停止自转，效果分别如图 177.1 的左图和右图所示。

主要代码如下：

```
public class MainActivity extends Activity {
    GLRenderer myGLRenderer;
    @Override
    protected void onCreate(Bundle savedInstanceState) {
        super.onCreate(savedInstanceState);
        setContentView(R.layout.activity_main);
        GLSurfaceView myGLSurfaceView =
                (GLSurfaceView) findViewById(R.id.myGLSurface);
        Bitmap myBitmap =
                BitmapFactory.decodeResource(getResources(), R.mipmap.myearth);
```

```
    myGLRenderer = new GLRenderer(myBitmap);
    myGLSurfaceView.setRenderer(myGLRenderer);
    myGLRenderer.isRotationEnabled = false;
}
//响应单击"开始自转"按钮
public void onClickBtn1(View v) { myGLRenderer.isRotationEnabled = true;}
//响应单击"停止自转"按钮
public void onClickBtn2(View v) { myGLRenderer.isRotationEnabled = false;}
}
```

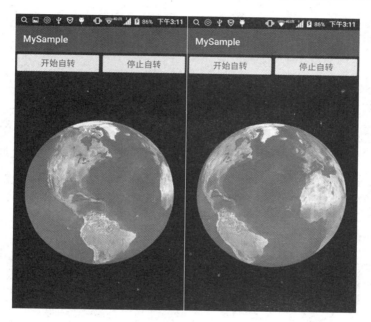

图 177.1

上面这段代码在 MyCode \ MySample779 \ app \ src \ main \ java \ com \ bin \ luo \ mysample \ MainActivity.java 文件中。在这段代码中，myGLRenderer = new GLRenderer(myBitmap) 中的 GLRenderer 是实现了 GLSurfaceView.Renderer 接口的自定义类，关于该自定义类的详细内容请参考源代码中的 MyCode\MySample779\app\src\main\java\com\bin\luo\ mysample\GLRenderer.java 文件。此实例的完整项目在 MyCode\MySample779 文件夹中。

第6章 音频和视频

178 使用 MediaPlayer 播放本地 mp3 音乐文件

此实例主要通过使用 MediaPlayer，实现播放 SD 卡上的 mp3 音乐文件。当实例运行之后，在"音乐文件："输入框中输入在 SD 卡上存储的音乐文件全路径，如"/storage /sdcard0/mymusic1.mp3"，然后单击"开始播放"按钮，则将播放指定的音乐文件，效果分别如图 178.1 的左图和右图所示。

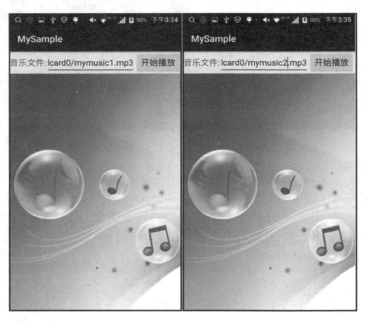

图 178.1

主要代码如下：

```
public void onClickmyBtn1(View v) {                                //响应单击"开始播放"按钮
  try {
    MediaPlayer myMediaPlayer = new MediaPlayer();
    myMediaPlayer.setDataSource(myEditText.getText().toString());  //音乐文件路径
    myMediaPlayer.prepare();
    myMediaPlayer.start();                                         //播放音乐
  } catch (Exception e) {                                          //注意：输入框的音乐文件必须在 SD
                                                                   //  卡中真实存在
```

```
Toast.makeText(getApplicationContext(),
                        e.getMessage(), Toast.LENGTH_SHORT).show();
}}
```

上面这段代码在 MyCode\MySample259\app\src\main\java\com\bin\luo\mysample\MainActivity.java 文件中。在这段代码中，myMediaPlayer.setDataSource(myEditText.getText().toString())用于设置 MediaPlayer 将要播放的音乐文件，myMediaPlayer.start()用于播放音乐。此外，读取在 SD 卡的文件需要在 AndroidManifest.xml 文件中添加< uses-permission android:name=" android.permission.READ_EXTERNAL_STORAGE"/>权限。此实例的完整项目在 MyCode\MySample259 文件夹中。

179　使用 MediaPlayer 播放本地 mp4 视频文件

此实例主要通过使用 MediaPlayer 和 SurfaceView，实现在 SD 卡上播放 mp4 视频文件。当实例运行之后，在"视频文件："输入框中输入 SD 卡上存储的视频文件全路径，然后单击"开始播放"按钮，则将播放指定的视频文件，如图 179.1 的左图和右图所示。

图　179.1

主要代码如下：

```
public class MainActivity extends Activity {
 EditText myEditText;
 private SurfaceView mySurfaceView;
 private SurfaceHolder mySurfaceHolder;
 private MediaPlayer myMediaPlayer = null;
 @Override
 protected void onCreate(Bundle savedInstanceState) {
  super.onCreate(savedInstanceState);
  setContentView(R.layout.activity_main);
```

```java
    myEditText = (EditText)findViewById(R.id.myEditText);
    myEditText.setText(Environment.getExternalStorageDirectory()
                            + "/myvideo1.mp4");

    myMediaPlayer = new MediaPlayer();
    mySurfaceView = (SurfaceView) findViewById(R.id.mySurfaceView);
    mySurfaceHolder = mySurfaceView.getHolder();
    mySurfaceHolder.setFixedSize(400, 800);
}
public void onClickmyBtn1(View v) {                            //响应单击"开始播放"按钮
    try {
        myMediaPlayer.setDataSource(myEditText.getText().toString());   //设置数据资源
        myMediaPlayer.setDisplay(mySurfaceHolder);              //设置视频播放窗口
        myMediaPlayer.prepare();
        myMediaPlayer.start();                                  //播放视频
    }
    catch (IOException e) {                                     //注意:输入框的视频文件必须在SD
                                                                //卡中真实存在
        Toast.makeText(getApplicationContext(),
                e.getMessage(), Toast.LENGTH_SHORT).show();
} } }
```

上面这段代码在 MyCode\MySample260\app\src\main\java\com\bin\luo\mysample\MainActivity.java 文件中。在这段代码中，myMediaPlayer.setDataSource(myEditText.getText().toString()) 用于设置 MediaPlayer 将要播放的视频文件。myMediaPlayer.setDisplay(mySurfaceHolder)用于设置视频播放窗口。MediaPlayer 通常用于播放音频，但是如果需要使用 MediaPlayer 播放视频，则需要使用 SurfaceView 设置视频显示窗口。SurfaceView 比普通的自定义 View 更有绘图上的优势，它支持完全的 OpenGL ES 库。此外，读取在 SD 卡上的视频文件需要在 AndroidManifest.xml 文件中添加＜uses-permission android：name＝ "android.permission.READ_EXTERNAL_STORAGE"/＞权限。此实例的完整项目在 MyCode\MySample260 文件夹中。

180 使用 MediaPlayer 播放指定网址的音乐文件

此实例主要通过使用 MediaPlayer 的 setOnBufferingUpdateListener()方法和 setDataSource()方法，实现缓冲并播放指定网址的网络音乐文件。当实例运行之后，在"歌曲网址："输入框中输入音乐文件的网络地址，然后单击"下载并播放此网络歌曲"按钮，则将立即缓冲网络数据，完成之后则播放此歌曲文件；单击右边的暂停按钮，则可以暂停播放，效果分别如图 180.1 的左图和右图所示。

主要代码如下：

```java
public class MainActivity extends Activity
                        implements SeekBar.OnSeekBarChangeListener {
    EditText myEditText;
    private MediaPlayer myMediaPlayer;
    TextView myTotalTime;
    ImageView myPlayPause;
    SeekBar mySeekBar;
    Handler myHandler = new Handler() {
        @Override
        public void handleMessage(Message msg) {
```

```java
            myCurrentTime.setText(getPlayerCurrentPositionToString(myMediaPlayer));
        }};
    private TextView myCurrentTime;
    @Override
    protected void onCreate(Bundle savedInstanceState) {
        super.onCreate(savedInstanceState);
        setContentView(R.layout.activity_main);
        myEditText = (EditText) findViewById(R.id.myEditText);
        myCurrentTime = (TextView) findViewById(R.id.myCurrentTime);
        myTotalTime = (TextView) findViewById(R.id.myTotalTime);
        myPlayPause = (ImageView) findViewById(R.id.myPlayPause);
        myPlayPause.setOnClickListener(new View.OnClickListener() {
            @Override
            public void onClick(View view) {
                if (myMediaPlayer == null) return;
                if (myMediaPlayer.isPlaying()) {                    //如果正在播放,则暂停
                    myMediaPlayer.pause();
                    myPlayPause.setImageResource(R.mipmap.myicon_play);
                } else {                                            //如果已经暂停,则播放
                    myMediaPlayer.start();
                    myPlayPause.setImageResource(R.mipmap.myicon_pause);
                }}});
        mySeekBar = (SeekBar) findViewById(R.id.mySeekBar);
    }
    @Override
    public void onProgressChanged(SeekBar seekBar, int i, boolean b) { }
    @Override
    public void onStartTrackingTouch(SeekBar seekBar) { }
    @Override
    public void onStopTrackingTouch(SeekBar seekBar) {
        myMediaPlayer.seekTo(seekBar.getProgress());
    }
    public void onClickmyBtn1(View v) {                 //响应单击"下载并播放此网络歌曲"按钮
        try {
            myMediaPlayer = new MediaPlayer();
            myMediaPlayer.setAudioStreamType(AudioManager.STREAM_MUSIC);
            myMediaPlayer.setDataSource(myEditText.getText().toString());
            myMediaPlayer.prepare();
            myMediaPlayer.start();
            myMediaPlayer.setOnBufferingUpdateListener(
                    new MediaPlayer.OnBufferingUpdateListener() {
                @Override
                public void onBufferingUpdate(MediaPlayer mediaPlayer, int i) {
                    mySeekBar.setSecondaryProgress((i / 100) *
                        myMediaPlayer.getDuration());           //显示缓冲进度

                    if (i == 100) {
                        Toast.makeText(MainActivity.this,
                                "音乐已下载(缓冲)完成!", Toast.LENGTH_SHORT).show();
                    }}});
            mySeekBar.setMax(myMediaPlayer.getDuration());
            myTotalTime.setText(getPlayerDurationToString(myMediaPlayer));
            myPlayPause.setImageResource(R.mipmap.myicon_pause);
```

```
        mySeekBar.setOnSeekBarChangeListener(this);
        final Timer myTimer = new Timer();
        myTimer.schedule(new TimerTask() {
         @Override
         public void run() {                                          //显示播放进度
           mySeekBar.setProgress(myMediaPlayer.getCurrentPosition());
           myHandler.sendEmptyMessage(1);
        } }, 0, 10);
      } catch (Exception e) { }
    }
    public String getPlayerDurationToString(MediaPlayer player) {    //获取音乐时长
      return player.getDuration()/1000/60 + ":" + (player.getDuration()/1000 % 60 < 10 ? "0" + player.
    getDuration()/1000 % 60 : player.getDuration()/1000 % 60);
    }
    public String getPlayerCurrentPositionToString(MediaPlayer player) {
      return player.getCurrentPosition()/1000/60 + ":" + (player.getCurrentPosition() /1000 % 60 < 10?"0"
    + player.getCurrentPosition()/1000 % 60 : player.getCurrentPosition() / 1000 % 60);
                                                            //获取歌曲当前播放位置
    } }
```

图 180.1

上面这段代码在 MyCode\MySample719\app\src\main\java\com\bin\luo\mysample\MainActivity.java 文件中。其中,myMediaPlayer.setOnBufferingUpdateListener()方法用于监听缓冲来自网络的音乐文件,可以在其参数中获取缓冲进度。myMediaPlayer.setDataSource(myEditText.getText().toString())用于在参数中设置网络音乐文件。setDataSource()方法设置播放源的常用调用形式如下。

(1) 播放当前应用的资源文件。如果把"test.mp3"音乐文件放在 res/raw/目录下,则调用形式如下:myMediaPlayer.setDataSource(this, Uri.parse("android.resource：//com.android.sim/" + R.raw.test))。com.android.sim 是当前应用包名,即 Uri.parse("android.resource://应用程序包名/" + R.raw.音乐文件名称)。如果把"test.mp3"音乐文件放在 assets 目录下,则调用形式如下:

```
AssetManager assetMg = this.getApplicationContext().getAssets();
AssetFileDescriptor fileDescriptor = assetMg.openFd("test.mp3");
myMediaPlayer.setDataSource(fileDescriptor.getFileDescriptor(),
        fileDescriptor.getStartOffset(), fileDescriptor.getLength());
```

（2）播放存储卡音乐文件，如 myMediaPlayer.setDataSource("/mnt/sdcard/test.mp3")。

（3）播放远程的资源文件，如 myMediaPlayer.setDataSource(Context, Uri.parse("http://**"))。

此外，访问网络需要在 AndroidManifest.xml 文件中添加< uses-permission android：name = "android.permission.INTERNET"/>权限。此实例的完整项目在 MyCode\MySample719 文件夹中。

181　使用滑块同步 MediaPlayer 播放音频的进度

此实例主要通过使用 SeekBar 和 MediaPlayer，实现在播放音乐时，SeekBar 的滑块始终跟随 MediaPlayer 的音乐播放进度。当实例运行之后，单击"选择音乐文件"按钮，则可以在打开的窗口中选择音乐文件，如图 181.1 的左图所示；选择的音乐文件的时长将显示在右端，如"5：6"，即 5 分 6 秒；单击下面的播放（或暂停）图标，则可以播放（或暂停）音乐；在播放音乐时，SeekBar 的滑块将自动跟随音乐播放进度向右滑动；向右或向左拖动滑块，则可调节音乐播放进度，如图 181.1 的右图所示。

图　181.1

主要代码如下：

```
public class MainActivity extends Activity implements View.OnClickListener,
SeekBar.OnSeekBarChangeListener {
    boolean isPlaying = false;
    private TextView myCurrentTime;
    private TextView myTotalTime;
    private ImageView myPlayPause;
    private MediaPlayer myPlayer;
    private SeekBar mySeekBar;
```

```java
Handler myHandler = new Handler() {
    @Override
    public void handleMessage(Message msg) {                    //显示当前播放位置
        myCurrentTime.setText(myPlayer.getCurrentPosition()/1000/60 + ":"
                + myPlayer.getCurrentPosition() / 1000 % 60);
    }
};
@Override
protected void onCreate(Bundle savedInstanceState) {
    super.onCreate(savedInstanceState);
    setContentView(R.layout.activity_main);
    myCurrentTime = (TextView) findViewById(R.id.myCurrentTime);
    myTotalTime = (TextView) findViewById(R.id.myTotalTime);
    Button myBtnFile = (Button) findViewById(R.id.myBtnFile);
    myPlayPause = (ImageView) findViewById(R.id.myPlayPause);
    myBtnFile.setOnClickListener(this);
    myPlayPause.setOnClickListener(this);
    mySeekBar = (SeekBar) findViewById(R.id.mySeekBar);
    mySeekBar.setOnSeekBarChangeListener(this);
}
@Override
public void onClick(View view) {
    if (view.getId() == R.id.myBtnFile) {                       //响应单击"选择音乐文件"按钮
        Intent myIntent = new Intent(Intent.ACTION_GET_CONTENT);
        myIntent.setType("audio/*");
        try {
            startActivityForResult(Intent.createChooser(myIntent, null), 2);
        } catch (android.content.ActivityNotFoundException ex) { }
    } else if (view.getId() == R.id.myPlayPause) {              //单击播放或暂停图标
        setPlayerStatus(!isPlaying);
    }
}
public void setPlayerStatus(boolean isplay) {
    if (isplay) {                                               //播放
        myPlayPause.setImageResource(R.mipmap.myicon_pause);
        isPlaying = true;
        myPlayer.start();
    } else {                                                    //暂停
        myPlayPause.setImageResource(R.mipmap.myicon_play);
        isPlaying = false;
        myPlayer.pause();
    }
}
public String getPath(Context context, Uri uri) {
    if ("content".equalsIgnoreCase(uri.getScheme())) {
        String[] myProjection = {"_data"};
        Cursor myCursor = null;
        try {
            myCursor = context.getContentResolver().query(uri,
                    myProjection, null, null, null);
            int myIndex = myCursor.getColumnIndexOrThrow("_data");
            if (myCursor.moveToFirst()) { return myCursor.getString(myIndex); }
        } catch (Exception e) { }
    } else if ("file".equalsIgnoreCase(uri.getScheme())) { return uri.getPath(); }
    return null;
}
```

```java
        @Override
    protected void onActivityResult(int requestCode, int resultCode, Intent data) {
      if (requestCode == 2) {
        if (resultCode == RESULT_OK) {
          Uri myUri = data.getData();
          String myPath = getPath(MainActivity.this, myUri);
          myPlayer = new MediaPlayer();
          try {
            myPlayer.setDataSource(myPath);
            myPlayer.prepare();                                    //加载音乐文件资源
            myTotalTime.setText(myPlayer.getDuration() / 1000 / 60 + ":"
                                + myPlayer.getDuration() / 1000 % 60);
            setPlayerStatus(true);
            mySeekBar.setMax(myPlayer.getDuration());
            new Timer().schedule(new TimerTask() {
              @Override
              public void run() {                                  //根据播放进度设置 SeekBar 位置
                mySeekBar.setProgress(myPlayer.getCurrentPosition());
                myHandler.sendEmptyMessage(1);
              } }, 0, 10);
          } catch (Exception e) { }
    } } }
    @Override
    public void onProgressChanged(SeekBar seekBar, int i, boolean b) { }
    @Override
    public void onStartTrackingTouch(SeekBar seekBar) { }
    @Override
    public void onStopTrackingTouch(SeekBar seekBar) {
      myPlayer.seekTo(seekBar.getProgress());                      //根据 SeekBar 位置设置播放进度
    }
```

上面这段代码在 MyCode \ MySample712 \ app \ src \ main \ java \ com \ bin \ luo \ mysample \ MainActivity.java 文件中。在这段代码中，mySeekBar.setMax(myPlayer.getDuration())用于根据音乐文件的时长设置 SeekBar 的最大值。mySeekBar.setProgress(myPlayer.getCurrentPosition())用于根据当前音乐播放进度设置 SeekBar 滑块的当前位置。myPlayer.seekTo(seekBar.getProgress())用于根据 SeekBar 滑块的当前位置(在被拖动之后)设置(重置)音乐播放进度。此外，读取在 SD 卡上的音乐文件需要在 AndroidManifest.xml 文件中添加< uses-permission android:name = "android.permission.READ_EXTERNAL_STORAGE"/>权限。此实例的完整项目在 MyCode\MySample712 文件夹中。

182 使用滑块同步 MediaPlayer 播放视频的进度

此实例主要通过使用 SeekBar 和 MediaPlayer，实现在播放视频时，SeekBar 的滑块始终跟随 MediaPlayer 的视频播放进度。当实例运行之后，单击"选择视频文件"按钮，则可以在打开的窗口中选择视频文件，如图 182.1 的左图所示；选择的视频文件的时长将显示在右端，如"5：53"，即 5 分 53 秒；单击右端的播放(或暂停)图标，则可以播放(或暂停)视频；在播放视频时，SeekBar 的滑块将自动跟随视频播放进度向右滑动；向右或向左拖动滑块，则可调节视频播放进度，如图 182.1 的右图所示。

图 182.1

主要代码如下：

```java
public class MainActivity extends Activity implements View.OnClickListener {
  boolean isPlaying = false;
  private SurfaceView mySurfaceView;
  private MediaPlayer myPlayer;
  private TextView myTotalTime;
  private ImageView myPlayPause;
  Handler myHandler = new Handler() {                              //更新当前播放位置
  @Override
  public void handleMessage(Message msg) {
   if (msg.what == 2) {
    String myCurrentPosition;
    if (myPlayer.getCurrentPosition() / 1000 % 60 < 10) {
     myCurrentPosition = "0" + myPlayer.getCurrentPosition() / 1000 % 60;
    } else { myCurrentPosition = myPlayer.getCurrentPosition()/1000 % 60 + "";}
    myCurrentTime.setText(myPlayer.getCurrentPosition() / 1000 / 60 + ":"
                                                     + myCurrentPosition);
  } } };
private TextView myCurrentTime;
private SeekBar mySeekBar;
 @Override
 protected void onCreate(Bundle savedInstanceState) {
  super.onCreate(savedInstanceState);
  setContentView(R.layout.activity_main);
  Button myBtnFile = (Button) findViewById(R.id.myBtnFile);
  myBtnFile.setOnClickListener(this);
  myPlayPause = (ImageView) findViewById(R.id.myPlayPause);
  mySurfaceView = (SurfaceView) findViewById(R.id.mySurfaceView);
  mySeekBar = (SeekBar) findViewById(R.id.mySeekBar);
  myCurrentTime = (TextView) findViewById(R.id.myCurrentTime);
```

```java
    myTotalTime = (TextView) findViewById(R.id.myTotalTime);
  }
  @Override
  public void onClick(View view) {                              //响应单击"选择视频文件"按钮
    if (view.getId() == R.id.myBtnFile) {
      Intent myIntent = new Intent(Intent.ACTION_GET_CONTENT);
      myIntent.setType("video/*");                              //选择视频类型文件
      myIntent.addCategory(Intent.CATEGORY_OPENABLE);
      try {
        startActivityForResult(Intent.createChooser(myIntent, null), 1);
      } catch (Exception ex) { }
    } }
  //响应单击播放(或暂停)按钮(ImageView控件)
  public void onClickPlayPause(View view) { setPlayerStatus(!isPlaying); }
  public void setPlayerStatus(boolean isplay) {
    if (isplay) {                                               //播放
      myPlayPause.setImageResource(R.mipmap.myicon_pause);
      isPlaying = true;
      myPlayer.start();
    }else{                                                      //暂停
      myPlayPause.setImageResource(R.mipmap.myicon_play);
      isPlaying = false;
      myPlayer.pause();
  } }
  @Override
  protected void onActivityResult(int requestCode, int resultCode, Intent data){
    if (requestCode == 1){
      if (resultCode == RESULT_OK){
        Uri myUri = data.getData();
        final String myPath = getPath(this, myUri);
        try {
          mySurfaceView.getHolder().setKeepScreenOn(true);
          mySurfaceView.getHolder().addCallback(new SurfaceHolder.Callback(){
            @Override
            public void surfaceCreated(SurfaceHolder surfaceHolder){
              try {
                myPlayer = new MediaPlayer();
                myPlayer.setDataSource(myPath);
                myPlayer.setDisplay(mySurfaceView.getHolder());
                myPlayer.prepare();
                setPlayerStatus(true);
                new Timer().schedule(new TimerTask(){
                  @Override
                  public void run(){                             //根据播放进度设置SeekBar的当前值
                    mySeekBar.setProgress(myPlayer.getCurrentPosition());
                    myHandler.sendEmptyMessage(2);
                  } }, 0, 10);
                myTotalTime.setText(myPlayer.getDuration() / 1000 / 60 + ":"
                                     + myPlayer.getDuration() / 1000 % 60);
                //根据视频文件的时长设置SeekBar控件的最大值
                mySeekBar.setMax(myPlayer.getDuration());
                mySeekBar.setOnSeekBarChangeListener(
                                      new SeekBar.OnSeekBarChangeListener() {
```

```java
        @Override
        public void onProgressChanged(SeekBar seekBar, int i, boolean b) {}
        @Override
        public void onStartTrackingTouch(SeekBar seekBar) {}
        @Override
        public void onStopTrackingTouch(SeekBar seekBar) {
          //根据 SeekBar 控件的当前值设置播放进度
          myPlayer.seekTo(seekBar.getProgress());
        } });
      } catch (IOException e) { e.printStackTrace(); }
    }
    @Override
    public void surfaceChanged(SurfaceHolder surfaceHolder,
                                        int i, int i1, int i2) { }

    @Override
    public void surfaceDestroyed(SurfaceHolder surfaceHolder) { } });
  } catch (Exception e){ e.printStackTrace(); }
} } }
public static String getPath(Context context, Uri uri) {        //获取视频文件全路径
if ("content".equalsIgnoreCase(uri.getScheme())) {
 String[] projection = {"_data"};
 Cursor cursor = null;
 try {
  cursor = context.getContentResolver().query(uri,
                                        projection, null, null, null);
  int column_index = cursor.getColumnIndexOrThrow("_data");
  if (cursor.moveToFirst()) { return cursor.getString(column_index); }
 } catch (Exception e) { }
} else if ("file".equalsIgnoreCase(uri.getScheme())) { return uri.getPath(); }
return null;
} }
```

上面这段代码在 MyCode \ MySample715 \ app \ src \ main \ java \ com \ bin \ luo \ mysample \ MainActivity.java 文件中。此实例的完整项目在 MyCode\MySample715 文件夹中。

183 使用 MediaController 创建视频播放控制栏

此实例主要通过使用 MediaController 对象作为 VideoView 的 setMediaController()方法的参数,实现在单击 VideoView 控件之后,在屏幕底部弹出视频播放控制栏。当实例运行之后,单击"浏览本地视频文件"按钮,然后在弹出的窗口中选择视频文件,即可在 VideoView 控件中自动播放此视频;单击正在播放的视频,则在屏幕底部弹出控制视频播放的控制栏 MediaController,在此控制栏上有播放、暂停等按钮,以及播放进度条,单击相关按钮即可控制视频播放,效果分别如图 183.1 的左图和右图所示。

主要代码如下:

```java
public class MainActivity extends Activity {
 VideoView myVideoView;
 Button myBtnBrowser;
 @Override
 protected void onCreate(Bundle savedInstanceState) {
```

```
super.onCreate(savedInstanceState);
setContentView(R.layout.activity_main);
myBtnBrowser = (Button) findViewById(R.id.myBtnBrowser);
myVideoView = (VideoView) findViewById(R.id.myVideoView);
myBtnBrowser.setOnClickListener(new View.OnClickListener() {
  @Override
  public void onClick(View v) {                           //浏览视频文件
    Intent myIntent = new Intent(Intent.ACTION_GET_CONTENT);
    myIntent.setType("video/mp4");
    startActivityForResult(myIntent, 0);
  } });
}
@Override
protected void onActivityResult(int requestCode, int resultCode, Intent data) {
  if(resultCode == RESULT_OK){
    Uri myUri = data.getData();                           //获取用户所选视频 Uri
    myVideoView.setVideoURI(myUri);                       //设置 Uri(视频文件全路径)
    myVideoView.start();                                  //播放视频
    //在单击视频之后,显示播放、暂停等按钮的控制栏,初始化 MediaController 对象
    MediaController myMediaController = new MediaController(this);
    //将该对象应用于 VideoView 控件
    myVideoView.setMediaController(myMediaController);
} } }
```

图 183.1

上面这段代码在 MyCode \ MySample814 \ app \ src \ main \ java \ com \ bin \ luo \ mysample \ MainActivity.java 文件中。在这段代码中,myVideoView.setMediaController(myMediaController)用于实现在单击视频之后,在屏幕底部显示播放控制栏。此实例的完整项目在 MyCode \ MySample814 文件夹中。

184　使用 MediaMetadataRetriever 实现视频截图

此实例主要通过使用 MediaMetadataRetriever 的 getFrameAtTime（）和 MediaPlayer 的 getCurrentPosition()方法，实现对当前正在播放的视频进行截图。当实例运行之后，在"视频文件："输入框中输入在 SD 卡上存储的视频文件名称，然后单击"开始播放视频"按钮，则将播放此视频，如图 184.1 的左图所示。单击"当前视频截图"按钮，则将把视频的当前图像保存在 SD 卡上，如图 184.1 的右图所示。

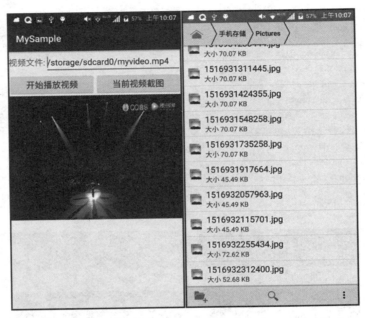

图　184.1

主要代码如下：

```java
public void onClickmyBtn2(View v) {                                    //响应单击"当前视频截图"按钮
    try {
        MediaMetadataRetriever myMediaMetadataRetriever =
                                    new MediaMetadataRetriever();
        myMediaMetadataRetriever.setDataSource(myVideoFile.getText().toString());
        Bitmap myBitmap = myMediaMetadataRetriever.getFrameAtTime(myMediaPlayer.getCurrentPosition() * 1000, MediaMetadataRetriever.OPTION_CLOSEST_SYNC);
        if (myBitmap != null) {
            Uri myImageUri = getContentResolver().insert(MediaStore.Images.Media.EXTERNAL_CONTENT_URI, new ContentValues());
            OutputStream myOutStream =
                        getContentResolver().openOutputStream(myImageUri);
            myBitmap.compress(Bitmap.CompressFormat.JPEG,
                90, myOutStream);                                      //将截图保存到存储卡
            Toast.makeText(MainActivity.this,
                        "截图已成功保存!",Toast.LENGTH_SHORT).show();
        }
    } catch (Exception e) { e.printStackTrace(); }
}
```

上面这段代码在 MyCode\MySample681\app\src\main\java\com\bin\luo\mysample\ MainActivity. java 文件中。在这段代码中，myBitmap = myMediaMetadataRetriever. getFrameAtTime(myMediaPlayer. getCurrentPosition() * 1000，MediaMetadataRetriever. OPTION_CLOSEST_SYNC)用于将当前视频播放位置（帧位置）的图像转化成位图，myMediaPlayer. getCurrentPosition()用于获取当前视频的播放位置（帧位置），由于 getFrameAtTime()方法的第一个参数代表的是微秒值，getCurrentPosition()方法的返回值是毫秒值，因此需要乘以 1000。使用 getFrameAtTime(long timeUs, int option)方法还需要注意的是：option：OPTION_CLOSEST 表示获取离 timeUs 最近的一帧图像，此种情况获取的帧是 Sync Frame，由于在 timeUs 此时间点不一定恰好有一个 Sync Frame，所以可能有一定的误差，在获取 Sync Frame 时可能有下列 3 种情况。

（1）在 timeUs 处恰好有一个 Sync Frame。
（2）获取 timeUs 前一个 Sync Frame。
（3）获取 timeUs 后一个 Sync Frame。

此外，在 SD 卡读写文件需要在 AndroidManifest. xml 文件中添加< uses-permission android：name＝"android. permission. READ_EXTERNAL_STORAGE"/>和< uses-permission android：name＝"android. permission. WRITE_EXTERNAL_STORAGE"/>权限。此实例的完整项目在 MyCode\MySample681 文件夹中。

185　使用 MediaMetadataRetriever 获取视频缩略图

此实例主要通过使用 MediaMetadataRetriever 的成员方法 getFrameAtTime()，实现根据指定的视频文件获取其视频缩略图。当实例运行之后，在"视频文件："输入框中输入在 SD 卡上存储的视频文件名称，然后单击"获取此视频文件的缩略图"按钮，则在下面显示此视频文件的缩略图，效果分别如图 185.1 的左图和右图所示。

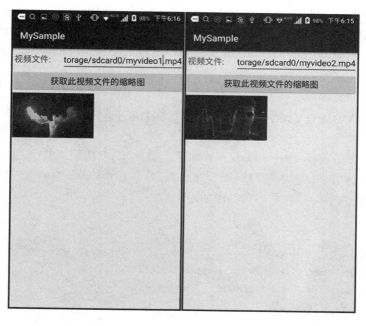

图　185.1

主要代码如下：

```java
public void onClickmyBtn1(View v) { //响应单击"获取此视频文件的缩略图"按钮
  String myPath = myEditText.getText().toString();
  Bitmap myBitmap = createVideoThumbnail(myPath,
         MediaStore.Images.Thumbnails.MINI_KIND);
  myImageView.setImageBitmap(myBitmap);
}
public Bitmap createVideoThumbnail(String myPath, int kind) {
  MediaMetadataRetriever myRetriever = new MediaMetadataRetriever();
  myRetriever.setDataSource(myPath);
  Bitmap myBmp = myRetriever.getFrameAtTime(-1);
  if (myBmp == null) return null;
  if (kind == MediaStore.Images.Thumbnails.MINI_KIND) {
   int width = myBmp.getWidth();
   int height = myBmp.getHeight();
   int max = Math.max(width, height);
   if (max > 512) {
    float scale = 512f / max;
    int w = Math.round(scale * width);
    int h = Math.round(scale * height);
    myBmp = Bitmap.createScaledBitmap(myBmp, w, h, true);
  } }
  return myBmp;
}
```

上面这段代码在 MyCode\MySample611\app\src\main\java\com\bin\luo\mysample\MainActivity.java 文件中。在这段代码中，myRetriever.setDataSource（myPath）用于设置 MediaMetadataRetriever 操作的视频文件。myBmp = myRetriever.getFrameAtTime(-1)用于获取该视频指定时间点的图像。myBmp = Bitmap.createScaledBitmap(myBmp, w, h, true)用于根据指定尺寸创建该图像的缩略图。此外，读取在 SD 卡上的文件需要在 AndroidManifest.xml 文件中添加 <uses-permission android:name="android.permission.READ_EXTERNAL_STORAGE"/>权限。此实例的完整项目在 MyCode\MySample611 文件夹中。

186 使用 VideoView 播放本地 mp4 视频文件

此实例主要通过使用 VideoView，实现在 SD 卡上播放 mp4 视频文件。当实例运行之后，在"视频文件："输入框中输入在 SD 卡上存储的视频文件路径，然后单击"开始播放"按钮，则将播放指定的视频文件，效果分别如图 186.1 的左图和右图所示。

主要代码如下：

```java
public void onClickmyBtn1(View v) {                    //响应单击"开始播放"按钮
  myVideoView.setMediaController(new MediaController(this));
  Uri myUri = Uri.parse(myEditText.getText().toString());
  myVideoView.setVideoURI(myUri);
  myVideoView.start();
  myVideoView.requestFocus();
}
```

图 186.1

上面这段代码在 MyCode\MySample261\app\src\main\java\com\bin\luo\mysample\MainActivity.java 文件中。在这段代码中，myVideoView.setVideoURI(myUri)用于以 Uri 的方式设置 VideoView 控件将要播放的视频文件，也可以直接使用 myVideoView.setVideoPath(myEditText.getText().toString())方法设置 VideoView 控件的视频文件。然后就可以调用 VideoView 控件的 start()方法来控制视频的播放。由于 VideoView 控件通过与 MediaController 类结合使用，因此可以不用自己控制视频的播放与暂停等操作。此外，读取在 SD 卡上的视频文件需要在 AndroidManifest.xml 文件中添加< uses-permission android:name= "android.permission.READ_EXTERNAL_STORAGE"/>权限。此实例的完整项目在 MyCode\MySample261 文件夹中。

187 使用 VideoView 播放指定网址的视频文件

此实例主要通过使用 VideoView 的 setVideoPath()方法设置视频文件的网络地址，实现播放存储在网络上的视频文件。当实例运行之后，在"视频网址："输入框中输入在网络上存储的视频文件网址（如"https://1251412368.vod2.myqcloud.com/vodtransgzp1251412368/4564972818539678338/v.f30.mp4"），然后单击"播放视频"按钮，则视频播放效果如图 187.1 的左图所示。在"视频网址："输入框中输入在网络上存储的其他视频文件网址（如"https://1251412368.vod2.myqcloud.com/vodtransgzp1251412368/4564972818871314502/v.f30.mp4"），然后单击"播放视频"按钮，则视频播放效果如图 187.1 的右图所示。

主要代码如下：

```
public void onClickBtn1(View v) {                         //响应单击"播放视频"按钮
    //初始化视频控制块
    MediaController myController = new MediaController(this);
    //将视频控制块与 VideoView 控件绑定
    myVideoView.setMediaController(myController);
    //设置网络视频的地址
```

```
myVideoView.setVideoPath(myEditText.getText().toString());
myVideoView.start();                                    //开始播放
}
```

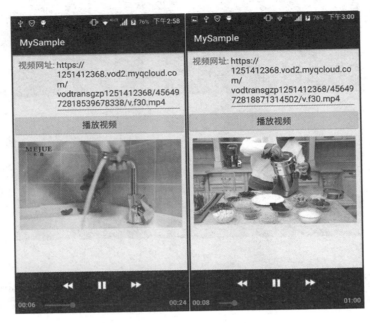

图 187.1

上面这段代码在 MyCode \ MySample816 \ app \ src \ main \ java \ com \ bin \ luo \ mysample \ MainActivity.java 文件中。由于 VideoView 控件能够通过与 MediaController 控件结合使用，因此可以不用自己控制视频的播放与暂停等操作。需要说明的是，播放网络视频需要在 AndroidManifest.xml 文件中添加< uses-permission android：name＝"android. permission. INTERNET"/>权限。此外，此实例在部分模拟器测试不成功，因此请直接在手机上测试。此实例的完整项目在 MyCode\ MySample816 文件夹中。

188　使用 MediaRecorder 录制音频文件

此实例主要通过使用录音机对象 MediaRecorder，实现使用手机录制音频文件。当实例运行之后，单击"开始录音"按钮，则实例自动根据当前时间生成一个文件记录音频数据，此时就可以对着手机讲话，内容自然就保存在此音频文件中，如图 188.1 的左图所示。单击"停止录音"按钮，则音频文件录制完成，此时即可以在 SD 卡找到此文件，然后进行播放，如图 188.1 的右图所示。

主要代码如下：

```
public void onClickmyBtn1(View v) {                     //响应单击"开始录音"按钮
    myFile = new File(Environment.getExternalStorageDirectory() + "/" +
        "myMediaRecorder" + new DateFormat().format("yyyyMMdd_hhmmss",
        Calendar.getInstance(Locale.CHINA)) + ".amr");  //将录音保存在 SD 卡
    myMediaRecorder = new MediaRecorder();              //创建录音对象
    myMediaRecorder.setAudioSource(
            MediaRecorder.AudioSource.DEFAULT);         //从话筒源进行录音
    myMediaRecorder.setOutputFormat(
```

```
                    MediaRecorder.OutputFormat.DEFAULT);        //设置输出格式
    myMediaRecorder.setAudioEncoder(
                    MediaRecorder.AudioEncoder.DEFAULT);         // 设置编码格式
    myMediaRecorder.setOutputFile(myFile.getAbsolutePath());     // 设置输出文件
    try {
      myMediaRecorder.prepare();                                 //准备录制
      myMediaRecorder.start();                                   //开始录制
      myTextView.setText(myFile.getAbsolutePath() + "文件正在录制中……");
    } catch (IllegalStateException e) { e.printStackTrace(); }
    catch (IOException e) { e.printStackTrace(); }
  }
  public void onClickmyBtn2(View v) {                            //响应单击"停止录音"按钮
    if (myMediaRecorder != null) {                               //如果正在录音,则可以停止录音
      myMediaRecorder.stop();
      myMediaRecorder.release();
      myMediaRecorder = null;
      myTextView.setText(myFile.getAbsolutePath() + "文件录制完毕");
    }
  }
```

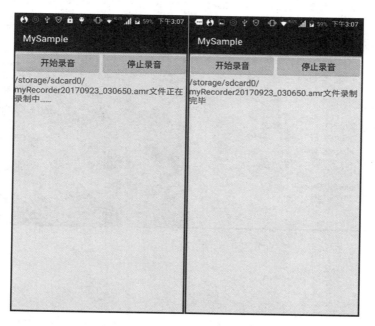

图 188.1

上面这段代码在 MyCode\MySample268\app\src\main\java\com\bin\luo\mysample\MainActivity.java 文件中。在这段代码中,myMedia Recorder= new MediaRecorder()用于创建一个录音机对象。myMedia Recorder.setOutputFile(myFile.getAbsolutePath())用于设置音频输出文件。myMedia Recorder.prepare()用于对录音机进行预处理,如果没有此操作,运行可能会报错。myMedia Recorder.start()表示开始录制音频。myMedia Recorder.stop()表示停止音频录制。此外,在 SD 卡写入文件需要在 AndroidManifest.xml 文件中添加< uses-permission android:name ="android.permission.RECORD _ AUDIO"/> 和 < uses-permission android:name = " android.permission.WRITE _ EXTERNAL _ STORAGE"/> 权限。此实例的完整项目在 MyCode\MySample268 文件夹中。

189　使用 RemoteViews 在通知栏上创建播放器

此实例主要通过使用 registerReceiver() 方法，实现以注册广播服务的方式在通知栏上创建播放器播放音乐。当实例运行之后，单击"选择音乐文件"按钮，则可以在打开的窗口中选择音乐文件，如"小品：梦幻家园_超清.mp3"，如图 189.1 的左图所示，同时自动播放此音乐文件；下拉滑出通知栏，则可以发现通知栏列表项有播放器正在播放"小品：梦幻家园_超清.mp3"，如图 189.1 的右图所示；单击右侧的按钮，则可以播放或暂停。

图　189.1

主要代码如下：

```
public class MainActivity extends Activity {
    boolean isPlaying = false;                              //播放暂停状态判断标志
    private MediaPlayer myMediaPlayer;
    private RemoteViews myRemoteViews;                      //通知对应的自定义视图对象
    private NotificationManager myNotificationManager;      //通知管理器对象
    private Notification myNotification;                    //通知对象
    @Override
    protected void onCreate(Bundle savedInstanceState) {
        super.onCreate(savedInstanceState);
        setContentView(R.layout.activity_main);
        myNotificationManager =
                (NotificationManager) getSystemService(NOTIFICATION_SERVICE);
        myNotification = new Notification(R.drawable.myplayer,
                                    null, System.currentTimeMillis());
        //自定义通知栏播放器布局
```

```java
    myRemoteViews = new RemoteViews(getPackageName(), R.layout.myplayerlayout);
    myNotification.contentView = myRemoteViews;
    registerReceiver(new BroadcastReceiver() {
      @Override
      public void onReceive(Context context, Intent intent) {
        if (intent.getAction() == "SwitchStatus") {
          if (isPlaying) {                                              //如果正在播放,则暂停
            myMediaPlayer.pause();
            isPlaying = false;
            myRemoteViews.setImageViewResource(R.id.myBtnPlayPause,
                                                          R.mipmap.myicon_play);
            myNotificationManager.notify(1, myNotification);
          } else {                                                      //如果已经暂停,则播放
            myMediaPlayer.start();
            isPlaying = true;
            myRemoteViews.setImageViewResource(R.id.myBtnPlayPause,
                                                          R.mipmap.myicon_pause);
            myNotificationManager.notify(1, myNotification);            //更新并重新推送通知
    } } }, new IntentFilter("SwitchStatus"));                           //注册广播,单击处理通知按钮事件
    //新建Intent,用于接收广播时过滤Intent信息
    Intent myIntentSwitchStatus = new Intent("SwitchStatus");
    PendingIntent myPendingIntent =
                    PendingIntent.getBroadcast(this, 0, myIntentSwitchStatus, 0);
    myRemoteViews.setOnClickPendingIntent(R.id.myBtnPlayPause, myPendingIntent);
    myNotification.flags = myNotification.FLAG_NO_CLEAR;
    myNotificationManager.notify(1, myNotification);
    Button myBtnFile = (Button) findViewById(R.id.myBtnFile);
    //浏览音乐文件并播放
    myBtnFile.setOnClickListener(new View.OnClickListener() {
      @Override
      public void onClick(View view) {
        Intent myIntent = new Intent(Intent.ACTION_GET_CONTENT);
        myIntent.setType("audio/*");
        try {
          startActivityForResult(Intent.createChooser(myIntent, null), 31415926);
        } catch (android.content.ActivityNotFoundException ex) { } } });
  }
  @Override
  protected void onActivityResult(int requestCode, int resultCode, Intent data){
    if (requestCode == 31415926) {
      if (resultCode == RESULT_OK) {
        Uri myUri = data.getData();
        String myPath = getFilePath(MainActivity.this, myUri);
        myMediaPlayer = new MediaPlayer();
        try {
          myMediaPlayer.setDataSource(myPath);
          myMediaPlayer.prepare();
          //在通知栏自定义播放器上显示音乐文件名称
          myRemoteViews.setTextViewText(R.id.myFileName,
                                  getFileName(MainActivity.this, myPath));
          myNotificationManager.notify(1, myNotification);
          myMediaPlayer.start();                                        //开始播放
          isPlaying = true;
```

```
        } catch (Exception e) { }
  } } }
  //获取音乐文件的完整路径,省略的部分请参考源代码
  public String getFilePath(Context context, Uri uri) { }
  //获取音乐文件的显示名称,省略的部分请参考源代码
  public static String getFileName(Context context, String path) { }
}
```

上面这段代码在 MyCode\MySample713\app\src\main\java\com\bin\luo\mysample\MainActivity.java 文件中。在这段代码中,myRemoteViews = new RemoteViews(getPackageName(),R.layout.myplayerlayout)用于根据布局文件 myplayerlayout.xml 在通知栏上创建音乐播放器。关于 myplayerlayout 布局的详细内容请参考源代码中的 MyCode\MySample713\app\src\main\res\layout\myplayerlayout.xml 文件。此外,读取 SD 卡文件需要在 AndroidManifest.xml 文件中添加 < uses-permission android:name="android.permission.READ_EXTERNAL_STORAGE"/>权限。此实例的完整项目在 MyCode\MySample713 文件夹中。

190　在使用 SurfaceView 播放视频时实现横屏显示

此实例主要通过使用 setRequestedOrientation()方法,实现在播放视频时进行横屏和竖屏切换。当实例运行之后,单击"选择视频文件"按钮,则可以在打开的窗口中选择视频文件,选择的视频文件的时长将显示在右端,如"5:53",即 5 分 53 秒,并且自动开始播放;单击"横屏竖屏切换"按钮,如果当前手机是竖屏,则切换为横屏;如果当前手机是横屏,则切换为竖屏,效果分别如图 190.1 的左图和右图所示。

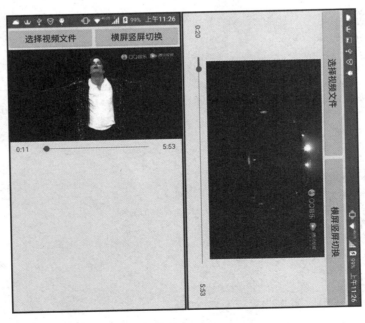

图　190.1

主要代码如下：

```java
public class MainActivity extends Activity implements SurfaceHolder.Callback {
    private MyUtils myUtils;
    private SurfaceView mySurfaceView;
    private SeekBar mySeekBar;
    Handler myHandler = new Handler() {
        @Override
        public void handleMessage(Message msg) {                    //显示已播放时间
            myCurrentTime.setText(myUtils.getMediaPlayerCurrentPositionByString());
        } };
    private TextView myCurrentTime, myTotalTime;
    private String myFilePath;
    @Override
    public void onConfigurationChanged(Configuration newConfig) {
        super.onConfigurationChanged(newConfig);
        if (newConfig.orientation == Configuration.ORIENTATION_LANDSCAPE) {
            setVideoParams(true);
        } else if (newConfig.orientation == Configuration.ORIENTATION_PORTRAIT){
            setVideoParams(false);
    } }
    public void setVideoParams(boolean isLandscape) {
        LinearLayout myMainLayout = (LinearLayout) findViewById(R.id.activity_main);
        ViewGroup.LayoutParams myMainLayoutParams = myMainLayout.getLayoutParams();
        ViewGroup.LayoutParams mySurfaceViewParams = mySurfaceView.getLayoutParams();
        float myScreenWidth = getResources().getDisplayMetrics().widthPixels;
        float myScreenHeight = getResources().getDisplayMetrics().heightPixels;
        if (isLandscape) {                                          //横屏显示模式
            //getWindow().addFlags(WindowManager.LayoutParams.FLAG_FULLSCREEN);
            //根据视频原始宽度和高度设置播放器的宽度和高度
            mySurfaceViewParams.width = myUtils.getMyPlayer().getVideoWidth();
            mySurfaceViewParams.height = myUtils.getMyPlayer().getVideoHeight();
        } else {                                                    //竖屏显示模式
            //getWindow().clearFlags(WindowManager.LayoutParams.FLAG_FULLSCREEN);
        }
        myMainLayoutParams.width = (int) myScreenWidth;
        myMainLayoutParams.height = (int) myScreenHeight;
        myMainLayout.setLayoutParams(myMainLayoutParams);
        mySurfaceView.setLayoutParams(mySurfaceViewParams);
    }
    @Override
    protected void onCreate(Bundle savedInstanceState) {
        super.onCreate(savedInstanceState);
        myUtils = new MyUtils();
        setContentView(R.layout.activity_main);
        mySurfaceView = (SurfaceView) findViewById(R.id.mySurfaceView);
        mySeekBar = (SeekBar) findViewById(R.id.mySeekBar);
        myCurrentTime = (TextView) findViewById(R.id.myCurrentTime);
        myTotalTime = (TextView) findViewById(R.id.myTotalTime);
    }
    public void onClickButton1(View view) {                         //响应单击"选择视频文件"按钮
        Intent myIntent = new Intent(Intent.ACTION_GET_CONTENT);
        myIntent.setType("video/*");
        myIntent.addCategory(Intent.CATEGORY_OPENABLE);
        try {
            startActivityForResult(Intent.createChooser(myIntent, null), 1);
        } catch (Exception ex) { }
```

```java
    }
    public void onClickButton2(View view) {                         //响应单击"横屏竖屏切换"按钮
        if (getResources().getConfiguration().orientation
                == Configuration.ORIENTATION_LANDSCAPE) {           //变成竖屏模式
            setRequestedOrientation(ActivityInfo.SCREEN_ORIENTATION_PORTRAIT);
        } else if (getResources().getConfiguration().orientation
                == Configuration.ORIENTATION_PORTRAIT) {            //变成横屏模式
            setRequestedOrientation(ActivityInfo.SCREEN_ORIENTATION_LANDSCAPE);
        }
    }
    @Override
    protected void onActivityResult(int requestCode, int resultCode, Intent data) {
        if (resultCode == RESULT_OK) {
            myFilePath = myUtils.getPath(MainActivity.this,
                    data.getData());                                 //获取视频文件的路径
            mySurfaceView.getHolder().setKeepScreenOn(true);
            mySurfaceView.getHolder().addCallback(MainActivity.this);
        }
    }
    @Override
    public void surfaceCreated(SurfaceHolder surfaceHolder) {
        try {
            myUtils.InitMediaPlayer(new TimerTask() {
                @Override
                public void run() {                                  //根据当前播放进度设置 SeekBar 的滑块位置
                    mySeekBar.setProgress(myUtils.getMediaPlayerCurrentPositionByInt());
                    myHandler.sendEmptyMessage(2);
                }
            }, myFilePath, mySurfaceView.getHolder());
            //显示视频时长
            myTotalTime.setText(myUtils.getMediaPlayerDurationByString());
            //根据视频时长设置滑块最大值
            mySeekBar.setMax(myUtils.getMediaPlayerDurationByInt());
            myUtils.InitSeekBarListener(mySeekBar);
        } catch (Exception e) { }
    }
    @Override
    public void surfaceChanged(SurfaceHolder surfaceHolder, int i, int i1, int i2){ }
    @Override
    public void surfaceDestroyed(SurfaceHolder surfaceHolder) { }
}
```

上面这段代码在 MyCode\MySample717\app\src\main\java\com\bin\luo\mysample\MainActivity.java 文件中。在这段代码中，setRequestedOrientation()方法用于设置屏幕显示方式，当该方法的参数为 ActivityInfo.SCREEN_ORIENTATION_LANDSCAPE 时，则实现横屏显示；当其参数为 ActivityInfo.SCREEN_ORIENTATION _PORTRAIT 时，则实现竖屏显示。getResources().getConfiguration().orientation==Configuration.ORIENTATION_LANDSCAPE 用于判断当前屏幕是否是横屏显示，如果值为 true，则表示横屏显示。getResources().getConfiguration().orientation==Configuration.ORIENTATION_PORTRAIT 用于判断当前屏幕是否是竖屏显示，如果值为 true，则表示竖屏显示。myFilePath=myUtils.getPath(MainActivity.this,data.getData())用于获取在选择文件窗口中选择的视频文件路径，myUtils 是自定义工具类 MyUtils 的实例，关于 MyUtils 类的详细内容请参考源代码中的 MyCode\MySample717\app\src\main\java\com\bin\luo\mysample\MyUtils.java 文件。需要说明的是，读取 SD 卡文件需要在 AndroidManifest.xml 文件中添加< uses-permission android:name="android.permission.READ_EXTERNAL_STORAGE"/>权限。此实例的完整项目在 MyCode\MySample717 文件夹中。

191 在选择音乐曲目窗口中选择音乐文件并播放

此实例主要通过使用 MediaStore.Audio.Media.EXTERNAL_CONTENT_URI 构造显示所有音乐文件的选择音乐曲目窗口,实现允许用户选择音乐文件并播放。当实例运行之后,单击"选择并播放音乐文件"按钮,则将显示"选择音乐曲目"窗口,如图 191.1 的左图所示。在音乐曲目列表中任意选择一个音乐文件,如"alarm01.wav",然后单击"确定"按钮,则将播放此音乐文件,如图 191.1 的右图所示。

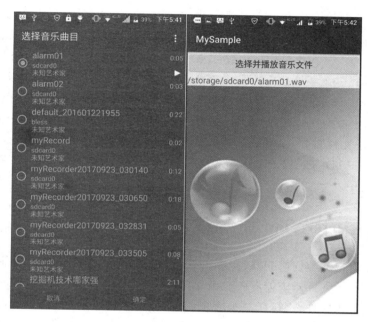

图 191.1

主要代码如下:

```
public void onClickmyBtn1(View v) {                               //响应单击"选择并播放音乐文件"按钮
    Intent myIntent = new Intent(Intent.ACTION_PICK, null);
    myIntent.setDataAndType(MediaStore.Audio.Media.EXTERNAL_CONTENT_URI, null);
    startActivityForResult(myIntent, 1);
}
@Override
protected void onActivityResult(int myRequestCode,
                int resultCode, Intent myIntent) {                //播放选择的音乐文件
    super.onActivityResult(myRequestCode, resultCode, myIntent);
    try {
      switch (myRequestCode) {
      case 1:
        if (resultCode == Activity.RESULT_OK) {
          Uri myUri = myIntent.getData();
          String[] myPathColumns = {MediaStore.Images.Media.DATA};
          Cursor myCursor =
             getContentResolver().query(myUri, myPathColumns, null, null, null);
          myCursor.moveToFirst();
          int myIndex = myCursor.getColumnIndex(myPathColumns[0]);
```

```
            String myFilePath = myCursor.getString(myIndex);
            MediaPlayer myMediaPlayer = new MediaPlayer();
            myMediaPlayer.setDataSource(myFilePath);            //设置音乐文件路径
            myMediaPlayer.prepare();
            myMediaPlayer.start();
            myTextView.setText(myFilePath);
            }
            break;
        }
    } catch (Exception e){ e.printStackTrace(); }
}
```

上面这段代码在 MyCode \ MySample380 \ app \ src \ main \ java \ com \ bin \ luo \ mysample \ MainActivity.java 文件中。在这段代码中，myIntent = new Intent(Intent.ACTION_PICK，null)用于创建可进行选择的 Intent。myIntent.setDataAndType(MediaStore.Audio.Media.EXTERNAL_CONTENT_URI，null)表示此 Intent 的数据类型为音乐文件。onActivityResult()用于处理用户在"选择音乐曲目"窗口中单击"确定"或"取消"按钮之后返回的信息，如音乐文件全路径信息等。此外，读取在 SD 卡上的音乐文件需要在 AndroidManifest.xml 文件中添加< uses-permission android:name="android.permission.READ_EXTERNAL_STORAGE"/>权限。此实例的完整项目在 MyCode\MySample380 文件夹中。

192　在 RecyclerView 中加载音乐文件并播放

此实例主要通过自定义数据适配器，实现在 RecyclerView 的列表中显示当前手机的所有音乐文件，选择任一列表项（音乐文件），则播放此音乐文件。当实例运行之后，将在 RecyclerView 的列表中显示当前手机的所有音乐文件，选择任一列表项（音乐文件），则立即播放音乐，在底部的 SeekBar 也将同步显示播放进度，单击右侧的暂停或播放按钮则可暂停或播放音乐，效果分别如图 192.1 的左图和右图所示。

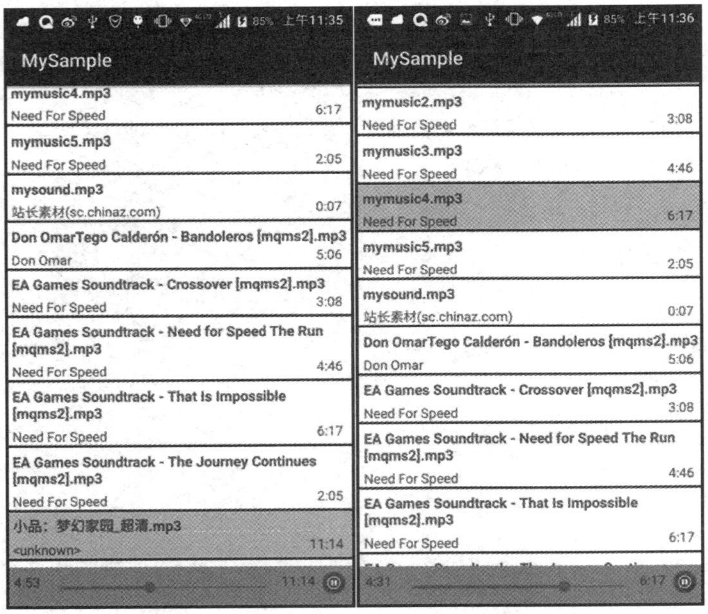

图　192.1

主要代码如下：

```java
public class MainActivity extends Activity {
  Handler myHandler = new Handler() {
    @Override
    public void handleMessage(Message msg) {                    //发送消息更新播放时间
      if (msg.what == 1) {
        String myCurrentSeconds = myMediaPlayer.getCurrentPosition() / 1000 / 60 + ":";
        if (myMediaPlayer.getCurrentPosition() / 1000 % 60 < 10) {
          myCurrentSeconds += "0" + myMediaPlayer.getCurrentPosition() / 1000 % 60;
        } else { myCurrentSeconds += myMediaPlayer.getCurrentPosition() / 1000 % 60; }
        myCurrentTime.setText(myCurrentSeconds);
  } } };
  MediaPlayer myMediaPlayer;
  Timer myTimer;
  ArrayList<MyMusicInfo> myData = new ArrayList<>();
  private TextView myCurrentTime;
  boolean isPlaying = false;
  @Override
  protected void onCreate(Bundle savedInstanceState) {
    super.onCreate(savedInstanceState);
    myMediaPlayer = new MediaPlayer();
    myTimer = new Timer();
    setContentView(R.layout.activity_main);
    RecyclerView myRecyclerView = (RecyclerView) findViewById(R.id.myRecyclerView);
    final TextView myTotalTime = (TextView) findViewById(R.id.myTotalTime);
    myCurrentTime = (TextView) findViewById(R.id.myCurrentTime);
    final SeekBar mySeekBar = (SeekBar) findViewById(R.id.mySeekBar);
    myRecyclerView.setLayoutManager(new LinearLayoutManager(this));
    final MyAdapter myAdapter = new MyAdapter(this);
    new MyUtils(this).getMusicInfo(myData);
    myAdapter.setMyData(myData);
    myRecyclerView.setAdapter(myAdapter);
    final ImageView myPlayPause = (ImageView) findViewById(R.id.myPlayPause);
    myAdapter.setMyListener(new MyAdapter.MyOnItemClickListener(){
      @Override
      public void onItemClick(View v, int position){
        try {
          if (myMediaPlayer != null || myTimer != null){
            myTimer.cancel();
            myTimer = null;
            myMediaPlayer.stop();
            myMediaPlayer = null;
            myPlayPause.setImageResource(R.mipmap.myicon_play);
          }
          myMediaPlayer = new MediaPlayer();
          myMediaPlayer.setDataSource(myData.get(position).myMusicPath);
          myMediaPlayer.prepare();
          myMediaPlayer.start();
          isPlaying = true;
          mySeekBar.setMax(myMediaPlayer.getDuration());
          mySeekBar.setOnSeekBarChangeListener(new SeekBar.OnSeekBarChangeListener(){
            @Override
```

```
    public void onProgressChanged(SeekBar seekBar, int i, boolean b) { }
    @Override
    public void onStartTrackingTouch(SeekBar seekBar) { }
    @Override
    public void onStopTrackingTouch(SeekBar seekBar) {
      myMediaPlayer.seekTo(seekBar.getProgress());
    } });
  String myDurations = myMediaPlayer.getDuration() / 1000 / 60 + ":";
  if (myMediaPlayer.getDuration() / 1000 % 60 < 10) {
    myDurations += "0" + myMediaPlayer.getDuration() / 1000 % 60;
  } else { myDurations += myMediaPlayer.getDuration() / 1000 % 60; }
  myTotalTime.setText(myDurations);
  myPlayPause.setImageResource(R.mipmap.myicon_pause);
  myTimer = new Timer();
  myTimer.schedule(new TimerTask() {
    @Override
    public void run() {
      mySeekBar.setProgress(myMediaPlayer.getCurrentPosition());
      myHandler.sendEmptyMessage(1);
    } }, 0, 10);
    myAdapter.notifyDataSetChanged();
  } catch (Exception e) { }
  } });
myPlayPause.setOnClickListener(new View.OnClickListener() {
  @Override
  public void onClick(View view) {
    if (isPlaying) {                                              //如果正在播放,则暂停
      isPlaying = false;
      myPlayPause.setImageResource(R.mipmap.myicon_play);
      myMediaPlayer.pause();
    } else {                                                       //如果已经暂停,则播放
      isPlaying = true;
      myPlayPause.setImageResource(R.mipmap.myicon_pause);
      myMediaPlayer.start();
} } });} }
```

上面这段代码在 MyCode \ MySample714 \ app \ src \ main \ java \ com \ bin \ luo \ mysample \ MainActivity.java 文件中。在这段代码中,myRecyclerView.setAdapter(myAdapter)表示使用自定义适配器 MyAdapter 的实例 myAdapter 作为 RecyclerView 的适配器,MyAdapter 自定义适配器的主要代码如下:

```
public class MyAdapter extends RecyclerView.Adapter< MyAdapter.MyViewHolder >{
  Context myContext;
  ArrayList< MyMusicInfo > myData = new ArrayList< MyMusicInfo >();
  MyOnItemClickListener myListener = null;
  public MyAdapter(Context context){ myContext = context; }
  public void setMyData(ArrayList< MyMusicInfo > data){
    myData = data;
    notifyDataSetChanged();
  }
  @Override
  public MyViewHolder onCreateViewHolder(ViewGroup parent, int viewType){
```

```java
        return new MyViewHolder(
            LayoutInflater.from(myContext).inflate(R.layout.myitem, parent, false));
    }
    public void setMyListener(MyOnItemClickListener listener){
        myListener = listener;
    }
    @Override
    public void onBindViewHolder(MyViewHolder holder, int position) {
        holder.myArtist.setText(myData.get(position).myArtist);
        holder.myDuration.setText(myData.get(position).myDuration);
        holder.myName.setText(myData.get(position).myName);
        if (myData.get(position).getSelected()) {
            holder.myItemView.setBackgroundColor(Color.LTGRAY);
        } else { holder.myItemView.setBackgroundColor(Color.WHITE); }
    }
    @Override
    public int getItemCount() { return myData.size(); }
    class MyViewHolder extends RecyclerView.ViewHolder {
        private final TextView myArtist;
        private final TextView myDuration;
        private final TextView myName;
        View myItemView;
        public MyViewHolder(View itemView) {
            super(itemView);
            myArtist = (TextView) itemView.findViewById(R.id.myArtist);      //艺术家
            myDuration = (TextView) itemView.findViewById(R.id.myDuration);  //音乐时长
            myName = (TextView) itemView.findViewById(R.id.myName);          //音乐名称
            myItemView = itemView;
            itemView.setOnClickListener(new View.OnClickListener() {
                @Override
                public void onClick(View view) {                              //响应单击列表项事件(即选择文件
                                                                              //  并播放)
                    for (int i = 0; i < myData.size(); i++) {myData.get(i).setSelected(false); }
                    myData.get(getAdapterPosition()).setSelected(true);
                    myListener.onItemClick(view, getAdapterPosition());
                } }); } }
        interface MyOnItemClickListener{ void onItemClick(View v, int position);}
}
```

上面这段代码在 MyCode\MySample714\app\src\main\java\com\bin\luo\mysample\MyAdapter.java 文件中。在这段代码中，new MyViewHolder(LayoutInflater.from(myContext).inflate(R.layout.myitem，parent，false))用于加载 myitem 布局以创建 MyAdapter 的列表项，即每个列表项的布局。关于 myitem 布局的内容请参考源代码中的 MyCode\MySample714\app\src\main\res\layout\myitem.xml 文件。在 MyAdapter.java 文件中，ArrayList＜MyMusicInfo＞myData = new ArrayList＜MyMusicInfo＞()用于创建一个数组，以保存音乐文件的 myName、myArtist、myDuration、myMusicPath、isSelected 等属性，从而决定是否播放或暂停。MyMusicInfo 是自定义类，关于 MyMusicInfo 类的详细内容请参考源代码中的 MyCode\MySample714\app\src\main\java\com\bin\luo\mysample\MyMusicInfo.java 文件。需要说明的是，使用此实例的相关类需要在 gradle 中引入 compile 'com.android.support:design:25.0.1'依赖项。此外，读取在 SD 卡上的音乐文件需要在 AndroidManifest.xml 文件中添加＜uses-permission android：name＝"android.

permission.READ_EXTERNAL_STORAGE"/>权限。此实例的完整项目在 MyCode \ MySample714 文件夹中。

193　依次播放在 RecyclerView 中的音乐文件

　　此实例主要通过在 MediaPlayer 的 setOnCompletionListener()方法中设置播放曲目,实现按照顺序播放 RecyclerView 加载的音乐文件。当实例运行后,将在 RecyclerView 的列表中显示当前手机的所有音乐文件,选择任一列表项(音乐文件),则立即播放音乐,在底部的 SeekBar 也将同步显示播放进度,单击右侧的暂停或播放按钮则可暂停或播放音乐;如果当前曲目播放完毕,则自动播放列表中的下一曲目,效果分别如图 193.1 的左图和右图所示。

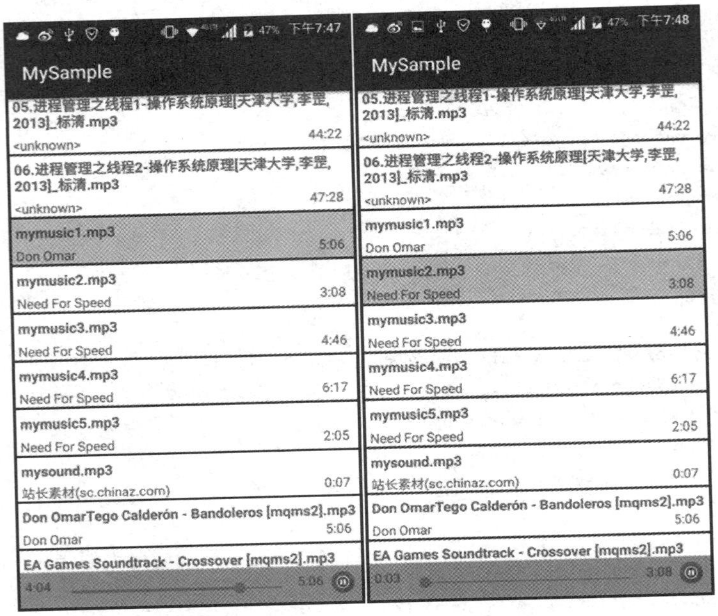

图　193.1

主要代码如下:

```java
public class MainActivity extends Activity implements View.OnClickListener,
SeekBar.OnSeekBarChangeListener {
 Handler myHandler = new Handler() {
  @Override
  public void handleMessage(Message msg) {
   myCurrentTime.setText(myUtils.getMediaPlayerCurrentPositionByString());
  } };
 private TextView myTotalTime, myCurrentTime;
 private MyUtils myUtils;
 private ImageView myPlayPause;
 private MyAdapter myAdapter;
 private SeekBar mySeekBar;
 int myIndex;
 @Override
 protected void onCreate(Bundle savedInstanceState) {
  super.onCreate(savedInstanceState);
```

```java
    myUtils = new MyUtils(this);
    setContentView(R.layout.activity_main);
    mySeekBar = (SeekBar) findViewById(R.id.mySeekBar);
    myTotalTime = (TextView) findViewById(R.id.myTotalTime);
    myPlayPause = (ImageView) findViewById(R.id.myPlayPause);
    myCurrentTime = (TextView) findViewById(R.id.myCurrentTime);
    myPlayPause.setOnClickListener(this);
    mySeekBar.setOnSeekBarChangeListener(this);
    RecyclerView myRecyclerView = (RecyclerView) findViewById(R.id.myRecyclerView);
    myRecyclerView.setLayoutManager(new LinearLayoutManager(this));
    myAdapter = new MyAdapter(this);
    myAdapter.setMyData(myUtils.getMyData());
    myRecyclerView.setAdapter(myAdapter);
    myAdapter.setMyListener(new MyAdapter.MyOnItemClickListener() {
        @Override
        public void onItemClick(View v, int position) {
            selectMusicItem(position);                              //选择曲目列表选项
        } }); }
public void selectMusicItem(int index) {
    ArrayList<MyMusicInfo> myData = myAdapter.getMyData();
    for (int i = 0; i < myData.size(); i++) { myData.get(i).setSelected(false); }
    myData.get(index).setSelected(true);
    myAdapter.setMyData(myData);
    try {
        myUtils.InitMyPlayer(myData.get(index).myPath, new TimerTask() {
            @Override
            public void run() {                                     //根据播放进度设置 SeekBar 滑块位置
                mySeekBar.setProgress(myUtils.getMediaPlayerCurrentPositionByInt());
                myHandler.sendEmptyMessage(1);
            } });
        mySeekBar.setMax(myUtils.getMyPlayer().getDuration());      //根据时长设置最大值
        myTotalTime.setText(myUtils.getMediaPlayerDurationByString());
        myPlayPause.setImageResource(R.mipmap.myicon_pause);
        myIndex = index;
        myUtils.getMyPlayer().setOnCompletionListener(
            new MediaPlayer.OnCompletionListener() {                //当播放完成之后继续播放下一曲目
                @Override
                public void onCompletion(MediaPlayer mediaPlayer) {
                    mediaPlayer.stop();
                    selectMusicItem(myIndex + 1 < myUtils.getMyData().size() ? myIndex + 1 : 0);
                } });
    } catch (Exception e) { }
    myAdapter.notifyDataSetChanged();
}
@Override
public void onClick(View view) {
    if (view.getId() == R.id.myPlayPause) {
        if (myUtils.getMyPlayer().isPlaying()) {                    //在暂停音乐时设置播放图标
            myPlayPause.setImageResource(R.mipmap.myicon_play);
            myUtils.getMyPlayer().pause();
        } else {                                                    //在播放音乐时设置暂停图标
            myPlayPause.setImageResource(R.mipmap.myicon_pause);
            myUtils.getMyPlayer().start();
```

```
        }}}
    @Override
    public void onProgressChanged(SeekBar seekBar, int i, boolean b) { }
    @Override
    public void onStartTrackingTouch(SeekBar seekBar) { }
    @Override
    public void onStopTrackingTouch(SeekBar seekBar) {
      myUtils.getMyPlayer().seekTo(seekBar.getProgress());
    }}
```

上面这段代码在 MyCode\MySample716\app\src\main\java\com\bin\luo\mysample\MainActivity.java 文件中。在这段代码中，myRecyclerView.setAdapter(myAdapter)表示使用自定义适配器 MyAdapter 的实例 myAdapter 作为 RecyclerView 的适配器，关于 MyAdapter 自定义适配器的详细内容请参考源代码中的 MyCode\MySample716\app\src\main\java\com\bin\luo\mysample\MyAdapter.java 文件。需要说明的是，使用此实例的相关类需要在 gradle 中引入 compile 'com.android.support:design:25.0.1' 依赖项。此外，读取在 SD 卡上的音乐文件需要在 AndroidManifest.xml 文件中添加< uses-permission android:name = "android.permission.READ_EXTERNAL_STORAGE"/>权限。此实例的完整项目在 MyCode\MySample716 文件夹中。

194　在 ListView 上加载手机外存的音乐文件

此实例主要是实现在 ListView 上加载手机外部存储介质上的所有音乐文件名称和大小信息。当实例运行后，将在 ListView 控件上显示手机外部存储上的所有音乐文件，单击在 ListView 控件上的任意音乐文件名称（列表项），将播放此音乐文件，效果分别如图 194.1 的左图和右图所示。

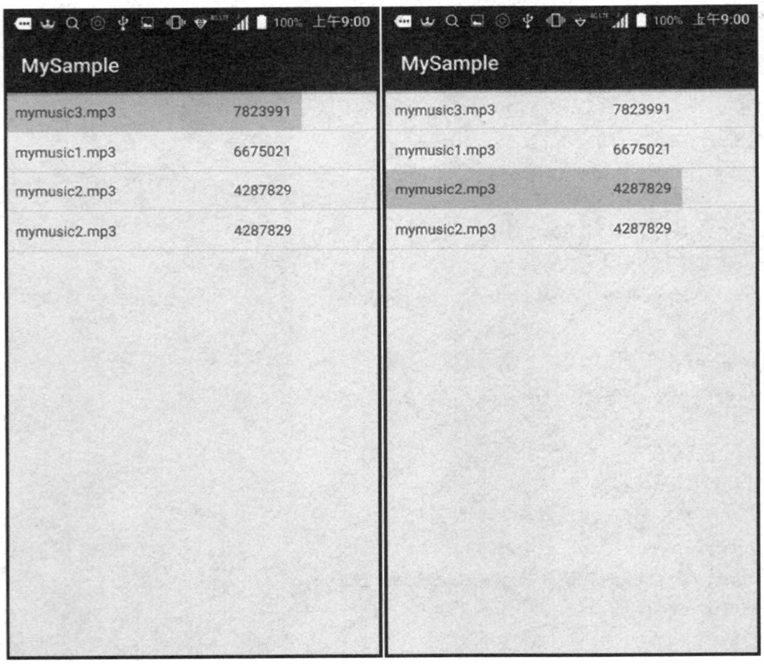

图　194.1

主要代码如下：

```java
public class MainActivity extends AppCompatActivity {
    private ListView myListView;
    @Override
    protected void onCreate(@Nullable Bundle savedInstanceState) {
        super.onCreate(savedInstanceState);
        setContentView(R.layout.activity_main);
        myListView = (ListView) findViewById(R.id.myListView);
        GetAudioFiles();
    }
    ArrayList myFiles = new ArrayList();
    ArrayList mySizes = new ArrayList();
    ArrayList<String> myPathData = new ArrayList();
    ArrayList<HashMap<String, Object>> myItems = new ArrayList<>();
    private void GetAudioFiles() {
        myFiles.clear();
        mySizes.clear();
        //通过 ContentResolver 查询所有外部存储音乐文件信息
        Cursor myCursor = getContentResolver().query(
            MediaStore.Audio.Media.EXTERNAL_CONTENT_URI, null, null, null, null);
        while (myCursor.moveToNext()) {
            String myName = myCursor.getString(myCursor.getColumnIndex(
                        MediaStore.Images.Media.DISPLAY_NAME));   //显示文件名称
            String mySize = myCursor.getString(myCursor.getColumnIndex(
                        MediaStore.Images.Media.SIZE));           //显示文件大小
            byte[] data = myCursor.getBlob(myCursor.getColumnIndex(
                    MediaStore.Images.Media.DATA));               //获取音乐文件的位置数据
            myFiles.add(myName);
            mySizes.add(mySize);
            String myPath = new String(data, 0,
                        data.length - 1);                         //将位置数据转换成字符串形式的路径
            myPathData.add(myPath);
        }
        for (int i = 0; i < myFiles.size(); i++) {
            HashMap<String, Object> myListItem = new HashMap<>();
            myListItem.put("name", myFiles.get(i));
            myListItem.put("size", mySizes.get(i));
            myItems.add(myListItem);
        }
        SimpleAdapter myAdapter = new SimpleAdapter(MainActivity.this,
                myItems, R.layout.myitem, new String[]{"name", "size"},
                new int[]{R.id.myName, R.id.mySize});
        myListView.setAdapter(myAdapter);
        myListView.setOnItemClickListener(new AdapterView.OnItemClickListener() {
            @Override
            public void onItemClick(AdapterView<?> parent, View view,
                        int position, long id) {                  //单击音乐文件名称(列表项)播放音乐
                try{
                    MediaPlayer myMediaPlayer = new MediaPlayer();
                    myMediaPlayer.setDataSource(myPathData.get(position));
                    myMediaPlayer.prepare();
                    myMediaPlayer.start();
                } catch (Exception e) { e.printStackTrace(); }
            }});
    }
}
```

上面这段代码在 MyCode\MySample669\app\src\main\java\com\bin\luo\mysample\MainActivity.java 文件中。在上面这段代码中，myAdapter = new SimpleAdapter(MainActivity.this, myItems, R.layout.myitem, new String[]{"name", "size"}, new int[] {R.id.myName, R.id.mySize})表示根据 myitem 布局的样式创建 ListView 的列表项，关于 myitem 布局的详细内容请参考源代码中的 MyCode\MySample669\app\src\main\res\layout\myitem.xml 文件中。此外，读取 SD 卡文件需要在 AndroidManifest.xml 文件中添加< uses-permission android:name = "android.permission.READ_EXTERNAL_STORAGE"/>权限。需要说明的是，使用此实例的相关类需要在 gradle 中引入 compile 'com.android.support:design:25.0.1'依赖项。此实例的完整项目在 MyCode\MySample669 文件夹中。

195 使用 SoundPool 播放较短的声音片段

此实例主要通过使用 SoundPool，实现从程序的资源中加载并播放较短的声音片段。当实例运行之后，单击"播放第一种声音两次"按钮，将播放声音片段两次；单击"播放第二种声音一次"按钮，将仅播放一次声音片段，效果分别如图 195.1 的左图和右图所示。

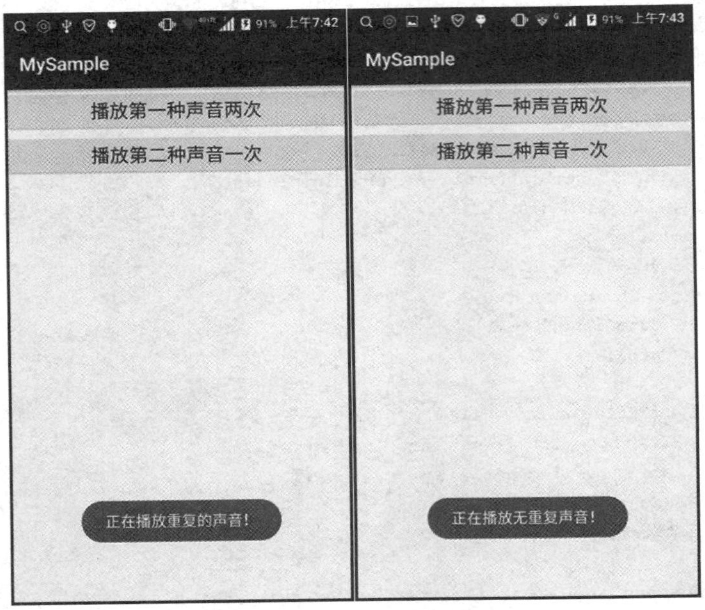

图 195.1

主要代码如下：

```java
public void onClickmyBtn1(View v) {                    //响应单击"播放第一种声音两次"按钮
    Toast.makeText(myContext,"正在播放重复的声音!",Toast.LENGTH_SHORT).show();
    playSound(1, 2);
}
public void onClickmyBtn2(View v) {                    //响应单击"播放第二种声音一次"按钮
    Toast.makeText(myContext,"正在播放无重复声音!",Toast.LENGTH_SHORT).show();
    playSound(2, 1);
}
public void InitSound() {                              //初始化声音
    mySoundPool = new SoundPool(5, AudioManager.STREAM_MUSIC, 0);
```

```
    myHashMap = new HashMap< Integer, Integer >();
    myHashMap.put(1, mySoundPool.load(this, R.raw.alarm01, 1));
    myHashMap.put(2, mySoundPool.load(this, R.raw.alarm02, 1));
    mySoundPool.setOnLoadCompleteListener(new SoundPool.OnLoadCompleteListener(){
     @Override
     public void onLoadComplete(SoundPool sound, int sampleId, int status){
      isLoaded = true;
      Toast.makeText(myContext,"音效加载完成!",Toast.LENGTH_SHORT).show();
}}); }
 public void playSound(int sound, int number) {              //播放声音
    AudioManager myAudioManager =
                (AudioManager) this.getSystemService(Context.AUDIO_SERVICE);
    float myMaxVolume =
                myAudioManager.getStreamMaxVolume(AudioManager.STREAM_MUSIC);
    float myCurVolume = myAudioManager.getStreamVolume(AudioManager.STREAM_MUSIC);
    float myRatio = myCurVolume/ myMaxVolume;
    mySoundPool.play(myHashMap.get(sound),
                    myRatio,               //左声道音量
                    myRatio,               //右声道音量
                    1,                     //优先级
                    number,                //循环播放次数
                    1);                    //回放速度,该值在 0.5～2.0 之间 1 为正常速度
 }
```

上面这段代码在 MyCode\MySample375\app\src\main\java\com\bin\luo\mysample\MainActivity.java 文件中。在这段代码中,mySoundPool = new SoundPool(5, AudioManager.STREAM_MUSIC, 0)用于创建一个 SoundPool 对象,SoundPool()构造函数的语法声明如下:

```
public SoundPool(int maxStream, int streamType, int srcQuality)
```

其中,参数 int maxStream 表示同时播放流的最大数量。参数 int streamType 表示流的类型,一般为 STREAM_MUSIC(具体在 AudioManager 类中列出)。参数 int srcQuality 表示采样率转化质量,使用 0 作为默认值。

mySoundPool.load(this,R.raw.alarm01,1)表示从资源 R.raw.alarm01 加载音效。如果使用 load()方法从文件中加载音效,则其语法为 load(String path, int priority)。

mySoundPool.play(myHashMap.get(sound),myRatio,myRatio,1,number,1)用于播放音效,play()方法的语法声明如下:

```
play(int soundID, float leftVolume, float rightVolume,
                        int priority, int loop, float rate)
```

其中,参数 int soundID 用于指定音效 ID,即在加载音效资源时设置的 ID 值。参数 float leftVolume 和参数 float rightVolume 表示左右声道音量。参数 int priority 表示优先级。参数 int loop 表示循环次数。参数 float rate 表示速率,速率最低 0.5,最高为 2,1 代表正常速度。

此实例的完整项目在 MyCode\MySample375 文件夹中。

196 使用 AudioManager 增大或减小音量

此实例主要通过使用 AudioManager 的 adjustStreamVolume()方法,实现直接调出系统音量控制条增大或减小音量。在实例运行之前,先使用任一音乐播放器播放歌曲;当实例运行之后,单击"减

小音量"按钮,则在顶部显示系统默认音量控制条并减小音量;单击"增大音量"按钮,也将在顶部显示系统默认音量控制条并增大音量,效果分别如图 196.1 的左图和右图所示。

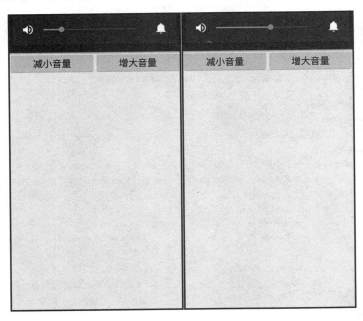

图 196.1

主要代码如下:

```
public void onClickButton1(View v) {                    //响应单击"减小音量"按钮
    AudioManager myAudioManager =
                (AudioManager)getSystemService(Context.AUDIO_SERVICE);
    myAudioManager.adjustStreamVolume(AudioManager.STREAM_MUSIC,
            AudioManager.ADJUST_LOWER,AudioManager.FX_FOCUS_NAVIGATION_UP);
}
public void onClickButton2(View v) {                    //响应单击"增大音量"按钮
    AudioManager myAudioManager =
                (AudioManager)getSystemService(Context.AUDIO_SERVICE);
    myAudioManager.adjustStreamVolume(AudioManager.STREAM_MUSIC,
            AudioManager.ADJUST_RAISE,AudioManager.FX_FOCUS_NAVIGATION_UP);
}
```

上面这段代码在 MyCode \ MySample064 \ app \ src \ main \ java \ com \ bin \ luo \ mysample \ MainActivity.java 文件中。在这段代码中,adjustStreamVolume(int streamType, int direction, int flags)方法用于以步长方式调节手机音量大小。其中,参数 streamType 表示声音类型,可取值为 STREAM_VOICE_CALL(通话)、STREAM_SYSTEM(系统声音)、STREAM_RING(铃声)、STREAM_MUSIC(音乐)、STREAM_ALARM(闹铃声);参数 direction 用于设置调整音量的方向,可取值为 ADJUST_LOWER(减小)、ADJUST_RAISE(增大)、ADJUST_SAME 等;参数 flags 表示可选标志位,当此值为 FX_FOCUS_NAVIGATION_UP 时,将在增大或减小音量时显示系统默认音量控制条,当此值为 FX_FOCUS_NAVIGATION_DOWN 时,将在增大或减小音量时隐藏系统默认音量控制条,此参数还支持 FX_KEY_CLICK、FX_KEYPRESS_STANDARD、FX_FOCUS_NAVIGATION_RIGHT、FX_FOCUS_NAVIGATION_LEFT、FX_KEYPRESS_SPACEBAR、FX_KEYPRESS_DELETE、FX_KEYPRESS_RETURN 等。此外,adjustVolume()方法也能实现与

adjustStreamVolume()方法类似的调节音量的功能。此实例的完整项目在 MyCode\ MySample064 文件夹中。

197　使用 AudioManager 播放系统预置的声音

此实例主要通过使用 AudioManager 的 playSoundEffect()方法,实现直接播放系统预置的删除、返回等音效。当实例运行后,单击"播放删除音效"按钮,则将播放删除音效;单击"播放返回音效"按钮,则将播放返回音效,效果分别如图197.1的左图和右图所示。

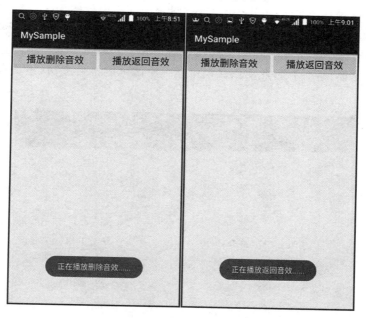

图　197.1

主要代码如下:

```java
public void onClickButton1(View v) {                    //响应单击"播放删除音效"按钮
    AudioManager myAudioManager =
            (AudioManager)getSystemService(Context.AUDIO_SERVICE);
    myAudioManager.playSoundEffect(AudioManager.FX_KEYPRESS_DELETE,1000);
    Toast.makeText(this,"正在播放删除音效……",Toast.LENGTH_SHORT).show();
}
public void onClickButton2(View v) {                    //响应单击"播放返回音效"按钮
    AudioManager myAudioManager =
            (AudioManager)getSystemService(Context.AUDIO_SERVICE);
    myAudioManager.playSoundEffect(AudioManager.FX_KEYPRESS_RETURN,1000);
    Toast.makeText(this,"正在播放返回音效……",Toast.LENGTH_SHORT).show();
}
```

上面这段代码在 MyCode\MySample065\app\src\main\java\com\bin\luo\mysample\MainActivity.java 文件中。在这段代码中,playSoundEffect(int effectType, float volume)方法用于播放系统预置的音效。其中,参数 int effectType 表示音效类型,可选值有 AudioManager.FX_KEY_CLICK、AudioManager.FX_KEYPRESS_RETURN、AudioManager.FX_FOCUS_NAVIGATION_DOWN、AudioManager.FX_FOCUS_NAVIGATION_LEFT、AudioManager.FX_FOCUS_

NAVIGATION_RIGHT、AudioManager.FX_KEYPRESS_STANDARD、AudioManager.FX_KEYPRESS_SPACEBAR、AudioManager.FX_KEYPRESS_DELETE、AudioManager.FX_FOCUS_NAVIGATION_UP 和 AudioManager.FX_KEYPRESS_INVALID 等。此实例的完整项目在 MyCode\MySample065 文件夹中。

198　使用 AudioManager 获取和设置铃声模式

此实例主要通过使用 AudioManager 的 getRingerMode()方法和 setRingerMode()方法，获取和设置当前手机的铃声模式。当实例运行之后，单击"设置当前铃声模式为振动模式"按钮，将设置当前手机铃声模式为振动模式，如图 198.1 的左图所示；单击"获取当前铃声模式"按钮，则在弹出的 Toast 中显示"当前手机铃声模式是：振动模式"，如图 198.1 的右图所示。单击其他按钮则会实现按钮标题所示的功能。需要说明的是，此实例在模拟器中测试可能不会成功，建议直接在 Android 手机中测试。

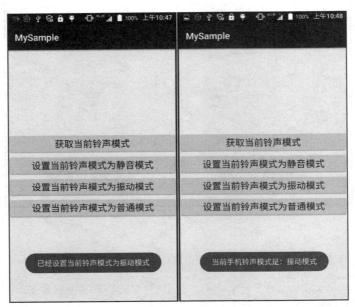

图　198.1

主要代码如下：

```
public class MainActivity extends Activity {
 public AudioManager myAudioManager;
 @Override
 protected void onCreate(Bundle savedInstanceState) {
  super.onCreate(savedInstanceState);
  setContentView(R.layout.activity_main);
  myAudioManager = (AudioManager) getSystemService(Context.AUDIO_SERVICE);
 }
 public void onClickmyBtnGetRingerMode(View v) {                    //响应单击"获取当前铃声模式"按钮
  int myMode = myAudioManager.getRingerMode();
  String myText = "当前手机铃声模式是:";
  switch (myMode) {
   case AudioManager.RINGER_MODE_NORMAL:
```

```
            myText += "普通模式";
            break;
        case AudioManager.RINGER_MODE_SILENT:
            myText += "静音模式";
            break;
        case AudioManager.RINGER_MODE_VIBRATE:
            myText += "振动模式";
            break;
        default:
            myText += "未知模式";
            break;
    }
    Toast.makeText(getApplicationContext(), myText, Toast.LENGTH_SHORT).show();
}
//响应单击"设置当前铃声模式为静音模式"按钮
public void onClickmyBtnSetRingerModeSilent(View v) {
    myAudioManager.setRingerMode(AudioManager.RINGER_MODE_SILENT);
    Toast.makeText(getApplicationContext(),
            "已经设置当前铃声模式为静音模式", Toast.LENGTH_SHORT).show();
}
//响应单击"设置当前铃声模式为振动模式"按钮
public void onClickmyBtnSetRingerModeVibrate(View v) {
    myAudioManager.setRingerMode(AudioManager.RINGER_MODE_VIBRATE);
    Toast.makeText(getApplicationContext(),
            "已经设置当前铃声模式为振动模式", Toast.LENGTH_SHORT).show();
}
//响应单击"设置当前铃声模式为普通模式"按钮
public void onClickmyBtnSetRingerModeNormal(View v) {
    myAudioManager.setRingerMode(AudioManager.RINGER_MODE_NORMAL);
    Toast.makeText(getApplicationContext(),
            "已经设置当前铃声模式为普通模式", Toast.LENGTH_SHORT).show();
}}
```

上面这段代码在 MyCode \ MySample066 \ app \ src \ main \ java \ com \ bin \ luo \ mysample \ MainActivity.java 文件中。在这段代码中，getRingerMode()方法用于获取当前手机的铃声模式，可以通过该方法的返回值来判断当前手机的铃声模式，如果返回值是 RINGER_MODE_NORMAL，则手机铃声是普通模式；如果返回值是 RINGER_MODE_SILENT，则手机铃声是静音模式、如果返回值是 RINGER_MODE_VIBRATE，则手机铃声是振动模式。setRingerMode(int Mode)方法用于设置手机铃声模式，该方法的参数 Mode 表示铃声模式，可能的取值包括：RINGER_MODE_NORMAL（普通模式）、RINGER_MODE_SILENT（静音模式）、RINGER_MODE_VIBRATE（振动模式）等。此实例的完整项目在 MyCode\ MySample066 文件夹中。

第 7 章

文件和数据

199 使用 JSONObject 解析 JSON 字符串

此实例主要通过使用 JSONTokener 和 JSONObject,实现解析 JSON 字符串的单个对象。当实例运行之后,单击"使用 JSONObject 解析 JSON 字符串的单个对象"按钮,则将把 JSON 字符串:{"myName":"Karli Watson","myBook":"C♯入门经典"},解析成如图 199.1 所示的效果。

图　199.1

主要代码如下:

```
//响应单击"使用JSONObject解析JSON字符串的单个对象"按钮
public void onClickmyBtn1(View v) {
  String myJsonData = "{\"myName\":\"Karli Watson\",\"myBook\":\"C♯入门经典\"}";
  try {
    String myInfo = "";
```

```
    JSONTokener myJSONTokener = new JSONTokener(myJsonData);
    JSONObject myJSONObject = (JSONObject) myJSONTokener.nextValue();
    myInfo += "作者:" + myJSONObject.getString("myName");
    myInfo += ",书名:" + myJSONObject.getString("myBook");
    Toast.makeText(getApplicationContext(),
                              myInfo, Toast.LENGTH_SHORT).show();
} catch (Exception e) {
    Toast.makeText(getApplicationContext(),
                       e.getMessage().toString(), Toast.LENGTH_SHORT).show();
}}
```

上面这段代码在 MyCode\MySample472\app\src\main\java\com\bin\luo\mysample\MainActivity.java 文件中。在这段代码中，JSONTokener myJSONTokener = new JSONTokener(myJsonData)用于根据 JSON 格式的字符串构造一个 JSONTokener 实例。JSONObject myJSONObject =（JSONObject) myJSONTokener.nextValue()用于将 JSON 字符串唯一的对象解析为 JSONObject 的实例。myJSONObject.getString("myName")用于从 JSON 对象中获取属性名称为 myName 的值。此实例的完整项目在 MyCode\MySample472 文件夹中。

200　使用 JSONArray 解析 JSON 字符串

此实例主要通过使用 JSONArray 和 JSONObject，实现解析 JSON 字符串的嵌套对象。当实例运行之后，单击"使用 JSONArray 解析 JSON 字符串的嵌套对象"按钮，则将把 JSON 字符串：[{"myName":"Karli Watson","myBook":{"bookName":"C♯入门经典","bookPrice":108,"bookStore":"京东商城"}},{"myName":"Bruce Eckel","myBook":{"bookName":"Java 编程思想","bookPrice":78,"bookStore":"天猫商城"}},{"myName":"Stephen Prata","myBook":{"bookName":"C++Primer Plus","bookPrice":95,"bookStore":"当当商城"}}]，解析成如图 200.1 所示的效果。

图　200.1

主要代码如下：

```java
//响应单击"使用JSONArray解析JSON字符串的嵌套对象"按钮
public void onClickmyBtn1(View v) {
    String myJsonData = "[{\"myName\":\"Karli Watson\",\"myBook\":{\"bookName\": \"C#入门经典\",\"bookPrice\":108,\"bookStore\":\"京东商城\"}},"
            + "{\"myName\":\"Bruce Eckel\",\"myBook\":{\"bookName\":\"Java 编程思想\", \"bookPrice\":78,\"bookStore\":\"天猫商城\"}},"
            + "{\"myName\":\"Stephen Prata\",\"myBook\":{\"bookName\":\"C++Primer Plus\",\"bookPrice\":95,\"bookStore\":\"当当商城\"}}" + "]";
    try {
        String myInfo = "";
        JSONArray myArray = new JSONArray(myJsonData);
        int myLength = myArray.length();
        for (int i = 0; i < myLength; i++) {
            JSONObject myJSONObject = myArray.getJSONObject(i);
            myInfo += "\n 作者:" + myJSONObject.getString("myName");
            myInfo += ",出版专著信息:" ;
            JSONObject myBooks = myJSONObject.getJSONObject("myBook");
            myInfo += ",书名:" + myBooks.getString("bookName");
            //myInfo += ",单价:" + myBooks.getInt("bookPrice");
            myInfo += ",单价:" + myBooks.getString("bookPrice");
            myInfo += ",代理书店:" + myBooks.getString("bookStore");
        }
        Toast.makeText(getApplicationContext(),
                                myInfo, Toast.LENGTH_SHORT).show();
    } catch (Exception e) {
        Toast.makeText(getApplicationContext(),
                    e.getMessage().toString(),Toast.LENGTH_SHORT).show();
    } }
```

上面这段代码在 MyCode\MySample476\app\src\main\java\com\bin\luo\mysample\MainActivity.java 文件中。在这段代码中，{"myName":"Stephen Prata","myBook"：{"bookName":"C++Primer Plus","bookPrice":95,"bookStore":"当当商城"}}有两对大括号"{...{...}}"，里面的大括号包含的对象是外面的大括号对象的子对象，因此两个对象在解析时有差异，myJSONObject = myArray.getJSONObject(i)是根据数组索引 i 解析的外面大括号对象，myBooks = myJSONObject.getJSONObject("myBook")是根据子对象名称 myBook 解析的里面大括号子对象。此实例的完整项目在 MyCode\MySample476 文件夹中。

201　使用 JSONTokener 解析 JSON 字符串

此实例主要通过使用 JSONTokener 的 nextString()方法，从而获取在 JSON 字符串中指定字符之前的部分文本。当实例运行之后，单击"使用 JSONTokener 获取 JSON 的部分文本"按钮，则在弹出的 Toast 中显示"大家系列:{"myName":"Karli Watson","myBook": "C#入门经典"}"字符串的"入"字符之前的所有文本，如图 201.1 所示。

图 201.1

主要代码如下:

```
//响应单击"使用JSONTokener获取JSON的部分文本"按钮
public void onClickmyBtn1(View v) {
  String myJsonData =
        "大家系列:{\"myName\":\"Karli Watson\",\"myBook\":\"C#入门经典\"}";
  try {
    String myInfo = "";
    JSONTokener myJSONTokener = new JSONTokener(myJsonData);
    myInfo += myJSONTokener.nextString('入');
    Toast.makeText(getApplicationContext(),
                            myInfo, Toast.LENGTH_SHORT).show();
  } catch (Exception e) {
    Toast.makeText(getApplicationContext(),
                  e.getMessage().toString(),Toast.LENGTH_SHORT).show();
  } }
```

上面这段代码在 MyCode\MySample478\app\src\main\java\com\bin\luo\mysample\MainActivity.java 文件中。在这段代码中，myJSONTokener.nextString('入')用于获取"大家系列：{"myName":"Karli Watson","myBook":"C#入门经典"}"字符串的"入"字符之前的所有文本"大家系列：{ "myName":"Karli Watson","myBook"："C#"。此实例的完整项目在 MyCode\MySample478 文件夹中。

202 使用 JsonReader 解析 JSON 字符串

此实例主要通过使用 JsonReader 的 beginArray()、beginObject()、nextName()和 nextString()等方法，实现将 JSON 格式的字符串以对象的方式解析出来。当实例运行之后，单击"使用

JsonReader解析JSON字符串"按钮，则将把JSON字符串：[{"myName":"Karli Watson","myBook":"C♯入门经典"},{"myName":"Bruce Eckel","myBook":"Java编程思想"},{"myName":"Stephen Prata","myBook":"C++Primer Plus"}]，解析成如图202.1所示的效果。

图　202.1

主要代码如下：

```java
//响应单击"使用JsonReader解析JSON字符串"按钮
public void onClickmyBtn1(View v) {
    String myJsonData =
        "[{\"myName\":\"Karli Watson\",\"myBook\":\"C♯入门经典\"}," +
        "{\"myName\":\"Bruce Eckel\",\"myBook\":\"Java编程思想\"}," +
        "{\"myName\":\"Stephen Prata\",\"myBook\":\"C++Primer Plus\"}]";
    try {
        JsonReader myJsonReader = new JsonReader(new StringReader(myJsonData));
        myJsonReader.beginArray();
        String myInfo = "";
        while (myJsonReader.hasNext()) {
            myJsonReader.beginObject();
            while (myJsonReader.hasNext()) {
                String myTagName = myJsonReader.nextName();
                if (myTagName.equals("myName")) {
                    myInfo += "\n作者:" + myJsonReader.nextString();
                } else if (myTagName.equals("myBook")) {
                    myInfo += ",书名:" + myJsonReader.nextString();
                }
            }
            myJsonReader.endObject();
        }
        myJsonReader.endArray();
        Toast.makeText(getApplicationContext(), myInfo, Toast.LENGTH_SHORT).show();
```

```
    } catch (Exception e) {
      Toast.makeText(getApplicationContext(),
              e.getMessage().toString(), Toast.LENGTH_SHORT).show();
    } }
```

上面这段代码在 MyCode\MySample468\app\src\main\java\com\bin\luo\mysample\MainActivity.java 文件中。在这段代码中，myJsonReader = new JsonReader(new StringReader(myJsonData))用于根据 JSON 字符串构造 JsonReader 实例。myJsonReader.beginArray()和 myJsonReader.beginObject()表示 JsonReader 开始执行解析动作。myJsonReader.nextName()用于获取 JSON 标签名称。myJsonReader.nextString()用于获取某个标签对应的内容，标签和内容之间用冒号连接，就像是一个键值对。此实例的完整项目在 MyCode\MySample468 文件夹中。

203 使用 JSONStringer 创建 JSON 字符串

此实例主要通过使用 JSONStringer，创建 JSON 格式的字符串。当实例运行之后，单击"使用 JSONStringer 创建 JSON 格式的字符串"按钮，则将创建包含三个对象的数组型的 JSON 字符串，如图 203.1 的 Toast 所示。

主要代码如下：

```
//响应单击"使用JSONStringer创建JSON格式的字符串"按钮
public void onClickmyBtn1(View v) {
    try {
        JSONStringer myJSONStringer = new JSONStringer();
        myJSONStringer.array();
        myJSONStringer.object();
        myJSONStringer.key("myName");
        myJSONStringer.value("Karli Watson");
        myJSONStringer.key("myBook");
        myJSONStringer.value("C#入门经典");
        myJSONStringer.endObject();
        myJSONStringer.object();
        myJSONStringer.key("myName");
        myJSONStringer.value("Bruce Eckel");
        myJSONStringer.key("myBook");
        myJSONStringer.value("Java编程思想");
        myJSONStringer.endObject();
        myJSONStringer.object();
        myJSONStringer.key("myName");
        myJSONStringer.value("Stephen Prata");
        myJSONStringer.key("myBook");
        myJSONStringer.value("C++Primer Plus");
        myJSONStringer.endObject();
        myJSONStringer.endArray();
        Toast.makeText(getApplicationContext(),
                    myJSONStringer.toString(), Toast.LENGTH_SHORT).show();
    } catch (Exception e) {
        Toast.makeText(getApplicationContext(),
                    e.getMessage().toString(), Toast.LENGTH_SHORT).show();
    } }
```

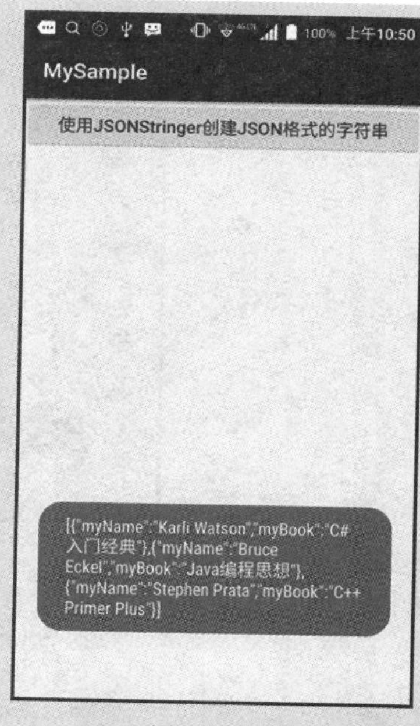

图 203.1

上面这段代码在 MyCode\MySample474\app\src\main\java\com\bin\luo\mysample\

MainActivity.java 文件中。在这段代码中，myJSONStringer.key("myName")用于创建键名 myName，myJSONStringer.value("Stephen Prata")用于创建键值 Stephen Prata，在 JSON 字符串中，键名和键值用 ":" 连接，每个键名键值对之间用 "," 连接。myJSONStringer.object() 和 myJSONStringer.endObject()用于构造一个对象，相当于 JSON 字符串的 "{" 和 "}"，每个对象之间也用 "," 连接。myJSONStringer.array() 和 myJSONStringer.endArray()用于构造对象数组，相当于 JSON 字符串的 "[" 和 "]"。myJSONStringer.toString()表示以 JSON 格式输出 myJSONStringer 的内容。此实例的完整项目在 MyCode\MySample474 文件夹中。

204　使用 JSONObject 根据 IP 显示所在城市

此实例主要通过使用 JSONObject 解析从 http://int.dpool.sina.com.cn 返回的信息，获取本机 IP 对应的城市信息。当实例运行之后，单击"查询本机 IP 地址"按钮，则在弹出的 Toast 中显示本机 IP 信息，如图 204.1 的左图所示；单击"查询本机位置"按钮，则在弹出的 Toast 中显示本机对应的城市信息，如图 204.1 的右图所示。

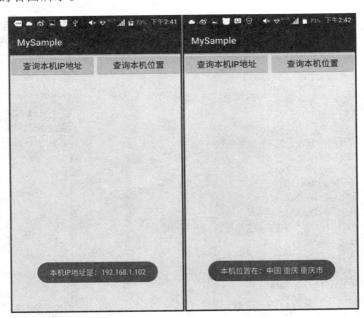

图　204.1

主要代码如下：

```
public class MainActivity extends Activity {
 Handler myHandler = new Handler() {
  @Override
  public void handleMessage(Message msg) {
   String myData = (String) msg.obj;
   try {
    JSONObject myJSONObject =
              new JSONObject(myData.substring(myData.indexOf("{")));
    String myCity = (String) myJSONObject.get("city");
    String myCoutry = (String) myJSONObject.get("country");
    String myProvince = (String) myJSONObject.get("province");
```

```java
            Toast.makeText(getApplicationContext(), "本机位置在:" + myCoutry + " "
                    + myProvince + " " + myCity + "市", Toast.LENGTH_LONG).show();
        } catch (Exception e) {
            Toast.makeText(getApplicationContext(),
                            "查询失败!", Toast.LENGTH_SHORT).show();
}}};
    private String myIPAddress;
    @Override
    protected void onCreate(Bundle savedInstanceState) {
        super.onCreate(savedInstanceState);
        setContentView(R.layout.activity_main);
    }
    public void onClickBtn1(View view) {                          //响应单击"查询本机 IP 地址"按钮
        ConnectivityManager myManager =
            (ConnectivityManager) getSystemService(Context.CONNECTIVITY_SERVICE);
        NetworkInfo myMobileNetInfo =
            myManager.getNetworkInfo(ConnectivityManager.TYPE_MOBILE);
        NetworkInfo myWifiNetInfo =
            myManager.getNetworkInfo(ConnectivityManager.TYPE_WIFI);
        if (!myMobileNetInfo.isConnected() && !myWifiNetInfo.isConnected()) {
            Toast.makeText(MainActivity.this,
                            "当前网络不可用!", Toast.LENGTH_SHORT).show();
        } else if (myMobileNetInfo.isConnected() && !myWifiNetInfo.isConnected()) {
            myIPAddress = getIPAddressByMobileNet();
        } else if (!myMobileNetInfo.isConnected() && myWifiNetInfo.isConnected()) {
            WifiManager myWifiManager =
                (WifiManager) getSystemService(Context.WIFI_SERVICE);
            WifiInfo myWifiInfo = myWifiManager.getConnectionInfo();
            myIPAddress = FormatIPAddress(myWifiInfo.getIpAddress());
        }
        Toast.makeText(getApplicationContext(),
                    "本机 IP 地址是:" + myIPAddress, Toast.LENGTH_LONG).show();
    }
//响应单击"查询本机位置"按钮
    public void onClickBtn2(View view) { getLocalInfo(); }
    public static String FormatIPAddress(int ipAddress) {
        StringBuilder myBuilder = new StringBuilder();
        myBuilder.append(ipAddress & 0xFF).append(".");
        myBuilder.append((ipAddress >> 8) & 0xFF).append(".");
        myBuilder.append((ipAddress >> 16) & 0xFF).append(".");
        myBuilder.append((ipAddress >> 24) & 0xFF);
        return myBuilder.toString();
    }
    public void getLocalInfo() {
        new Thread() {
            @Override
            public void run() {
                try {
                    URL myUrl =
                        new URL("http://int.dpool.sina.com.cn/iplookup/iplookup.php?format = js");
                    URLConnection myConnection = myUrl.openConnection();
                    BufferedReader myReader = new BufferedReader(
                            new InputStreamReader(myConnection.getInputStream(), "GBK"));
```

```
            String myLength = null;
            StringBuffer myBuffer = new StringBuffer();
            while ((myLength = myReader.readLine()) != null) {
             myBuffer.append(myLength);
            }
            myReader.close();
            String myResult = myBuffer.toString();
            Message myMessage = new Message();
            myMessage.obj = myResult;
            myHandler.sendMessage(myMessage);
        } catch (Exception e) { } }
    }.start();
}
public static String getIPAddressByMobileNet() {
    try {
        for (Enumeration< NetworkInterface > myEnum =
            NetworkInterface.getNetworkInterfaces(); myEnum.hasMoreElements(); ) {
            NetworkInterface myNetInterface = myEnum.nextElement();
            for (Enumeration< InetAddress > myIPAddress =
                myNetInterface.getInetAddresses(); myIPAddress.hasMoreElements(); ) {
                InetAddress myAddress = myIPAddress.nextElement();
                if (!myAddress.isLoopbackAddress()&& myAddress instanceof Inet4Address) {
                    return myAddress.getHostAddress().toString();
                } } }
    } catch (Exception e) {
    }
    return null;
} }
```

上面这段代码在 MyCode\MySample740\app\src\main\java\com\bin\luo\mysample\ MainActivity.java 文件中。在这段代码中，myUrl = new URL("http://int.dpool.sina.com.cn/iplookup/iplookup.php?format=js")用于创建 URL，该 URL 在网络连接成功（有可能失败）之后，将以 JSON 字符串的形式返回查询者本身的位置信息，如图 204.2 所示。当使用 myJSONObject = new JSONObject(myData.substring(myData.indexOf("{")))构造 JSON 对象之后，即可用 myCity = (String) myJSONObject.get("city")的形式进行查询。

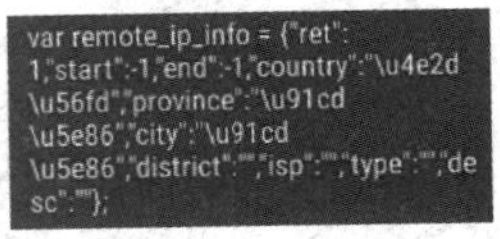

图 204.2

此外，访问网络需要在 AndroidManifest.xml 文件中添加< uses-permission android:name="android.permission.ACCESS_WIFI_STATE"/>、< uses-permission android:name="android.permission.INTERNET"/>和< uses-permission android:name="android.permission.ACCESS_NETWORK_STATE"/>权限。此实例的完整项目在 MyCode\MySample740 文件夹中。

205 使用 Gson 将数组转换成 JSON 字符串

此实例主要通过使用 Gson 的 toJson()方法，实现将数组对象转换成 JSON 字符串。当实例运行之后，单击"使用 Gson 将数组对象转换成 JSON 字符串"按钮，将把包含三个对象的数组解析成如图 205.1 所示的 JSON 字符串。

主要代码如下:

```java
public class MainActivity extends Activity {
    @Override
    protected void onCreate(Bundle savedInstanceState) {
        super.onCreate(savedInstanceState);
        setContentView(R.layout.activity_main);
    }
    //响应单击"使用Gson将数组对象转换成JSON字符串"按钮
    public void onClickmyBtn1(View v) {
        //创建数组对象并添加数据
        ArrayList<User> myArray = new ArrayList<>();
        User myUser1 = new User();
        myUser1.setMyName("Karli Watson");
        myUser1.setMyBook("C#入门经典");
        User myUser2 = new User();
        myUser2.setMyName("Bruce Eckel");
        myUser2.setMyBook("Java 编程思想");
        User myUser3 = new User();
        myUser3.setMyName("Stephen Prata");
        myUser3.setMyBook("C++Primer Plus");
        myArray.add(myUser1);
        myArray.add(myUser2);
        myArray.add(myUser3);
        try {
            Gson myGson = new Gson();
            String myInfo = myGson.toJson(myArray);
            Toast.makeText(getApplicationContext(), myInfo, Toast.LENGTH_SHORT).show();
        } catch (Exception e) {
            Toast.makeText(getApplicationContext(),
                    e.getMessage().toString(), Toast.LENGTH_SHORT).show();
        } } }
class User{ //User 类定义
    public String getMyName() { return myName; }
    public void setMyName(String myName) { this.myName = myName; }
    public String getMyBook() { return myBook; }
    public void setMyBook(String myBook) { this.myBook = myBook; }
    String myName;
    String myBook;
} }
```

图　205.1

上面这段代码在 MyCode\MySample471\app\src\main\java\com\bin\luo\mysample\MainActivity.java 文件中。在这段代码中,myInfo=myGson.toJson(myArray)用于将 myArray 数组中的三个对象以标准格式的 JSON 字符串输出到 myInfo 中。此外,在 Android Studio 中使用 Gson 需要在 gradle 中引入 compile 'com.google.code.gson:gson:2.8.1'依赖项。此实例的完整项目在 MyCode\MySample471 文件夹中。

206　使用 Gson 解析 JSON 字符串

此实例主要通过使用 Gson 的 fromJson()方法,实现将 JSON 格式的字符串以对象的方式解析出来。当实例运行之后,单击"使用 Gson 解析 JSON 格式的字符串"按钮,则将把 JSON 字符串: [{"myName":"Karli Watson","myBook":"C#入门经典"},{"myName":"Bruce Eckel","myBook":"Java 编程思想"},{"myName":"Stephen Prata","myBook":"C++Primer Plus"}],解析

成如图 206.1 所示的效果。

图 206.1

主要代码如下：

```java
public class MainActivity extends Activity {
 @Override
 protected void onCreate(Bundle savedInstanceState) {
  super.onCreate(savedInstanceState);
  setContentView(R.layout.activity_main);
 }
 //响应单击"使用Gson解析JSON格式的字符串"按钮
 public void onClickmyBtn1(View v) {
  String myJsonData = "[{\"myName\":\"Karli Watson\",\"myBook\":\"C#入门经典\"},"
            + "{\"myName\":\"Bruce Eckel\",\"myBook\":\"Java 编程思想\"},"
            + "{\"myName\":\"Stephen Prata\",\"myBook\":\"C++Primer Plus\"}]";
  try {
   Type myType = new TypeToken<LinkedList<User>>(){}.getType();
   Gson myGson = new Gson();
   LinkedList<User> myUsers = myGson.fromJson(myJsonData, myType);
   String myInfo = "";
   for (Iterator myIterator = myUsers.iterator(); myIterator.hasNext();) {
    User myUser = (User) myIterator.next();
    myInfo += "\n 作者:" + myUser.getMyName();
    myInfo += ", 书名:" + myUser.getMyBook();
   }
   Toast.makeText(getApplicationContext(), myInfo, Toast.LENGTH_SHORT).show();
  } catch (Exception e) {
   Toast.makeText(getApplicationContext(),
            e.getMessage().toString(), Toast.LENGTH_SHORT).show();
  }
 }
```

```
//User 类定义
class User{
    public String getMyName() { return myName; }
    public void setMyName(String myName) { this.myName = myName; }
    public String getMyBook() { return myBook; }
    public void setMyBook(String myBook) { this.myBook = myBook; }
    String myName;
    String myBook;
}
```

上面这段代码在 MyCode \ MySample469 \ app \ src \ main \ java \ com \ bin \ luo \ mysample \ MainActivity.java 文件中。在这段代码中，myType = new TypeToken < LinkedList < User >>(){}.getType() 用于获取自定义类 User 的类型定义。LinkedList < User > myUsers = myGson.fromJson(myJsonData, myType) 用于根据 User 的类型定义 myType 和 Json 字符串 myJsonData 的内容获取对象集合。myIterator = myUsers.iterator() 用于获取集合枚举器。myUser =（User）myIterator.next() 用于从集合枚举器中获取单个对象。此外，在 Android Studio 中使用 Gson 需要在 gradle 中引入 compile 'com.google.code.gson:gson:2.8.1' 依赖项。此实例的完整项目在 MyCode\MySample469 文件夹中。

207　使用 XmlPullParser 解析城市天气数据

此实例主要通过调用聚合数据提供的天气预报 API 接口，查询指定城市的天气信息并返回 XML 格式数据，然后使用 XmlPullParser 解析这些城市的天气数据，从而获取指定城市的天气信息。当实例运行之后，在"城市："输入框中输入城市名称，如"重庆"，然后单击"查询当地今天的天气"按钮，则在下面显示该地今天的天气信息；输入其他城市名称查询天气信息将获得相应的结果，效果分别如图 207.1 的左图和右图所示。

图　207.1

主要代码如下：

```java
public class MainActivity extends Activity {
 EditText myEditCity;
 Handler myHandler = new Handler() {
  @Override
  public void handleMessage(Message msg) {
   String myData = (String) msg.obj;
   if (msg.what == 2) {
    //将字符串转换为输入流形式
    InputStream myInputStream = new ByteArrayInputStream(myData.getBytes());
    ParseWeatherInfo(myInputStream);                //使用 XmlPullParser 对 XML 数据进行解析
    myTemperature.setText(myWeatherList.get(0).TemperatureInfo);
    myWind.setText(myWeatherList.get(0).WindInfo);
    myWeatherText.setText(myWeatherList.get(0).WeatherInfo);
    myLayout.setVisibility(View.VISIBLE);           //显示天气信息
    myLoading.setVisibility(GONE);
}}};
 private ArrayList<Weather> myWeatherList;
 private Weather myWeather;
 private TextView myTemperature;
 private TextView myWind;
 private TextView myWeatherText;
 private TextView myLoading;
 private LinearLayout myLayout;
 @Override
 protected void onCreate(Bundle savedInstanceState) {
  super.onCreate(savedInstanceState);
  setContentView(R.layout.activity_main);
  myEditCity = (EditText) findViewById(R.id.myEditCity);
  myTemperature = (TextView) findViewById(R.id.myTemperature);
  myWind = (TextView) findViewById(R.id.myWind);
  myWeatherText = (TextView) findViewById(R.id.myWeather);
  myLoading = (TextView) findViewById(R.id.myLoading);
  myLayout = (LinearLayout) findViewById(R.id.myLayout);
 }
 public void onClickmyBtn1(View v) {                //响应单击"查询当地今天的天气"按钮
  myLayout.setVisibility(GONE);
  myLoading.setVisibility(View.VISIBLE);
  getWeatherInfo(myEditCity.getText().toString());
 }
 //这里使用的是聚合数据的 API 接口,调用时请更换为自己申请的 APPKEY 值
 public void getWeatherInfo(final String myCityName) {
  new Thread() {
   @Override
   public void run() {
    try {
     URL myUrl = new URL("http://v.juhe.cn/weather/index?format=2&cityname=" +
             myCityName + "&key=94e8b799a9d12a50bacb61c5282e603e&dtype=xml");
     URLConnection myConnection = myUrl.openConnection();
     BufferedReader myReader = new BufferedReader(new InputStreamReader(
             myConnection.getInputStream(), "UTF-8"));
     String myLength = null;
```

```java
            StringBuffer myBuffer = new StringBuffer();
            while ((myLength = myReader.readLine())!= null){ myBuffer.append(myLength);}
            myReader.close();
            String myResult = myBuffer.toString();
            Message myMessage = new Message();
            myMessage.obj = myResult;
            myMessage.what = 2;
            myHandler.sendMessage(myMessage);
        } catch (Exception e) { e.printStackTrace(); }
        }
    }.start();
}
public void ParseWeatherInfo(InputStream inputStream) {
    try {
        XmlPullParser myParser = Xml.newPullParser();
        myParser.setInput(inputStream, "UTF-8");              //传入需要解析的 XML 数据流
        int myEvent = myParser.getEventType();                //获取事件类型
        while (myEvent != XmlPullParser.END_DOCUMENT) {
            String myNodeName = myParser.getName();
            switch (myEvent) {
                case XmlPullParser.START_DOCUMENT:            //文档开始
                    myWeatherList = new ArrayList<Weather>();
                    break;
                case XmlPullParser.START_TAG:                 //标签开始
                    if ("today".equals(myNodeName)) { myWeather = new Weather(); }
                    if ("temperature".equals(myNodeName)) {
                        String myTemperatureInfo = myParser.nextText();
                        myWeather.setTemperatureInfo(myTemperatureInfo);
                    }
                    if ("weather".equals(myNodeName)) {
                        String myWeatherInfo = myParser.nextText();
                        myWeather.setWeatherInfo(myWeatherInfo);
                    }
                    if ("wind".equals(myNodeName)) {
                        String myWindInfo = myParser.nextText();
                        myWeather.setWindInfo(myWindInfo);
                    }
                    break;
                case XmlPullParser.END_TAG:                   //标签结束
                    if ("today".equals(myNodeName)) {
                        myWeatherList.add(myWeather);
                        myWeather = null;
                    }
                    break;
            }
            myEvent = myParser.next();                        //下一个标签
        }
    } catch (Exception e) {
        e.printStackTrace();
    }
} }
```

上面这段代码在 MyCode\MySample753\app\src\main\java\com\bin\luo\mysample\MainActivity.java 文件中。在这段代码中，myData =（String）msg.obj 返回的天气信息原本是一个

XML 格式的文本，如下所示：

```
<?xml version = "1.0" encoding = "utf-8"?>
<root>
<resultcode>200</resultcode>
<reason>successed!</reason>
<result><sk><temp>9</temp>
<wind_direction>西北风</wind_direction>
<wind_strength>2级</wind_strength>
<humidity>95%</humidity>
<time>14:01</time></sk>
<today><temperature>9℃～11℃</temperature>
<weather>小雨</weather>
<weather_id><fa>07</fa><fb>07</fb></weather_id>
<wind>持续无风向微风</wind>
<week>星期三</week>
<city>重庆</city>
<date_y>2018年02月21日</date_y>
<dressing_index>较冷</dressing_index>
<dressing_advice>建议着厚外套加毛衣等服装.年老体弱者宜着大衣、呢外套加羊毛衫.</dressing_advice>
<uv_index>最弱</uv_index>
<comfort_index></comfort_index>
<wash_index>不宜</wash_index>
<travel_index>较不宜</travel_index>
<exercise_index>较不宜</exercise_index>
<drying_index></drying_index></today></result><error_code>0</error_code></root>
```

ParseWeatherInfo()方法则使用 XmlPullParser 根据上述文本的各个 XML 标签进行指定内容的解析，Pull 解析 XML 是基于事件驱动方式的解析，Pull 开始解析时，可以先通过 getEventType()方法获取当前解析事件类型，并且通过 next()方法获取下一个解析事件类型。Pull 解析器提供了 START_DOCUMENT（开始文档）、END_DOCUMENT（结束文档）、START_TAG（开始标签）、END_TAG（结束标签）四种事件解析类型。当处于某个元素时，可以调用 getAttributeValue()方法获取属性的值，也可以通过 nextText()方法获取本节点的文本值。

此外，访问网络需要在 AndroidManifest.xml 文件中添加< uses-permission android：name = "android.permission.INTERNET"/>权限。此实例的完整项目在 MyCode\MySample753 文件夹中。

208 采用 SAX 方式解析 XML 文件内容

此实例主要通过使用 SAXParserFactory 等 SAX 操作类，实现以对象的形式读取 XML 文件的内容。当实例运行后，单击"采用 SAX 方式解析 XML 文件内容"按钮，则在 ListView 控件中显示 personsax.xml 文件的内容，效果分别如图 208.1 的左图和右图所示。

主要代码如下：

```
public void onClickmyBtn1(View v) {                    //响应单击"采用 SAX 方式解析 XML 文件内容"按钮
  try {
    myArray = ReadXMLFileBySAX();
  } catch (Exception e) { e.printStackTrace(); }
  myAdapter = new ArrayAdapter<Person>(MainActivity.this,
            android.R.layout.simple_expandable_list_item_1, myArray);
```

```
    myListView.setAdapter(myAdapter);
}
private ArrayList<Person> ReadXMLFileBySAX() throws Exception {
    InputStream myInputStream =
            getAssets().open("personsax.xml");          //根据 XML 文件创建输入流对象
    SaxHelper mySaxHelper = new SaxHelper();            //创建 XML 解析处理器
    SAXParserFactory myFactory =
            SAXParserFactory.newInstance();             //得到 SAX 解析工厂
    SAXParser myParser = myFactory.newSAXParser();      //创建 SAX 解析器
    myParser.parse(myInputStream, mySaxHelper);         //对文档进行解析
    myInputStream.close();
    return mySaxHelper.getPersonArrayList();
}
```

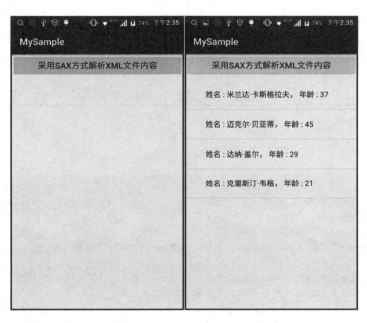

图　208.1

上面这段代码在 MyCode\MySample332\app\src\main\java\com\bin\luo\mysample\MainActivity.java 文件中。在这段代码中，mySaxHelper = new SaxHelper()用于创建 SaxHelper 类的实例，以解析 XML 文件的标签。关于 SaxHelper 类的定义请参考源代码中的 MyCode\MySample332\app\src\main\java\com\bin\luo\mysample\SaxHelper.java 文件。SAX，全称 Simple API for XML，SAX 不同于 DOM 解析，它逐行扫描文档，一边扫描一边解析。由于应用只是在读取时检查数据，因此不需要将全部数据加载在内存中，这对于大型文档的解析是个巨大优势。此实例的完整项目在 MyCode\MySample332 文件夹中。

209　使用 Pattern 根据正则表达式校验手机号码

此实例主要通过使用 Pattern 的 compile() 和 matcher() 方法，实现通过正则表达式校验字符串格式的手机号码是否正确。当实例运行之后，如果在"手机号码："输入框中输入"13996060872-1"，然后单击"使用正则表达式检验手机号码"按钮，则在弹出的 Toast 中提示此号码错误，如图 209.1 的左图所示。如果在"手机号码："输入框中输入"13996060872"，然后单击"使用正则表达式检验手机号码"

按钮,则在弹出的 Toast 中提示此号码正确,如图 209.1 的右图所示。

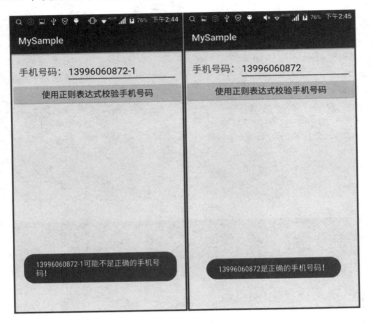

图 209.1

主要代码如下:

```java
public void onClickmyBtn1(View v) {                    //响应单击"使用正则表达式校验电话号码"按钮
    String myPhone = myEditText.getText().toString();
    if(!checkPhone(myPhone)){
        Toast.makeText(getApplicationContext(),
                myPhone + "可能不是正确的手机号码!", Toast.LENGTH_SHORT).show();
    }else{
        Toast.makeText(getApplicationContext(),
                myPhone + "是正确的手机号码!", Toast.LENGTH_SHORT).show();
    } }
public static boolean checkPhone(String myPhone){   //判断最新的中国电话号码
    Pattern myPattern = Pattern.compile(
            "^((13[0-9])|(14[5|7])|(15([0-3]|[5-9]))|(18[0,5-9]))\\d{8}$",
            Pattern.CASE_INSENSITIVE);
    Matcher myMatcher = myPattern.matcher(myPhone);
    return myMatcher.matches();
}
```

上面这段代码在 MyCode\MySample372\app\src\main\java\com\bin\luo\mysample\MainActivity.java 文件中。在这段代码中,myPattern = Pattern.compile("^((13[0-9])|(14[5|7])|(15([0-3]|[5-9]))|(18[0,5-9]))\d{8}$", Pattern.CASE_INSENSITIVE)用于根据字符串指定的规则创建一个正则表达式的实例,当前这个字符串主要用于匹配中国大陆地区的手机号码,如下:

* 移动号码段:139、138、137、136、135、134、150、151、152、157、158、159、182、183、187、188、147。
* 联通号码段:130、131、132、136、185、186、145。
* 电信号码段:133、153、180、189。

myMatcher=myPattern.matcher(myPhone)用于校验 myPhone 字符串是否符合 myPattern 正

则表达式的要求,如果myMatcher.matches()返回值为true,则表示符合要求;否则不符合要求。此实例的完整项目在MyCode\MySample372文件夹中。

210　使用SharedPreferences保存账户和密码

此实例主要通过使用SharedPreferences,实现以键值对的形式在手机中保存账户和密码信息。当实例运行之后,在账户和密码输入框中输入内容,选中"记住账户和密码信息"复选框,然后单击"登录"按钮,则将执行登录操作,并将账户和密码信息保存在手机中,否则提示操作错误,效果分别如图210.1的左图和右图所示。在Android系统中,SharedPreferences保存的数据文件一般存储在/data/data/com.bin.luo.mysample(your packagename)/shared_prefs目录下。此实例保存的SharedPreferences数据文件名称是com.bin.luo.mysample_preferences.xml,内容如下:

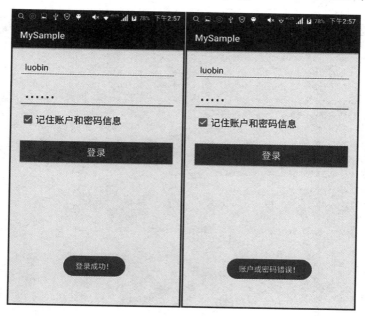

图　210.1

```
<?xml version = '1.0' encoding = 'utf-8' standalone = 'yes' ?>
<map><string name = "account">luobin</string>
    <string name = "password">123456</string>
    <boolean name = "remember_password" value = "true" /></map>
```

该功能的具体实现代码如下:

```java
public class MainActivity extends Activity {
  public static final String REMEMBER_PASSWORD = "remember_password";
  public static final String ACCOUNT = "account";
  public static final String PASSWORD = "password";
  private EditText myEditAccount, myEditPassword;
  private CheckBox myCheckbox;
  private SharedPreferences mySharedPreferences;
  private SharedPreferences.Editor myEditor;
  private boolean isRemember;
```

```java
@Override
protected void onCreate(Bundle savedInstanceState) {
    super.onCreate(savedInstanceState);
    setContentView(R.layout.activity_main);
    myEditAccount = (EditText) findViewById(R.id.myAccount);
    myEditPassword = (EditText) findViewById(R.id.myPassword);
    myCheckbox = (CheckBox) findViewById(R.id.myCheckbox);
    //获取 SharedPreferences 对象
    mySharedPreferences = PreferenceManager.getDefaultSharedPreferences(this);
    isRemember = mySharedPreferences.getBoolean(REMEMBER_PASSWORD, false);
    if (isRemember) {
        //将 SharedPreferences 保存的账户和密码设置到 TextView 中
        myEditAccount.setText(mySharedPreferences.getString(ACCOUNT, ""));
        myEditPassword.setText(mySharedPreferences.getString(PASSWORD, ""));
        myCheckbox.setChecked(true);
    }
}
public void onClickmyBtn1(View v) {                      //响应单击"登录"按钮
    String account = myEditAccount.getText().toString();
    String password = myEditPassword.getText().toString();
    //如果账户是:luobin,密码是:123456,则认为登录成功
    if (account.equals("luobin") && password.equals("123456")) {
        myEditor = mySharedPreferences.edit();           //获取 SharedPreferences.Editor 对象
        if (myCheckbox.isChecked()) {
            //如果选中复选框,则向 SharedPreferences.Editor 对象保存账户和密码信息
            myEditor.putBoolean(REMEMBER_PASSWORD, true);
            myEditor.putString(ACCOUNT, account);
            myEditor.putString(PASSWORD, password);
        } else { myEditor.clear(); }
        myEditor.apply();                                //提交数据
        //进行跳转,并 finish 当前 activity
        //startActivity(new Intent(MainActivity.this, OtherActivity.class));
        //finish();
        Toast.makeText(MainActivity.this, "登录成功!", Toast.LENGTH_SHORT).show();
    } else {
        Toast.makeText(MainActivity.this,
                "账户或密码错误!", Toast.LENGTH_SHORT).show();
    }
}
```

上面这段代码在 MyCode\MySample319\app\src\main\java\com\bin\luo\mysample\MainActivity.java 文件中。在这段代码中,mySharedPreferences = PreferenceManager.getDefaultSharedPreferences(this)用于获取 SharedPreferences 对象,并自动使用当前应用的包名作为前缀来命名 SharedPreferences 文件。mySharedPreferences.getString(ACCOUNT, "")用于获取 ACCOUNT 键的信息。myEditor = mySharedPreferences.edit()用于获取一个 SharedPreferences.Editor 对象。myEditor.putString(ACCOUNT, account)用于设置 ACCOUNT 键的信息。myEditor.apply()则用于提交(保存)设置的(所有)键值数据。此实例的完整项目在 MyCode\MySample319 文件夹中。

211 使用 ListPreference 读写单选按钮值

此实例主要通过使用 ListPreference,实现在存储卡上保存用户在单选按钮中的选择结果或者从存储卡上读取用户在单选按钮中的选择结果。当实例运行之后,单击"年度精品图书"列表选项,则将显示"年度精品图书"对话框,在该对话框中任意选中一个单选按钮,然后返回,如"改善既有代码的设

计"单选按钮,如图211.1的左图所示。然后退出此应用,再重新启动此应用,则在弹出的 Toast 中显示"刚才选择了图书:改善既有代码的设计",如图211.1的右图所示。主要代码如下:

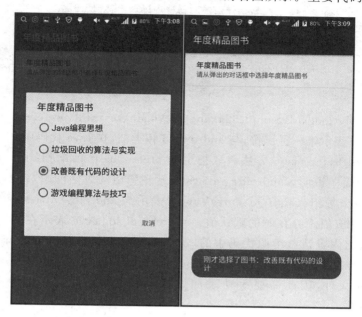

图 211.1

```
public class MainActivity extends PreferenceActivity {
@Override
protected void onCreate(Bundle savedInstanceState) {
  super.onCreate(savedInstanceState);
  addPreferencesFromResource(R.xml.preferences);
  PreferenceManager myPreferenceManager = getPreferenceManager();
  // 根据 android:key 中指定的名称(相当于 ID)获取首选项
  ListPreference myListPreference = (ListPreference)
          myPreferenceManager.findPreference("myListPreference");
  Toast.makeText(MainActivity.this, "刚才选择了图书:" +
          myListPreference.getValue(),Toast.LENGTH_SHORT).show();
} }
```

上面这段代码在 MyCode \ MySample324 \ app \ src \ main \ java \ com \ bin \ luo \ mysample \ MainActivity.java 文件中。在这段代码中,PreferenceManager myPreferenceManager = getPreferenceManager()用于保存用户的设置,它根据包名和布局文件创建新文件来实现在存储卡上保存内容,因此在关闭应用,再重新启动应用之后,能够显示此前用户在对话框的操作结果。myListPreference=(ListPreference)myPreferenceManager.findPreference("myListPreference")用于在 XML 文件中查找 key 为 myListPreference 的 ListPreference。myListPreference.getValue()用于获取 ListPreference 值。addPreferencesFromResource(R.xml.preferences)表示 MainActivity 加载的视图是 xml 目录下的 preferences 文件。preferences 文件的主要内容如下:

```
<?xml version = "1.0" encoding = "UTF - 8"?>
< PreferenceScreen xmlns:android = "http://schemas.android.com/apk/res/android"
                android:key = "screen_list"
                android:title = "年度精品图书"
                android:summary = "请从弹出的对话框中选择年度精品图书" >
```

```xml
<ListPreference  android:key = "myListPreference"
                 android:title = "年度精品图书"
                 android:summary = "请从弹出的对话框中选择年度精品图书"
                 android:entries = "@array/list_entries"
                 android:entryValues = "@array/list_entries_value"
                 android:dialogTitle = "年度精品图书"
                 android:defaultValue = "@array/list_entries_value2"></ListPreference>
</PreferenceScreen>
```

上面这段代码在 MyCode\MySample324\app\src\main\res\xml\preferences.xml 文件中。在这段代码中，android:key 表示唯一标识符，与 android:id 相类似，在 Java 代码中，PreferenceManager 可以以其为参数通过 findPreference()方法获取指定的 preference。android:title 表示标题。android:summary 表示选项的简单说明。android:entries 表示在弹出的对话框中，列表项显示的文本内容，注意，这里指定的是一个数组。android:entryValues 表示与 android:entries 相对应的值。android:defaultValue 表示当对应值不存在时的默认值。android:dialogTitle 表示在弹出的对话框中的标题信息。上述三个数组的内容请参考源代码中的 MyCode\MySample324\app\src\main\res\values\arrays.xml 文件。需要说明的是，默认情况下，Android Studio 创建的工程在 res 目录下没有 xml 目录及 preferences.xml 文件，因此需要手动添加该目录和文件。此实例的完整项目在 MyCode\MySample324 文件夹中。

212　在代码中获取 CheckBoxPreference 值

此实例主要通过在 onSharedPreferenceChanged()方法中获取 CheckBoxPreference 的 key 值，实现在 Java 代码中判断用户是否选中了 CheckBoxPreference 控件。当实例运行之后，如果选择了"是否开启定期更新功能"CheckBoxPreference，则在弹出的 Toast 中显示"刚才选择了定期更新功能！"，如图 212.1 的左图所示。如果未选择"是否开启定期更新功能"CheckBoxPreference，则在弹出的 Toast 中显示"刚才取消了定期更新功能！"，如图 212.1 的右图所示。注意：此实例应在 Android 手机中进行测试，在部分模拟器测试时发现存在不能正常显示 Toast 的漏洞。

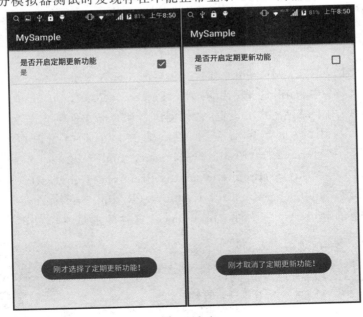

图　212.1

主要代码如下：

```java
public class MainActivity extends PreferenceActivity {
    SharedPreferences mySharedPreferences;
    @Override
    protected void onCreate(Bundle savedInstanceState) {
        super.onCreate(savedInstanceState);
        addPreferencesFromResource(R.xml.preferences);
        mySharedPreferences = PreferenceManager.getDefaultSharedPreferences(this);
        mySharedPreferences.registerOnSharedPreferenceChangeListener(
                new SharedPreferences.OnSharedPreferenceChangeListener() {
            @Override
            public void onSharedPreferenceChanged(
                    SharedPreferences sharedPreferences, String key) {
                if (key.equals("myUpdate")) {
                    Boolean myValue = sharedPreferences.getBoolean("myUpdate", false);
                    if (myValue) {
                        Toast.makeText(MainActivity.this,
                                "刚才选择了定期更新功能!", Toast.LENGTH_SHORT).show();
                    } else {
                        Toast.makeText(MainActivity.this,
                                "刚才取消了定期更新功能!", Toast.LENGTH_SHORT).show();
                    }}}});
    }
}
```

上面这段代码在 MyCode\MySample323\app\src\main\java\com\bin\luo\mysample\MainActivity.java 文件中。在这段代码中，mySharedPreferences = PreferenceManager.getDefaultSharedPreferences(this) 用于保存 PreferenceActivity 中的设置，根据包名和 PreferenceActivity 的布局文件创建一个键名保存，因此当前应用的 CheckBoxPreference 如果是选中状态，在关闭应用，再重新启动应用之后，该 CheckBoxPreference 仍然是选中状态；如果当前应用的 CheckBoxPreference 是未选状态，在关闭应用，再重新启动应用之后，该 CheckBoxPreference 仍然是未选状态。myValue = sharedPreferences.getBoolean("myUpdate", false) 用于获取 key 为 myUpdate 的 CheckBoxPreference 的布尔值。addPreferencesFromResource(R.xml.preferences) 表示 MainActivity 加载的视图是 xml 目录下的 preferences 文件。preferences 文件的主要内容如下：

```xml
<?xml version = "1.0" encoding = "UTF-8"?>
<PreferenceScreen xmlns:android = "http://schemas.android.com/apk/res/android">
    <CheckBoxPreference android:key = "myUpdate"
                        android:title = "是否开启定期更新功能"
                        android:summaryOn = "是"
                        android:summaryOff = "否"
                        android:defaultValue = "true"/>
</PreferenceScreen>
```

上面这段代码在 MyCode\MySample323\app\src\main\res\xml\preferences.xml 文件中。此实例的完整项目在 MyCode\MySample323 文件夹中。

213 通过 PreferenceScreen 跳转到 WiFi 设置

此实例主要通过使用 PreferenceScreen，实现从当前应用跳转到手机的 WiFi 设置界面。当实例运行之后，将显示 PreferenceScreen 列表项"WiFi 设置"，如图 213.1 的左图所示。单击列表项"WiFi

设置",则将跳转到当前手机的有效 WiFi 列表,如图 213.1 的右图所示。

图 213.1

主要代码如下:

```
public class MainActivity extends PreferenceActivity {
 @Override
 protected void onCreate(Bundle savedInstanceState) {
  super.onCreate(savedInstanceState);
  addPreferencesFromResource(R.xml.preferences);
} }
```

上面这段代码在 MyCode\MySample321\app\src\main\java\com\bin\luo\mysample\MainActivity.java 文件中。在这段代码中,addPreferencesFromResource(R.xml.preferences)表示 MainActivity 加载的视图是 xml 目录下的 preferences 文件,而不是默认的 setContentView(R.layout.activity_main),因为此处不再需要默认的 activity_main 布局。preferences 文件的主要内容如下:

```
<?xml version = "1.0" encoding = "UTF-8"?>
<PreferenceScreen xmlns:android = "http://schemas.android.com/apk/res/android">
<PreferenceScreen  android:key = "wifi_settings"
                  android:title = "WiFi 设置"
                  android:summary = "单击即可跳转到手机 WiFi 设置">
  <intent  android:action = "android.intent.action.MAIN"
           android:targetPackage = "com.android.settings"
           android:targetClass = "com.android.settings.wifi.WifiSettings" />
 </PreferenceScreen>
</PreferenceScreen>
```

上面这段代码在 MyCode\MySample321\app\src\main\res\xml\preferences.xml 文件中。此实例的完整项目在 MyCode\MySample321 文件夹中。

214　使用 Intent 实现在 Activity 之间传递小图像

此实例主要通过使用 Intent 的 getParcelableExtra()方法，实现在两个 Activity 之间传递小图像。当实例运行之后，将在 MainActivity 的 ImageView 控件中显示大拇指图像，如图 214.1 的左图所示；单击"将下面的图像传递到第二个 Activity"按钮，则将跳转到 SecondActivity，此时 SecondActivity 的 ImageView 控件是一片空白；单击"接收从第一个 Activity 传递的图像"按钮，则在 ImageView 控件中显示从 MainActivity 传送的大拇指图像，如图 214.1 的右图所示。

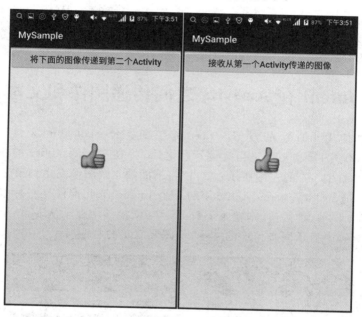

图　214.1

主要代码如下：

```
//响应单击"将下面的图像传递到第二个 Activity"按钮
public void onClickmyBtnSend(View v) {//注意:大尺寸图像无法通过测试
    Bitmap myBitmap = ((BitmapDrawable) myImageMain.getDrawable()).getBitmap();
    Intent myIntent = new Intent(MainActivity.this, SecondActivity.class);
    myIntent.putExtra("myBitmap", myBitmap);
    startActivity(myIntent);
}
```

上面这段代码在 MyCode\MySample442\app\src\main\java\com\bin\luo\mysample\MainActivity.java 文件中。在这段代码中，myBitmap = ((BitmapDrawable) myImageMain.getDrawable()).getBitmap()表示从 myImageMain 控件中获取图像 Bitmap。myIntent = new Intent(MainActivity.this, SecondActivity.class)表示从 MainActivity 跳转到 SecondActivity。myIntent.putExtra("myBitmap", myBitmap)表示在从 MainActivity 跳转到 SecondActivity 时附加图像 myBitmap。SecondActivity 的主要代码如下：

```
//响应单击"接收从第一个 Activity 传递的图像"按钮
public void onClickmyBtnReceive(View v) {
```

```
    Intent myIntent = getIntent();
    if(myIntent!= null){
     Bitmap myBitmap = myIntent.getParcelableExtra("myBitmap");
     myImageSecond.setImageBitmap(myBitmap);
} }
```

上面这段代码在 MyCode\MySample442\app\src\main\java\com\bin\luo\mysample\SecondActivity.java 文件中。在这段代码中，myIntent = getIntent()用于获取传递的 Intent，myBitmap＝myIntent.getParcelableExtra("myBitmap")用于从 myIntent 中获取 myBitmap 代表的图像数据。此外，当新增了 SecondActivity 时，则需要在 AndroidManifest.xml 文件中注册该 Activity，如< activity android:name = ".SecondActivity"/>。此实例的完整项目在 MyCode\MySample442 文件夹中。

215　使用 Intent 在 Activity 之间传递图像和文本

此实例主要通过使用 Intent 的 putExtra()方法和 getSerializableExtra()方法，实现在两个 Activity 之间传递多幅小图像和文本。当实例运行之后，将在 MainActivity 的 4 个 ImageView 控件中显示 4 幅小图像，如图 215.1 的左图所示；单击"将下面的 4 幅图像发送到第二个 Activity"按钮，则将跳转到 SecondActivity，此时 SecondActivity 的 4 个 ImageView 控件是一片空白；单击"接收从第一个 Activity 发送的 4 幅图像"按钮，则在 4 个 ImageView 控件中显示从 MainActivity 传来的 4 幅小图像，并在弹出的 Toast 中显示同步传递的文本，如图 215.1 的右图所示。

图 215.1

主要代码如下：

```
//响应单击"将下面的 4 幅图像发送到第二个 Activity"按钮
public void onClickmyBtnSend(View v) {
  try{
```

```
    Bitmap myBitmap1 = ((BitmapDrawable) myImageViewA1.getDrawable()).getBitmap();
    Bitmap myBitmap2 = ((BitmapDrawable) myImageViewA2.getDrawable()).getBitmap();
    Bitmap myBitmap3 = ((BitmapDrawable) myImageViewA3.getDrawable()).getBitmap();
    Bitmap myBitmap4 = ((BitmapDrawable) myImageViewA4.getDrawable()).getBitmap();
    ArrayList<HashMap<String,Object>> myArray = new ArrayList<>();
    HashMap<String,Object> myItem = new HashMap<>();
    myItem.put("myImage",myBitmap1);
    myItem.put("myName","夺命三头鲨");
    myArray.add(myItem);
    myItem = new HashMap<>();
    myItem.put("myImage",myBitmap2);
    myItem.put("myName","天启");
    myArray.add(myItem);
    myItem = new HashMap<>();
    myItem.put("myImage",myBitmap3);
    myItem.put("myName","变形金刚");
    myArray.add(myItem);
    myItem = new HashMap<>();
    myItem.put("myImage",myBitmap4);
    myItem.put("myName","速度与激情");
    myArray.add(myItem);
    Intent myIntent = new Intent(MainActivity.this,SecondActivity.class);
    myIntent.putExtra("arrayList", myArray);
    startActivity(myIntent);
    }catch (Exception e){
    Toast.makeText(getApplicationContext(),
                   e.getMessage().toString(), Toast.LENGTH_SHORT).show();
    } }
```

上面这段代码在 MyCode\MySample463\app\src\main\java\com\bin\luo\mysample\MainActivity.java 文件中。在这段代码中，ArrayList<HashMap<String, Object>> myArray = new ArrayList<>()用于创建存放键值对的数组列表。HashMap<String, Object> myItem = new HashMap<>()用于创建一个哈希键值对 myItem。myItem.put("myImage",myBitmap1)用于将 myBitmap1 保存在哈希键值对 myItem 中，并取键名为 myImage。myItem.put("myName","夺命三头鲨")用于将文本"夺命三头鲨"保存在哈希键值对 myItem 中，并取键名为 myName。myArray.add(myItem)用于将哈希键值对 myItem 添加到数组列表 myArray 中。myIntent = new Intent(MainActivity.this，SecondActivity.class)表示 myIntent 用于从 MainActivity 跳转到 SecondActivity。myIntent.putExtra("arrayList",myArray)表示 myIntent 将要传递的图像文本数组列表 myArray。SecondActivity 的主要代码如下：

```
//响应单击"接收从第一个 Activity 发送的 4 幅图像"按钮
public void onClickmyBtnReceive(View v) {
  try{
   Intent myIntent = getIntent();
   if(myIntent!= null){
    String myName1,myName2,myName3,myName4;
    ArrayList<HashMap<String, Object>> myArray = (ArrayList<HashMap<String,
            Object>>) myIntent.getSerializableExtra("arrayList");
    myImageViewB1.setImageBitmap((Bitmap) myArray.get(0).get("myImage"));
    myImageViewB2.setImageBitmap((Bitmap) myArray.get(1).get("myImage"));
    myImageViewB3.setImageBitmap((Bitmap) myArray.get(2).get("myImage"));
    myImageViewB4.setImageBitmap((Bitmap) myArray.get(3).get("myImage"));
    myName1 = (String) myArray.get(0).get("myName");
```

```
            myName2 = (String) myArray.get(1).get("myName");
            myName3 = (String) myArray.get(2).get("myName");
            myName4 = (String) myArray.get(3).get("myName");
            Toast.makeText(getApplicationContext(),"成功接收的电影海报分别是:"
            + myName1 + "、" + myName2 + "、" + myName3 + "、" + myName4, Toast.LENGTH_SHORT).show();
        } }catch (Exception e){
            Toast.makeText(getApplicationContext(),
                    e.getMessage().toString(),Toast.LENGTH_SHORT).show();
        } }
```

上面这段代码在 MyCode\MySample463\app\src\main\java\com\bin\luo\mysample\SecondActivity.java 文件中。在这段代码中，myIntent = getIntent()用于获取传递的 Intent，ArrayList < HashMap < String, Object >> myArray =（ArrayList < HashMap < String, Object >>) myIntent.getSerializableExtra("arrayList")用于从 Intent 中获取名称为 arrayList 代表的图像文本数组列表。myImageViewB1.setImageBitmap((Bitmap) myArray.get(0).get("myImage"))用于在 myImageViewB1 控件中设置从图像文本数组列表传递的图像。myName1 = (String) myArray.get(0).get("myName")用于获取从图像文本数组列表中传递的文本。注意：在接收 Intent 传递的图像或文本时，需要进行数据类型强制转换；此外，图像过大或过多可能会导致应用崩溃，在实际应用中需要考虑这一问题。需要注意的是，当新增了 SecondActivity，则需要在 AndroidManifest.xml 文件中注册该 Activity，如< activity android:name = ".SecondActivity"/>。此实例的完整项目在 MyCode\MySample463 文件夹中。

216 使用 Intent 在 Activity 之间传递集合数据

此实例主要通过使用 putSerializable()方法和 getSerializableExtra()方法，实现在两个 Activity 之间传递集合数据。当实例运行之后，在 MainActivity 的输入框中输入两本书名，如图 216.1 的左图所示，单击"发送数据"按钮，则将跳转到 SecondActivity。在 SecondActivity 中单击"接收数据"按钮，则在弹出的 Toast 中显示在 MainActivity 的输入框中输入的两本书名，如图 216.1 的右图所示。

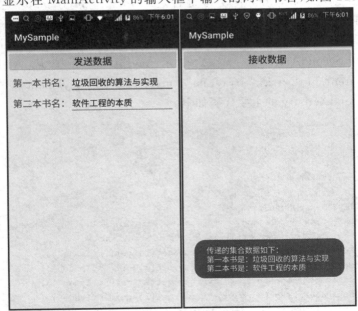

图 216.1

主要代码如下:

```java
public void onClickmyBtnSend(View v) {                    //响应单击"发送数据"按钮
    Intent myIntent = new Intent(MainActivity.this,SecondActivity.class);
    String myBook1 = myEditText1.getText().toString();
    String myBook2 = myEditText2.getText().toString();
    myList.add(myBook1);
    myList.add(myBook2);
    Bundle myBundle = new Bundle();
    myBundle.putSerializable("myBook",(Serializable)myList);
    myIntent.putExtras(myBundle);
    startActivity(myIntent);
}
```

上面这段代码在 MyCode \ MySample810 \ app \ src \ main \ java \ com \ bin \ luo \ mysample \ MainActivity.java 文件中。在这段代码中,myBundle.putSerializable("myBook",(Serializable)myList)表示将 myList 中的集合数据序列化之后存放在 myBundle 中。myIntent.putExtras(myBundle)表示将 myBundle 附加在 myIntent 上,当执行 startActivity(myIntent)实现从 MainActivity 跳转到 SecondActivity 时,集合数据将同时被传递。SecondActivity 的主要代码如下:

```java
public void onClickmyBtnReceive(View v) {                 //响应单击"接收数据"按钮
    String myInfo = "传递的集合数据如下:\n";
    Intent myIntent = getIntent();
    myList = (List<String>)myIntent.getSerializableExtra("myBook");
    myInfo += "第一本书是:" + myList.get(0) + "\n";
    myInfo += "第二本书是:" + myList.get(1);
    Toast.makeText(getApplicationContext(), myInfo, Toast.LENGTH_SHORT).show();
}
```

上面这段代码在 MyCode \ MySample810 \ app \ src \ main \ java \ com \ bin \ luo \ mysample \ SecondActivity.java 文件中。在这段代码中,myList = (List < String >)myIntent.getSerializableExtra("myBook")表示从 myIntent 获取集合数据,myBook 是数据名称,必须与发送方的名称一致。此外,当新增了 SecondActivity,则需要在 AndroidManifest.xml 文件中注册该 Activity,如< activity android:name = ".SecondActivity"/>。此实例的完整项目在 MyCode \ MySample810 文件夹中。

217 在 Intent 传递数据时使用 Bundle 携带数组

此实例主要演示了当使用 Intent 在两个 Activity 之间传递数据时,使用 Bundle 携带数组数据。当实例运行之后,在 MainActivity 中单击"获取并发送所有应用包名数据"按钮,如图 217.1 的左图所示,则将跳转到 SecondActivity。在 SecondActivity 中单击"接收并显示所有应用包名数据"按钮,则在 TextView 控件中显示在 MainActivity 中获取的所有应用包名信息,如图 217.1 的右图所示。

主要代码如下:

```java
//响应单击"获取并发送所有应用包名数据"按钮
public void onClickmyBtnSend(View v) {
    final PackageManager packageManager =
        getApplicationContext().getPackageManager();          //获取所有应用包名数据
```

```
List<PackageInfo> packageInfos = packageManager.getInstalledPackages(0);
ArrayList<String> packageNames = new ArrayList<String>();
if (packageInfos != null) {
  for (int i = 0; i < packageInfos.size(); i++) {
    String myPackageName = packageInfos.get(i).packageName;
    packageNames.add(myPackageName);
  }
}
//发送所有应用包名数据
Intent myIntent = new Intent(MainActivity.this,SecondActivity.class);
Bundle myBundle = new Bundle();
myBundle.putStringArrayList("myNames",packageNames);
myIntent.putExtras(myBundle);
startActivity(myIntent);
}
```

图 217.1

上面这段代码在 MyCode\MySample446\app\src\main\java\com\bin\luo\mysample\MainActivity.java 文件中。在这段代码中，Bundle 是一个简单的数据携带包，当使用 Bundle 携带字符串类型的数组数据时，应使用 Bundle 的 putStringArrayList()方法。putStringArrayList()方法的语法格式如下：

```
putStringArrayList(@Nullable String key, @Nullable ArrayList<String> value)
```

其中，参数 String key 表示数据名称。参数 ArrayList<String> value 表示数组类型的字符串数组。当将字符串数组通过 Bundle 的 putStringArrayList()方法附加到 Bundle 对象之后，还需要将 Bundle 对象通过 Intent 的 putExtras()方法附加到 Intent。当使用 startActivity(myIntent)方法从 MainActivity 跳转到 SecondActivity 时，myIntent 包含的 Bundle 数组就传递到了 SecondActivity。SecondActivity 的主要代码如下：

```
//响应单击"接收并显示所有应用包名数据"按钮
public void onClickmyBtnReceive(View v) {
  String myText = "当前手机安装的应用包名如下:";
  Intent myIntent = getIntent();
  Bundle myBundle = myIntent.getExtras();
  ArrayList<String> packageNames = myBundle.getStringArrayList("myNames");
  if (packageNames != null) {
    for (int i = 0; i < packageNames.size(); i++) {
      String myName = packageNames.get(i).toString();
      myText += "\n" + myName;
    } }
  myTextView.setText(myText);
}
```

上面这段代码在 MyCode\MySample446\app\src\main\java\com\bin\luo\mysample\SecondActivity.java 文件中。在这段代码中，myBundle = myIntent.getExtras()用于从 Intent 取得 Bundle 对象。packageNames = myBundle.getStringArrayList("myNames")用于从 Bundle 对象中获取数据名称为 myNames 的字符串数组。需要说明的是，putStringArrayList()和 getStringArrayList()方法仅支持字符串类型的数组数据传递，其他类型的数组数据传递应使用 Bundle 提供的其他方法，如传递 int 类型的数组数据应使用 putIntegerArrayList()、getIntegerArrayList()等。此外，当新增了 SecondActivity，则需要在 AndroidManifest.xml 文件中注册该 Activity，如<activity android:name = ".SecondActivity"/>。此实例的完整项目在 MyCode\MySample446 文件夹中。

218 使用 Intent 在 Service 和 Activity 之间传递数据

此实例主要通过在自定义 Service 中采集时间并发送广播，实现使用 Intent 以广播的方式在 Service 和 Activity 之间传递数据。当实例运行之后，单击"启动服务"按钮，则自定义 Service 采集的时间数据将以广播的形式发送，并被 IntentFilter 过滤接收，即逐秒显示当前时间，如图 218.1 的左图所示；单击"停止服务"按钮，则服务停止，显示的当前时间不再更新，如图 218.1 的右图所示。

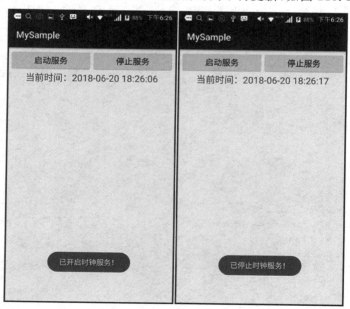

图 218.1

主要代码如下：

```java
public class MainActivity extends Activity {
  TextView myTextView;
  @Override
  protected void onCreate(Bundle savedInstanceState) {
    super.onCreate(savedInstanceState);
    setContentView(R.layout.activity_main);
    myTextView = (TextView) findViewById(R.id.myTextView);
    IntentFilter myIntentFilter = new IntentFilter();
    myIntentFilter.addAction("TimeService");                              //设置过滤 Action 字符串
    registerReceiver(new ServiceReceiver(), myIntentFilter);              //动态注册广播
  }
  public void onClickBtn1(View v) {                                        //响应单击"启动服务"按钮
    Intent myIntent = new Intent(MainActivity.this, TimeService.class);
    startService(myIntent);                                                //通过 Intent 启动服务
    Toast.makeText(MainActivity.this,
                   "已开启时钟服务!", Toast.LENGTH_SHORT).show();
  }
  public void onClickBtn2(View v) {                                        //响应单击"停止服务"按钮
    Intent myIntent = new Intent(MainActivity.this, TimeService.class);
    stopService(myIntent);                                                 //通过 Intent 停止服务
    Toast.makeText(MainActivity.this,
                   "已停止时钟服务!", Toast.LENGTH_SHORT).show();
  }
  //接收从服务发送数据的广播
  public class ServiceReceiver extends BroadcastReceiver {
    @Override
    public void onReceive(Context context, Intent intent) {
      myTextView.setText("当前时间:" +
          intent.getStringExtra("myTime"));                                //通过 TextView 控件显示当前时间
} } }
```

上面这段代码在 MyCode\MySample911\app\src\main\java\com\bin\luo\mysample\MainActivity.java 文件中。在这段代码中，Intent myIntent = new Intent(MainActivity.this, TimeService.class)用于根据指定的 MainActivity 和 TimeService 创建 Intent，TimeService 是获取当前时间的自定义 Service，TimeService 类的主要代码如下：

```java
public class TimeService extends Service{
  Timer myTimer;TimerTask myTimerTask;
  @Override
  public void onCreate(){
    super.onCreate();
    myTimer = new Timer();
    myTimerTask = new TimerTask(){
      @Override
      public void run(){
        Intent myIntent = new Intent();
        SimpleDateFormat mySimpleDateFormat =
            new SimpleDateFormat("yyyy-MM-dd HH:mm:ss");                   //格式化日期
        myIntent.putExtra("myTime",mySimpleDateFormat.format(new Date()));
        myIntent.setAction("TimeService");
        sendBroadcast(myIntent);                                            //通过广播传递指定数据
      } };
    myTimer.schedule(myTimerTask,0,1000);                                   //每隔1秒发送一次广播
```

```
    }
    @Override
    public IBinder onBind(Intent intent){return null;}
    @Override
    public void onDestroy(){                     //停止并销毁定时器及定时任务对象
      super.onDestroy();
      myTimerTask.cancel();
      myTimerTask = null;
      myTimer.cancel();
    } }
```

上面这段代码在 MyCode\MySample911\app\src\main\java\com\bin\luo\mysample\TimeService.java 文件中。一般情况下，当在项目中添加 Service 之后，通常应该在 AndroidManifest.xml 文件中进行注册，主要代码如下：

```
<service android:name=".TimeService"
         android:enabled="true"
         android:exported="true"></service>
```

上面这段代码在 MyCode\MySample911\app\src\main\AndroidManifest.xml 文件中。此实例的完整项目在 MyCode\MySample911 文件夹中。

219 使用 FileInputStream 和 FileOutputStream 读取和保存文本文件

此实例主要通过使用 FileInputStream 和 FileOutputStream，实现在应用中读取和保存文本文件。当实例运行之后，在"文件名称："输入框中输入文件名称，在"文件内容："输入框中输入内容，单击"保存文本文件"按钮，则将根据指定的文件名称保存文本内容到"data\data\应用包名\files"中。在"文件名称："输入框中输入文件名称，单击"读取文本文件"按钮，则将从"data\data\应用包名\files"中读取指定文件的文本内容到"文件内容："输入框中，如图 219.1 的左图（修改前）和右图（修改后）所示。

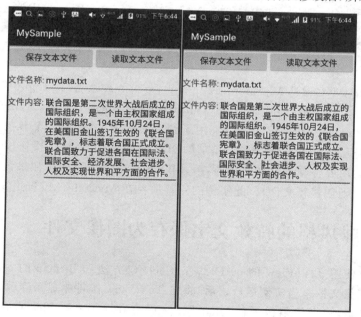

图 219.1

主要代码如下：

```java
public void onClickmyBtn1(View v) {                //响应单击"保存文本文件"按钮
  FileOutputStream myOutStream = null;
  BufferedWriter myBufferedWriter = null;
  try {
    myOutStream = openFileOutput(myEditName.getText().toString(),
                                  Context.MODE_PRIVATE);
    myBufferedWriter = new BufferedWriter(new OutputStreamWriter(myOutStream));
    myBufferedWriter.write(myEditText.getText().toString());
  } catch (IOException e) { e.printStackTrace();}
  finally {
    try {
      if (myBufferedWriter != null) { myBufferedWriter.close(); }
    } catch (IOException e) { e.printStackTrace(); }
  }
  Toast.makeText(getApplicationContext(),
                  "已经成功保存文件", Toast.LENGTH_SHORT).show();
}
public void onClickmyBtn2(View v) {                //响应单击"读取文本文件"按钮
  FileInputStream myInputStream = null;
  BufferedReader myBufferedReader = null;
  StringBuilder myText = new StringBuilder();
  try {
    myInputStream = openFileInput(myEditName.getText().toString());
    myBufferedReader = new BufferedReader(new InputStreamReader(myInputStream));
    String myLines = "";
    while ((myLines = myBufferedReader.readLine())!= null){
       myText.append(myLines);
    } } catch (IOException e) { e.printStackTrace();}
    finally {
      if (myBufferedReader != null) {
        try {
        myBufferedReader.close();
        } catch (IOException e) { e.printStackTrace(); }
    } }
    myEditText.setText(myText.toString());
}
```

上面这段代码在 MyCode \ MySample420 \ app \ src \ main \ java \ com \ bin \ luo \ mysample \ MainActivity.java 文件中。在这段代码中，myOutStream = openFileOutput(myEditName. getText().toString(), Context.MODE_PRIVATE)用于获取文件输出流，第一个参数是文件名称，第二个参数指定文件的操作模式；文件的操作模式有：MODE_PRIVATE（默认操作模式，表示当指定同名文件的时候，所写入的内容将覆盖原文件的内容）、MODE_APPEND（表示如果该文件已存在，就往文件里追加内容，不存在则创建新文件）。此实例的完整项目在 MyCode\MySample420 文件夹中。

220 将浮雕风格的特效文字保存为图像文件

此实例主要通过使用 TextView 的 getDrawingCache()方法和 Bitmap 的 compress()方法，实现将特效文字保存为图像文件。当实例运行之后，将在 TextView 控件中显示浮雕文字"炫酷"，单击"将文字保存为图像"按钮，将把浮雕效果的"炫酷"二字保存在 SD 卡的 Pictures 目录中，效果分别如

图 220.1 的左图和右图所示。

图 220.1

主要代码如下：

```
void OnClickButton1(View view) {                    //响应单击"将文字保存为图像"按钮
  myTextView.setDrawingCacheEnabled(true);
  myTextView.buildDrawingCache();
  Bitmap myBitmap = myTextView.getDrawingCache();
  try {
   Uri myImageUri = getContentResolver().insert(
          MediaStore.Images.Media.EXTERNAL_CONTENT_URI, new ContentValues());
   OutputStream myOutStream = getContentResolver().openOutputStream(myImageUri);
   myBitmap.compress(Bitmap.CompressFormat.JPEG, 90, myOutStream);
   Toast.makeText(MainActivity.this,
              "将文字保存为图像操作成功!", Toast.LENGTH_SHORT).show();
  } catch (Exception e) { e.printStackTrace(); }
}
```

上面这段代码在 MyCode\MySample682\app\src\main\java\com\bin\luo\mysample\MainActivity.java 文件中。在这段代码中，myBitmap = myTextView.getDrawingCache()用于将 TextView 控件绘制的内容（浮雕文字）转换成 Bitmap。myBitmap.compress（Bitmap.CompressFormat.JPEG，90，myOutStream)用于将 Bitmap 的内容保存为图像文件。此外，访问 SD 卡需要在 AndroidManifest.xml 文件中添加< uses-permission android:name= "android.permission.WRITE_EXTERNAL_STORAGE"/>权限。此实例的完整项目在 MyCode\MySample682 文件夹中。

221 在 SD 卡上将 Bitmap 保存为 PNG 图像文件

此实例主要通过使用 Bitmap 类的 compress()方法，实现以文件流方式将 Bitmap 图像在 SD 卡上保存为 PNG 格式的图像文件。当实例运行之后，在"图像文件："输入框中输入图像文件的保存路径,如图221.1的左图所示,单击"保存图像"按钮,将把下面显示的图像在SD卡上保存为PNG格式的图像文件,如图221.1的右图所示。

图 221.1

主要代码如下:

```
public void onClickmyBtn1(View v) {                    //响应单击"保存图像"按钮
  try {
    FileOutputStream myOutStream =
            new FileOutputStream(myEditText.getText().toString());
    Bitmap myBitmap = ((BitmapDrawable)myImageView.getDrawable()).getBitmap();
    myBitmap.compress(Bitmap.CompressFormat.PNG, 100, myOutStream);
    myOutStream.flush();
    myOutStream.close();
    Toast.makeText(this,"保存成功!",Toast.LENGTH_SHORT).show();
  } catch (Exception e) {
    Toast.makeText(this,e.getMessage(),Toast.LENGTH_SHORT).show();
} }
```

上面这段代码在 MyCode \ MySample222 \ app \ src \ main \ java \ com \ bin \ luo \ mysample \ MainActivity.java文件中。在这段代码中,myOutStream = new FileOutputStream(myEditText.getText().toString())用于根据指定的文件路径创建文件输出流。myBitmap=((BitmapDrawable)myImageView.getDrawable()).getBitmap()用于获取 ImageView 控件的图像。myBitmap.compress(Bitmap.CompressFormat.PNG,100,myOutStream)用于根据文件输出流和 Bitmap.CompressFormat.PNG 参数将图像保存为 PNG 格式文件。此外,在 SD 卡写入文件需要在AndroidManifest.xml文件中添加< uses-permission android:name="android.permission.WRITE_EXTERNAL_STORAGE"/>权限。此实例的完整项目在 MyCode\MySample222 文件夹中。

222 从手机相册中选择图像文件并裁剪头像

此实例主要通过使用 startActivityForResult() 方法调用相册和裁剪 Activity，实现从手机相册中选择图像文件并裁剪头像。当实例运行之后，单击"从图库中选择图像文件"按钮，则将显示相册 Activity，如图 222.1 的左图所示。在相册 Activity 中选择一个图像文件，则将显示裁剪 Activity；在裁剪 Activity 中移动网格线剪裁图像，如图 222.1 的右图所示，单击左上角的"保存"按钮，则在主 Activity 中显示裁剪的头像。

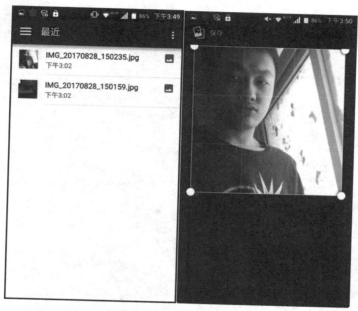

图 222.1

主要代码如下：

```java
public class MainActivity extends Activity {
    /* 请求识别码 */
    private static final int CODE_GALLERY_REQUEST = 0xa0;
    private static final int CODE_RESULT_REQUEST = 0xa2;
    //裁剪后图像的宽 X 和高 Y,480 * 480 的正方形.
    private static int output_X = 480;
    private static int output_Y = 480;
    private ImageView myImageView = null;
    @Override
    protected void onCreate(Bundle savedInstanceState) {
        super.onCreate(savedInstanceState);
        setContentView(R.layout.activity_main);
        myImageView = (ImageView) findViewById(R.id.myImageView);
    }
    public void onClickmyBtn1(View v) {                        //响应单击"从图库中选择图像文件"按钮
        Intent myIntent = new Intent();
        myIntent.setType("image/*");
        myIntent.setAction(Intent.ACTION_GET_CONTENT);
        startActivityForResult(myIntent, CODE_GALLERY_REQUEST);
```

```java
}
@Override
protected void onActivityResult(int requestCode, int resultCode,Intent intent) {
    if (resultCode == RESULT_CANCELED) {                //如果用户没有进行有效操作,则直接返回
        Toast.makeText(getApplication(), "取消", Toast.LENGTH_LONG).show();
        return;
    }
    switch (requestCode) {
        case CODE_GALLERY_REQUEST:
            cropRawPhoto(intent.getData());
            break;
        case CODE_RESULT_REQUEST:
            if (intent != null) { setImageToHeadView(intent); }
            break;
    }
    super.onActivityResult(requestCode, resultCode, intent);
}
public void cropRawPhoto(Uri uri) {                     //裁剪原始图像
    Intent intent = new Intent("com.android.camera.action.CROP");
    intent.setDataAndType(uri, "image/*");
    intent.putExtra("crop", "true");
    intent.putExtra("aspectX", 1);
    intent.putExtra("aspectY", 1);
    intent.putExtra("outputX", output_X);               //裁剪图像的宽和高
    intent.putExtra("outputY", output_Y);
    intent.putExtra("return-data", true);
    startActivityForResult(intent, CODE_RESULT_REQUEST);
}
private void setImageToHeadView(Intent intent){         //保存裁剪图像,并设置头像
    Bundle extras = intent.getExtras();
    if (extras != null) {
        Bitmap myBmp = extras.getParcelable("data");
        myImageView.setImageBitmap(myBmp);
    } } }
```

上面这段代码在 MyCode\MySample186\app\src\main\java\com\bin\luo\mysample\MainActivity.java 文件中。在这段代码中,startActivityForResult(myIntent, CODE_GALLERY_REQUEST)用于根据 myIntent 请求码访问相册,并在 onActivityResult(int requestCode, int resultCode,Intent intent)方法中处理访问结果。由于每个 Activity 都可以启动任意的子 Activity 并等待结果,而结果处理方法只有一个 onActivityResult(int requestCode, int resultCode, Intent intent),因此为了区别请求的 Activity 是谁,Android 将每个请求设定为一个大于等于 0 的值,这就是 requestCode。由此,在 onActivityResult()方法中即可利用 requestCode 区别不同的 Activity 并返回结果。在此实例中,requestCode 主要有 CODE_GALLERY_REQUEST 和 CODE_RESULT_REQUEST,它表明 onActivityResult()至少执行了两次,一次是获取选择的图像文件,第二次是获取裁剪之后的头像。此实例的完整项目在 MyCode\MySample186 文件夹中。

223 在 ListView 上加载手机外存的图像文件

此实例主要实现了在 ListView 上加载手机外部存储介质上的所有图像文件。当实例运行之后,将在 ListView 控件上显示手机外部存储上的所有图像文件的名称和大小,如图 223.1 的左图所示;单击在 ListView 控件上的任意图像文件名称(列表项),则在弹出的对话框中显示该图像,如图 223.1

的右图所示。

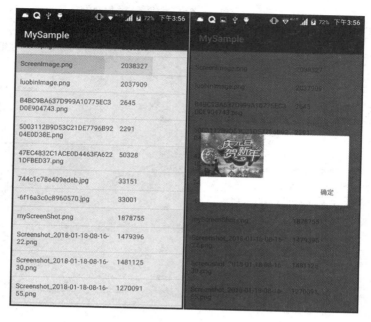

图 223.1

主要代码如下：

```
public class MainActivity extends AppCompatActivity {
  private ListView myListView;
  @Override
  protected void onCreate(@Nullable Bundle savedInstanceState) {
    super.onCreate(savedInstanceState);
    setContentView(R.layout.activity_main);
    myListView = (ListView) findViewById(R.id.myListView);
    GetImageFiles();
  }
  ArrayList myFiles = new ArrayList();
  ArrayList mySizes = new ArrayList();
  ArrayList<String> myPathData = new ArrayList();
  ArrayList<HashMap<String, Object>> myItems = new ArrayList<>();
  private void GetImageFiles() {
    myFiles.clear();
    mySizes.clear();
    //通过 ContentResolver 查询外部存储介质上的所有图像文件信息
    Cursor myCursor = getContentResolver().query(MediaStore.Images.Media.EXTERNAL_CONTENT_URI, null, null, null, null);
    while (myCursor.moveToNext()) {
      String myName = myCursor.getString(myCursor.getColumnIndex(
MediaStore.Images.Media.DISPLAY_NAME));              //显示名称
      String mySize = myCursor.getString(myCursor.getColumnIndex(
MediaStore.Images.Media.SIZE));                       //显示文件大小
      byte[] data = myCursor.getBlob(myCursor.getColumnIndex(
MediaStore.Images.Media.DATA));                       //获取图像的位置数据
      myFiles.add(myName);
      mySizes.add(mySize);
```

```
    String myPath = new String(data, 0,
            data.length - 1);                    //将位置数据转换成字符串形式的路径
    myPathData.add(myPath);
}
for (int i = 0; i < myFiles.size(); i++) {
    HashMap<String, Object> myListItem = new HashMap<>();
    myListItem.put("name", myFiles.get(i));
    myListItem.put("size", mySizes.get(i));
    myItems.add(myListItem);
}
SimpleAdapter myAdapter = new SimpleAdapter(MainActivity.this, myItems,
        R.layout.myitem, new String[]{"name", "size"},
        new int[]{R.id.myName, R.id.mySize});
myListView.setAdapter(myAdapter);
//单击图像文件名称(列表项)显示图像
myListView.setOnItemClickListener(new AdapterView.OnItemClickListener() {
    @Override
    public void onItemClick(AdapterView<?> parent, View view, int position, long id) {
        View myDialog = getLayoutInflater().inflate(R.layout.mydlg, null);
        ImageView myImageView = (ImageView) myDialog.findViewById(R.id.myImageView);
        //获取对应列表项的指定图像
        Bitmap myBitmap = BitmapFactory.decodeFile(myPathData.get(position));
        myImageView.setImageBitmap(myBitmap);
        new AlertDialog.Builder(MainActivity.this).setView(myDialog)
                .setPositiveButton("确定", null).show();
    }
}); } }
```

上面这段代码在 MyCode\MySample668\app\src\main\java\com\bin\luo\mysample\MainActivity.java 文件中。在这段代码中,SimpleAdapter myAdapter = new SimpleAdapter(MainActivity.this, myItems, R.layout.myitem, new String[]{"name", "size"}, new int[] {R.id.myName, R.id.mySize}) 表示根据 myitem 布局的样式创建 ListView 的列表项,关于 myitem 布局的详细内容请参考源代码中的 MyCode\MySample668\app\src\main\res\layout\myitem.xml 文件。myDialog = getLayoutInflater().inflate (R.layout.mydlg, null) 表示加载 mydlg 布局作为显示图像的对话框。关于 mydlg 布局的详细内容请参考源代码中的 MyCode\MySample668\app\src\main\res\layout\mydlg.xml 文件。myCursor = getContentResolver().query(MediaStore.Images.Media.EXTERNAL_CONTENT_URI, null, null, null, null) 表示查找外部存储介质上的所有图像文件。此外,当访问外部存储介质时,需要在 AndroidManifest.xml 文件中添加< uses-permission android:name="android.permission.READ_EXTERNAL_STORAGE"/>权限。需要说明的是,使用此实例的相关类需要在 gradle 中引入 compile 'com.android.support:design:25.0.1'依赖项。此实例的完整项目在 MyCode\MySample668 文件夹中。

224 使用 DownloadManager 下载网络文件

此实例主要实现了使用 Android 内置的 DownloadManager 下载网络文件。当实例运行之后,在 "下载网址:" 输入框中输入网络地址,如 "http://gdown.baidu.com/data/wisegame/0904344dee4a2d92/QQ_718.apk",如图 224.1 的左图所示;然后单击"下载"按钮,则将执行下载操作,下载进度将显示在通知栏上,如图 224.1 的右图所示。

图 224.1

主要代码如下：

```
public void onClickButton1(View v){                      //响应单击"下载"按钮
    DownloadManager myDownloadManager =
        (DownloadManager)getSystemService(DOWNLOAD_SERVICE);  //获取下载管理器
    String myUrl = myEditText.getText().toString();
    DownloadManager.Request myRequest =
        new DownloadManager.Request(Uri.parse(myUrl));        //向Url发送下载请求
    myRequest.setDestinationInExternalPublicDir("","QQ.apk"); //设置下载保存位置
    myRequest.setNotificationVisibility(DownloadManager.Request.
        VISIBILITY_VISIBLE_NOTIFY_COMPLETED);                 //在通知栏显示下载进度
    myDownloadManager.enqueue(myRequest);                     //将请求加入下载队列
}
```

上面这段代码在 MyCode\MySample892\app\src\main\java\com\bin\luo\mysample\ MainActivity.java 文件中。在这段代码中，myRequest.setNotificationVisibility（DownloadManager.Request.VISIBILITY_VISIBLE_NOTIFY_COMPLETED)用于在通知栏上显示下载进度,此参数可以使得下载进度在下载过程中和下载完成后均可见。如果是 myRequest.setNotificationVisibility(DownloadManager.Request.VISIBILITY_VISIBLE),则在下载过程中在通知栏上显示下载进度,下载完成之后自动消失。DownloadManager 是专用于处理耗时较长的 HTTP 文件下载的系统服务,它在后台进行下载,并自动处理网络连接变化,失败重试。DownloadManager 可以在自己的应用中提交下载请求,在通知栏中自动显示下载进度,可以指定下载文件的保存位置,并实时获取下载进度,监听下载结果。DownloadManager 中有两个重要的内部类。

（1）DownloadManager.Request,该类封装一个下载请求添加到系统下载器队列中。

（2）DownloadManager.Query,该类查询下载任务,可实时获取下载进度,下载结果。

此外,访问网络和读写 SD 卡上的文件需要在 AndroidManifest.xml 文件中添加< uses-permission android: name = " android. permission. INTERNET "/> 和 < uses-permission android: name="android.permission.WRITE_EXTERNAL_STORAGE"/>权限。此实例的完整项目在

MyCode\MySample892 文件夹中。

225　使用 RandomAccessFile 实现断点续传下载

此实例主要通过使用 RandomAccessFile 的 seek()、write() 等方法，实现在存储卡上写入下载文件，并在下载网络文件的过程中支持断点续传功能。当实例运行之后，在"下载地址："输入框中输入网络地址，如"http://gdown.baidu.com/data/wisegame/ 0904344dee4a2d92/QQ_718.apk"，然后单击"开始下载"按钮，执行下载操作，进度条同时显示下载进度；单击"暂停下载"按钮，则暂停下载操作，同时暂停进度条下载进度，如图 225.1 的左图所示；再次单击"开始下载"按钮，则在上次暂停的位置继续执行下载操作，直到下载完成；由于实例下载的是应用安装包，因此在下载完成之后将自动执行安装操作，如图 225.1 的右图所示。

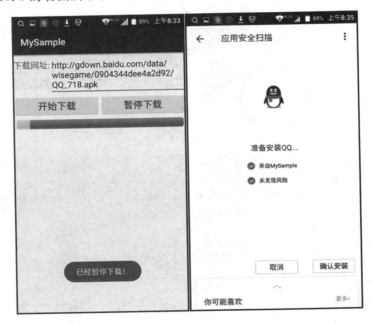

图　225.1

主要代码如下：

```
public class MainActivity extends Activity {
 ProgressBar myProgressBar;
 EditText myEditText;
 int myPausedPosition = 0;
 boolean isPaused = false;
 Handler myHandler = new Handler() {
  @Override
  public void handleMessage(Message msg) {
   if (msg.what == 0) {
    int myFileSize = msg.arg1;
    downloadFile(myFileSize);                    //开始异步下载文件
   } else if (msg.what == 1) {
    myProgressBar.setProgress(msg.arg1);         //实时显示当前下载进度
   } else if (msg.what == 2) {
    Toast.makeText(MainActivity.this,
```

```java
                            "下载完成!即将开始安装...", Toast.LENGTH_SHORT).show();
    Intent myAPKIntent = new Intent(Intent.ACTION_VIEW);
    myAPKIntent.setDataAndType(Uri.fromFile(new File(
            Environment.getExternalStorageDirectory().getAbsolutePath()
            + "/QQ.apk")), "application/vnd.android.package-archive");
    startActivity(myAPKIntent);                          //下载完成后跳转至APK安装界面
  } else if (msg.what == 3) {
    Toast.makeText(MainActivity.this, "开始下载!", Toast.LENGTH_SHORT).show();
  } } };
@Override
protected void onCreate(Bundle savedInstanceState) {
  super.onCreate(savedInstanceState);
  setContentView(R.layout.activity_main);
  myProgressBar = (ProgressBar) findViewById(R.id.myProgressBar);
  myEditText = (EditText) findViewById(R.id.myEditText);
}
public void onClickButton1(View v) {                     //响应单击"开始下载"按钮
  if (isPaused) isPaused = false;
  calcFileSize();
}
public void onClickButton2(View v) {                     //响应单击"暂停下载"按钮
  isPaused = true;
  Toast.makeText(this, "已经暂停下载!", Toast.LENGTH_SHORT).show();
}
public void calcFileSize() {                             //计算下载文件大小
  new Thread(new Runnable() {
    @Override
    public void run() {
      try {
        URL myUrl = new URL(myEditText.getText().toString());
        HttpURLConnection myConnection = (HttpURLConnection) myUrl.openConnection();
        myConnection.setRequestMethod("GET");
        myConnection.setConnectTimeout(5000);
        if (myConnection.getResponseCode() == 200) {
          Message myFileSizeMessage = new Message();
          myFileSizeMessage.what = 0;
          myFileSizeMessage.arg1 = myConnection.getContentLength();
          //获取文件大小,并发送至Handler
          myHandler.sendMessage(myFileSizeMessage);
        } } catch (Exception e) {
          Log.e("mytag", e.toString());
      } } }).start();
}
public void downloadFile(final int fileSize) {
  new Thread(new Runnable() {
    @Override
    public void run() {
      try {
        URL myUrl = new URL(myEditText.getText().toString());
        HttpURLConnection myConnection =
                          (HttpURLConnection) myUrl.openConnection();
        myConnection.setRequestMethod("GET");
        myConnection.setConnectTimeout(5000);
```

```
            myConnection.setRequestProperty("Range", "bytes = "
                    + myPausedPosition + " - " + fileSize);    //指定下载的区间范围
            if (myConnection.getResponseCode() == 206) {
                myHandler.sendEmptyMessage(3);
                InputStream myInputStream = myConnection.getInputStream();
                File myFile = new File(Environment.getExternalStorageDirectory().
                        getAbsolutePath() + "/QQ.apk");        //设置下载之后的保存位置
                RandomAccessFile myRandomAccessFile = new RandomAccessFile(myFile, "rwd");
                myRandomAccessFile.seek(myPausedPosition);     //设置文件写入点
                byte[] myBuffer = new byte[1024];
                int myCount = 0, myLength;
                while ((myLength = myInputStream.read(myBuffer)) != -1) {
                    if (!isPaused) {                           //若未处于暂停状态,则正常进行下载操作
                        myRandomAccessFile.write(myBuffer, 0, myLength);
                        myCount += myLength;
                    } else {                                   //若处于暂停状态,则停止文件写入操作
                        myPausedPosition = myCount;
                        myRandomAccessFile.close();
                    }
                    Message myProgressMessage = new Message();
                    myProgressMessage.what = 1;
                    myProgressMessage.arg1 = (int) ((((isPaused ? 0 : myPausedPosition) +
                            myCount) / (float) fileSize) * 100);
                    //向 Handler 发送消息并更新进度
                    myHandler.sendMessage(myProgressMessage);
                }
                myRandomAccessFile.close();
                if (myProgressBar.getProgress() == 100) myHandler.sendEmptyMessage(2);
            } } catch (Exception e) {
                Log.e("mytag", e.toString());
            } } }).start();
        }
    }
```

上面这段代码在 MyCode\MySample881\app\src\main\java\com\bin\luo\mysample\MainActivity.java 文件中。在这段代码中,myRandomAccessFile.seek(myPausedPosition)用于根据指定参数设置下载文件的写入位置。myRandomAccessFile.write(myBuffer,0,myLength)表示以 0 为起始位置,从 myBuffer 写入 myLength 的数据。RandomAccessFile 是用来支持随机存取文件的类,随机存取文件的行为就像存取在文件系统里一个很大的字节数组一样。为了方便存取文件,此字节数组提供了"文件指针"概念:类似于游标和下标,在此实例中,则使用该指针记录暂停的下载位置,控制下载文件的读写,从而实现断点续传的功能。此外,访问网络和读写 SD 卡需要在 AndroidManifest.xml 文件中添加< uses-permission android:name = " android.permission. INTERNET"/>和< uses-permission android:name = " android.permission.WRITE_EXTERNAL_STORAGE"/>权限。此实例的完整项目在 MyCode\MySample881 文件夹中。

226 使用 HttpURLConnection 下载图像文件

此实例主要通过使用 BitmapFactory 的 decodeStream()方法解析从指定网络地址获取的图像数据流,实现在 ImageView 控件中显示指定网址的图像。当实例运行之后,在"图像地址:"输入框中输

入代表一个图像的网络地址,如"https://p1.ssl.qhmsg.com/t01992db65fb4747713.jpg",然后单击"显示图像"按钮,则在下面显示该网址代表的图像,效果分别如图 226.1 的左图和右图所示。

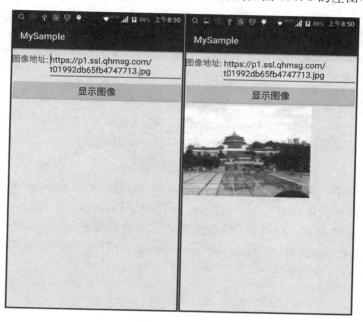

图 226.1

主要代码如下：

```
public class MainActivity extends Activity {
  EditText myEditText;
  ImageView myImageView;
  private Handler myHandler = new Handler() {              //在主线程中定义一个 Handler 对象
    public void handleMessage(Message msg) {
      Bitmap myBmp = (Bitmap) msg.obj;
      myImageView.setImageBitmap(myBmp);                  //把 Bitmap 显示到 ImageView 控件上
    }
  };
  @Override
  protected void onCreate(Bundle savedInstanceState) {
    super.onCreate(savedInstanceState);
    setContentView(R.layout.activity_main);
    myEditText = (EditText) findViewById(R.id.myEditText);
    myImageView = (ImageView) findViewById(R.id.myImageView);
  }
  //响应单击"显示图像"按钮
  public void onClickmyBtn1(View v) throws IOException {
    new Thread() {                                         //创建一个子线程
      public void run() {
        try {
          String myPath = myEditText.getText().toString().trim();
          URL myUrl = new URL(myPath);
          HttpURLConnection myHttpURLConnection =
                            (HttpURLConnection) myUrl.openConnection();
          myHttpURLConnection.setConnectTimeout(5000);
          int myCode = myHttpURLConnection.getResponseCode();   //获取服务器返回状态码
          if (myCode == 200) {
```

```
            //获取图像数据,不管是什么数据都是以流的形式返回
            InputStream myStream = myHttpURLConnection.getInputStream();
            Bitmap myBmp = BitmapFactory.decodeStream(myStream);
            Message myMsg = Message.obtain();
            myMsg.obj = myBmp;                          //发消息把 Bitmap 显示到 ImageView 控件
            myHandler.sendMessage(myMsg);
        } } catch (Exception e) {
        Toast.makeText(getApplicationContext(),
                    e.getMessage().toString(), Toast.LENGTH_SHORT).show();
        } } }.start();
    } }
```

上面这段代码在 MyCode \ MySample353 \ app \ src \ main \ java \ com \ bin \ luo \ mysample \ MainActivity.java 文件中。在这段代码中,myStream = myHttpURLConnection.getInputStream()用于根据指定的网址获取图像的数据流,myBmp = BitmapFactory.decodeStream(myStream)用于将图像数据流解析为位图格式。此外,访问指定网址的图像需要在 AndroidManifest.xml 文件中添加< uses-permission android:name="android.permission.INTERNET"/>权限。此实例的完整项目在 MyCode\MySample353 文件夹中。

系统和设备

227　使用 QuickContactBadge 访问联系人

此实例主要通过使用 QuickContactBadge 控件的 assignContactFromPhone()方法,实现快速联络联系人。当实例运行之后,单击电话图标,如图 227.1 的左图所示,如果电话号码指定的联系人在通讯录中已经存在,则直接显示联系人信息;否则弹出新增电话号码窗口,如图 227.1 的右图所示。

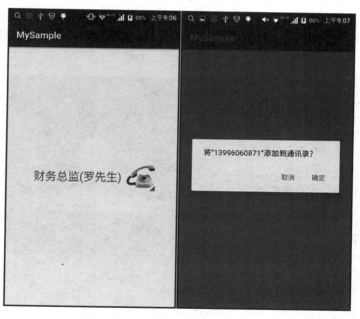

图　227.1

主要代码如下:

```
public class MainActivity extends Activity {
    @Override
    protected void onCreate(Bundle savedInstanceState) {
        super.onCreate(savedInstanceState);
        setContentView(R.layout.activity_main);
        QuickContactBadge myQuickContactBadge =
            (QuickContactBadge) findViewById(R.id.myQuickContactBadge);
```

```
        myQuickContactBadge.assignContactFromPhone("13996060871",
                              false); //为联系人指定一个电话号码
}}
```

上面这段代码在 MyCode\MySample067\app\src\main\java\com\bin\luo\mysample\MainActivity.java 文件中。在这段代码中,assignContactFromPhone(String phoneNumber, boolean lazyLookup)用于为联系人指定电话号码,其中,参数 phoneNumber 表示联系人的电话号码;参数 lazyLookup 如果为 true,将不会立即查找这个电话号码,直到 View 被点击;否则会在通讯录中立即查找这个电话号码。此外,还可以使用 QuickContactBadge 的 assignContactFromEmail(String emailAddress, boolean lazyLookup)方法指定联系人的电子邮箱地址,该方法的 emailAddress 参数表示联系人的电子邮箱地址,该方法的 lazyLookup 参数如果为 true,将不会立即查找这个邮箱地址,直到 View 被单击;否则会立即查找这个邮箱地址。此实例的完整项目在 MyCode\MySample067 文件夹中。

228 使用 ContentProviderOperation 增加联系人

此实例主要通过使用 ContentProviderOperation.newInsert()方法,实现在当前手机的通讯录中增加联系人。当实例运行之后,在 3 个输入框中分别输入"联系人姓名""联系人电话号码"和"联系人电子邮箱"等信息,然后单击"在当前手机的通讯录中增加联系人"按钮,如果增加成功,则在弹出的 Toast 中显示增加成功的信息,如图 228.1 的左图所示。此时也可以在手机的通讯录中查看刚才增加的联系人信息,如图 228.1 的右图所示。

图 228.1

主要代码如下:

```
//响应单击"在当前手机的通讯录中增加联系人"按钮
public void onClickButton1(View v) {
```

```java
    try {
      Contact myContact = new Contact();
      myContact.setName(myEditName.getText().toString());
      myContact.setEmail(myEditEmail.getText().toString());
      myContact.setNumber(myEditPhone.getText().toString());
      addContact(myContact);
    } catch (Exception e) {
      Toast.makeText(MainActivity.this,
                     e.getMessage().toString(), Toast.LENGTH_SHORT).show();
    } }
  public String getContactID(String myName) {                    //根据联系人姓名查询 ID
    String myID = "0";
    Cursor myCursor = getContentResolver().query(
          android.provider.ContactsContract.Contacts.CONTENT_URI,
          new String[]{android.provider.ContactsContract.Contacts._ID},
          android.provider.ContactsContract.Contacts.DISPLAY_NAME
                 + " = '" + myName + "'", null, null);
    if(myCursor.moveToNext()) {
      myID = myCursor.getString(myCursor.getColumnIndex(
            android.provider.ContactsContract.Contacts._ID));
    }
    return myID;
  }
  public void addContact(Contact myContact) {                    //新增联系人
    ArrayList<ContentProviderOperation> myData =
                       new ArrayList<ContentProviderOperation>();
    String myID = getContactID(myContact.getName());
    if(!myID.equals("0")) {
      Toast.makeText(MainActivity.this,
                     "联系人已经存在", Toast.LENGTH_SHORT).show();
      return;
    } else if(myContact.getName().trim().equals("")){
      Toast.makeText(MainActivity.this,
                     "联系人名字不能为空", Toast.LENGTH_SHORT).show();
      return;
    } else {
      myData.add(ContentProviderOperation.newInsert(
            ContactsContract.RawContacts.CONTENT_URI)
        .withValue(ContactsContract.RawContacts.ACCOUNT_TYPE, null)
        .withValue(ContactsContract.RawContacts.ACCOUNT_NAME, null).build());
      myData.add(ContentProviderOperation.newInsert(
            ContactsContract.Data.CONTENT_URI)
        .withValueBackReference(COLUMN_RAW_CONTACT_ID, 0)
        .withValue(COLUMN_MIMETYPE, MIMETYPE_STRING_NAME)
        .withValue(COLUMN_NAME, myContact.getName()).build());       //增加姓名
      if(!myContact.getNumber().trim().equals("")) {
        myData.add(ContentProviderOperation.newInsert(
              ContactsContract.Data.CONTENT_URI)
          .withValueBackReference(COLUMN_RAW_CONTACT_ID, 0)
          .withValue(COLUMN_MIMETYPE, MIMETYPE_STRING_PHONE)
          .withValue(COLUMN_NUMBER, myContact.getNumber()).build());  //增加电话
      }
      if(!myContact.getEmail().trim().equals("")) {
        myData.add(ContentProviderOperation.newInsert(
              ContactsContract.Data.CONTENT_URI)
          .withValueBackReference(COLUMN_RAW_CONTACT_ID, 0)
          .withValue(COLUMN_MIMETYPE, MIMETYPE_STRING_EMAIL)
```

```
                .withValue(COLUMN_EMAIL, myContact.getEmail()).build());        //增加邮箱
        }
        try {
            ContentResolver myContentResolver = this.getContentResolver();
            myContentResolver.applyBatch(ContactsContract.AUTHORITY, myData);
            Toast.makeText(MainActivity.this,
                    "成功增加联系人" + myContact.name, Toast.LENGTH_SHORT).show();
        } catch (Exception e) {
            Toast.makeText(MainActivity.this,
                    e.getMessage().toString(), Toast.LENGTH_SHORT).show();
        } }
}
```

上面这段代码在 MyCode\MySample229\app\src\main\java\com\bin\luo\mysample\ MainActivity. java 文件中。在这段代码中,增加联系人的整个操作由 4 个 ContentProviderOperation. newInsert 操作组合而成,从第二个操作开始,每个操作都有一个 withValueBackReference(COLUMN_RAW_CONTACT_ID, 0)步骤,这是因为它参照了第一个操作新添加联系人的 ID,因为是批处理,插入数据前并不知道 ID 值,在进行批处理插入数据时,它会重新引用新 ID 值,因此不会影响最终的结果。此外,增加手机联系人需要在 AndroidManifest. xml 文件中添加< uses-permission android:name="android. permission. READ_CONTACTS"/>权限和< uses-permission android:name="android. permission. WRITE_CONTACTS"/>权限。此实例的完整项目在 MyCode\MySample229 文件夹中。

229　使用 ContentProviderOperation 修改联系人

此实例主要通过使用 ContentProviderOperation. newUpdate()方法,实现在当前手机的通讯录中根据联系人姓名修改电话号码和电子邮箱。当实例运行之后,在三个输入框中分别输入"联系人姓名""电话号码"和"电子邮箱"等信息,然后单击"修改信息"按钮,如果操作成功,则在弹出的 Toast 中显示操作结果,如图 229.1 的左图所示。此时也可以在手机的通讯录中查看刚才修改的联系人信息,如图 229.1 的右图所示。

图　229.1

主要代码如下：

```java
public void onClickmyBtn1(View v) {                    //响应单击"修改信息"按钮
    Contact oldContact = new Contact();
    oldContact.setName(myEditName.getText().toString());
    Contact newContact = new Contact(oldContact);
    newContact.setName(myEditName.getText().toString());
    newContact.setNumber(myEditPhone.getText().toString());
    newContact.setEmail(myEditEmail.getText().toString());
    updateContact(oldContact, newContact);
}
public String getContactID(String myName) {            //根据联系人姓名查询ID
    String myID = "0";
    Cursor myCursor = getContentResolver().query(
            android.provider.ContactsContract.Contacts.CONTENT_URI,
            new String[]{android.provider.ContactsContract.Contacts._ID},
            android.provider.ContactsContract.Contacts.DISPLAY_NAME +
            " = '" + myName + "'", null, null);
    if (myCursor.moveToNext()) {
        myID = myCursor.getString(myCursor.getColumnIndex(
                    android.provider.ContactsContract.Contacts._ID));
    }
    return myID;
}
//根据联系人姓名修改其电话号码和电子邮箱地址
public void updateContact(Contact oldContact, Contact newContact) {
    String myID = getContactID(oldContact.getName());
    if(myID.equals("0")) {
        Toast.makeText(MainActivity.this,
                oldContact.getName() + "不存在!", Toast.LENGTH_SHORT).show();
    } else if(newContact.getName().trim().equals("")){
        Toast.makeText(MainActivity.this,
                "联系人姓名不能是空白!", Toast.LENGTH_SHORT).show();
    } else {
        ArrayList<ContentProviderOperation> myData =
                                    new ArrayList<ContentProviderOperation>();
        //修改联系人姓名
        myData.add(ContentProviderOperation.newUpdate(
            ContactsContract.Data.CONTENT_URI).withSelection(COLUMN_CONTACT_ID
            + " = ? AND " + COLUMN_MIMETYPE + " = ?",new String[]{myID, MIMETYPE_STRING_NAME})
            .withValue(COLUMN_NAME, newContact.getName()).build());
        if(!newContact.getNumber().trim().equals("")) {
            //修改电话号码
            myData.add(ContentProviderOperation.newUpdate(
                ContactsContract.Data.CONTENT_URI).withSelection(COLUMN_CONTACT_ID +
                " = ? AND " + COLUMN_MIMETYPE + " = ?",
                new String[]{myID, MIMETYPE_STRING_PHONE})
                .withValue(COLUMN_NUMBER, newContact.getNumber()).build());
        }
        if(!newContact.getEmail().trim().equals("")) {
            //修改电子邮箱地址
            myData.add(ContentProviderOperation.newUpdate(
                ContactsContract.Data.CONTENT_URI).withSelection(COLUMN_CONTACT_ID
                + " = ? AND " + COLUMN_MIMETYPE + " = ?",new String[]{myID, MIMETYPE_STRING_EMAIL})
                .withValue(COLUMN_EMAIL, newContact.getEmail()).build());
        }
        try {
```

```
        getContentResolver().applyBatch(ContactsContract.AUTHORITY, myData);
        Toast.makeText(MainActivity.this,
                "修改联系人信息操作成功!", Toast.LENGTH_SHORT).show();
    } catch (Exception e) {
        Toast.makeText(MainActivity.this,
                e.getMessage().toString(), Toast.LENGTH_SHORT).show();
    }
  }
}
```

上面这段代码在 MyCode\MySample231\app\src\main\java\com\bin\luo\mysample\ MainActivity. java 文件中。在这段代码中,ContentProviderOperation. newUpdate (ContactsContract. Data. CONTENT_ URI). withSelection(COLUMN_CONTACT_ID + "=? AND " + COLUMN_MIMETYPE + "=?",new String[]{myID, MIMETYPE_STRING_PHONE}). withValue(COLUMN_NUMBER, newContact. getNumber()). build()用于根据联系人的 myID 修改电话号码,联系人的 myID 是联系人在通讯录中的唯一标识,此处则是通过 myID = getContactID(oldContact. getName())获取。此外,修改手机联系人信息需要在 AndroidManifest. xml 文件中添加< uses-permission android:name = "android. permission. READ_CONTACTS"/>权限和< uses-permission android:name = "android. permission. WRITE_CONTACTS"/>权限。此实例的完整项目在 MyCode\MySample231 文件夹中。

230 使用 ContentProviderOperation 删除联系人

此实例主要通过使用 ContentProviderOperation. newDelete()方法,实现在当前手机的通讯录中根据姓名删除此联系人。当实例运行之后,在"联系人姓名:"输入框中输入姓名,如"罗斌",然后单击"删除联系人"按钮,如果操作成功,则在弹出的 Toast 中显示操作结果,否则显示联系人不存在,效果分别如图 230.1 的左图和右图所示。此时也可以在手机的通讯录中查看刚才的操作结果。

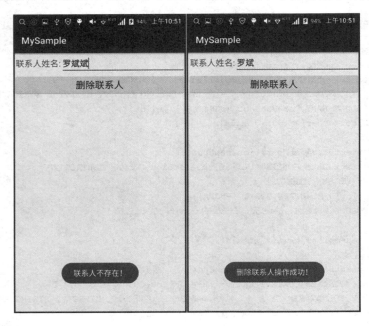

图 230.1

主要代码如下：

```java
public void onClickmyBtn1(View v) {                //响应单击"删除联系人"按钮
    Contact myContact = new Contact();
    myContact.setName(myEditName.getText().toString());
    deleteContact(myContact);
}
public String getContactID(String myName) {        //根据联系人姓名查询ID
    String myID = "0";
    Cursor myCursor = getContentResolver().query(
            android.provider.ContactsContract.Contacts.CONTENT_URI,
            new String[]{android.provider.ContactsContract.Contacts._ID},
            android.provider.ContactsContract.Contacts.DISPLAY_NAME
                    + " = '" + myName + "'", null, null);
    if (myCursor.moveToNext()) {
        myID = myCursor.getString(myCursor.getColumnIndex(
                android.provider.ContactsContract.Contacts._ID));
    }
    return myID;
}
public void deleteContact(Contact myContact) {     //根据姓名删除该联系人
    ArrayList<ContentProviderOperation> myData =
                            new ArrayList<ContentProviderOperation>();
    String myID = getContactID(myContact.getName());
    if(myID == "0"){
        Toast.makeText(MainActivity.this,
                "联系人不存在!", Toast.LENGTH_SHORT).show();
        return;
    }
    myData.add(ContentProviderOperation.newDelete(
        ContactsContract.RawContacts.CONTENT_URI)
        .withSelection(ContactsContract.RawContacts.CONTACT_ID
        + " = " + myID, null).build());
    try {
        getContentResolver().applyBatch(ContactsContract.AUTHORITY, myData);
        Toast.makeText(MainActivity.this,
                "删除联系人操作成功!", Toast.LENGTH_SHORT).show();
    } catch (Exception e) {
        Toast.makeText(MainActivity.this,
                e.getMessage().toString(), Toast.LENGTH_SHORT).show();
    }
}
```

上面这段代码在MyCode\MySample232\app\src\main\java\com\bin\luo\mysample\ MainActivity.java文件中。在这段代码中，ContentProviderOperation.newDelete（ContactsContract.RawContacts.CONTENT_URI).withSelection(ContactsContract.RawContacts.CONTACT_ID+"="+myID, null).build()用于根据联系人的myID删除联系人，联系人的myID是联系人在通讯录中的唯一标识，此处则是通过myID = getContactID（myContact.getName()）获取。此外，删除手机联系人需要在AndroidManifest.xml文件中添加<uses-permission android:name="android.permission.READ_CONTACTS"/>权限和<uses-permission android:name=" android.permission.WRITE_CONTACTS"/>权限。此实例的完整项目在MyCode\MySample232文件夹中。

231 使用 ContentResolver 检测飞行模式的状态

此实例主要通过在 Settings.System.getInt()方法的参数中使用 ContentResolver 参数和 Settings.System.AIRPLANE_MODE_ON 参数,实现检测当前手机的飞行模式设置情况。当实例运行之后,单击"检测飞行模式的使用状态"按钮,如果在手机设置中启用飞行模式,如图 231.1 的左图所示,则在弹出的 Toast 中提示"飞行模式已启用!",如图 231.1 的右图所示,否则在弹出的 Toast 中提示"飞行模式已禁用!"。

图 231.1

主要代码如下:

```java
public void onClickBtn1(View v) {                    //响应单击"检测飞行模式的使用状态"按钮
    ContentResolver myContentResolver = getContentResolver();
    //通过 ContentResolver 获取飞行模式启用状态标志
    int myFlag = Settings.System.getInt(myContentResolver,
                        Settings.System.AIRPLANE_MODE_ON, 0);

    //通过标志判断飞行模式是否启用
    if (myFlag == 1) Toast.makeText(MainActivity.this,
                "飞行模式已启用!", Toast.LENGTH_SHORT).show();
    else if (myFlag == 0)
      Toast.makeText(MainActivity.this,
                "飞行模式已禁用!", Toast.LENGTH_SHORT).show();
}
```

上面这段代码在 MyCode\MySample896\app\src\main\java\com\bin\luo\mysample\MainActivity.java 文件中。此实例的完整项目在 MyCode\MySample896 文件夹中。

232 使用 ContentResolver 检测手机的时间格式

此实例主要通过在 android.provider.Settings.System.getString()方法的参数中使用 ContentResolver 参数和 android.provider.Settings.System.TIME_12_24 参数,实现检测当前手机的时间格式的设置情况。当实例运行后,单击"检测当前手机的时间格式"按钮,如果在手机设置中设置的时间格式是 24 小时格式,如图 232.1 的左图所示,则在弹出的 Toast 中提示"当前手机的时间格式是:24",如图 232.1 的右图所示,否则在弹出的 Toast 中提示"当前手机的时间格式是:12"。

图 232.1

主要代码如下:

```
public void onClickBtn1(View v) {                    //响应单击"检测当前手机的时间格式"按钮
    ContentResolver myContentResolver = getContentResolver();
    //通过 ContentResolver 获取当前手机的时间格式
    String myFormat = android.provider.Settings.System.getString(
            myContentResolver, android.provider.Settings.System.TIME_12_24);
    Toast.makeText(MainActivity.this,
            "当前手机的时间格式是:" + myFormat, Toast.LENGTH_SHORT).show();
}
```

上面这段代码在 MyCode\MySample902\app\src\main\java\com\bin\luo\mysample\MainActivity.java 文件中。此实例的完整项目在 MyCode\MySample902 文件夹中。

233 使用 ContentResolver 获取所有短信

此实例主要通过使用 getContentResolver()的 query()方法,实现查询当前手机的所有短信内容。当实例运行之后,单击"获取当前手机的所有短信内容"按钮,则在下面的列表中显示当前手机中的所有短信内容,效果分别如图 233.1 的左图和右图所示。

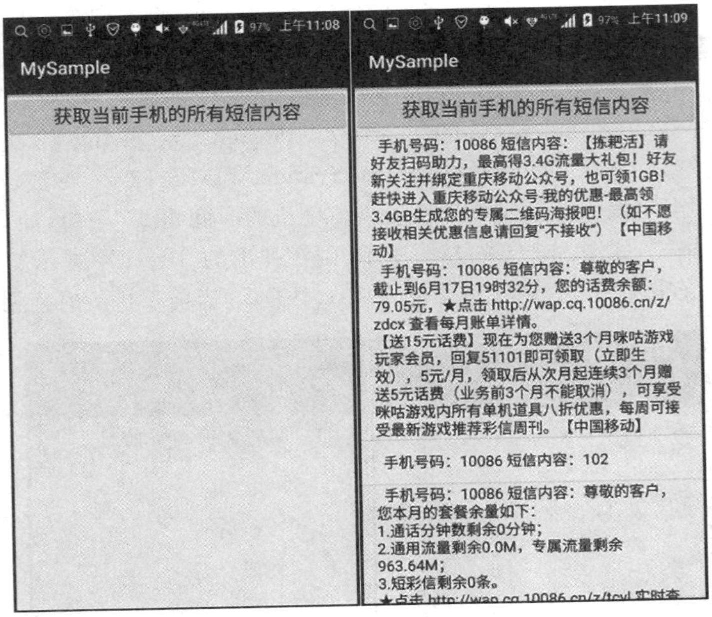

图 233.1

主要代码如下:

```java
public void onClickmyBtn1(View v) {                    //响应单击"获取当前手机的所有短信内容"按钮
    try {
        ArrayList myArray = new ArrayList<>();
        ContentResolver myContentResolver = getContentResolver();
        Cursor myCursor = myContentResolver.query(Uri.parse("content://sms/"),
                new String[]{"address", "body"}, null, null, "date desc");
        if (myCursor != null) {
            while (myCursor.moveToNext()) {
                String myPhone = myCursor.getString(myCursor.getColumnIndex("address"));
                String myBody = myCursor.getString(myCursor.getColumnIndex("body"));
                String myInfo = " 手机号码:" + myPhone + " 短信内容:" + myBody;
                myArray.add(myInfo);
            }
            myCursor.close();
        }
        myListView.setAdapter(new ArrayAdapter(this,
                android.R.layout.simple_list_item_1, myArray));
    } catch (Exception e) {
        Toast.makeText(MainActivity.this,
                e.getMessage().toString(), Toast.LENGTH_SHORT).show();
    }
}
```

上面这段代码在 MyCode \ MySample228 \ app \ src \ main \ java \ com \ bin \ luo \ mysample \ MainActivity.java 文件中。在这段代码中,myContentResolver().query(Uri.parse("content://sms/"), new String[]{"address", "body"}, null, null, "date desc")用于查询当前手机的短信信息,参数 Uri.parse("content://sms/")表示从内容提供者中查询手机的短信;参数 new String[]{"address", "body"}表示在查询结果中返回短信的内容和电话号码;第三个参数名为 selection,表示设置查询条件,相当于 SQL 语句中的 where,null 表示不进行筛选;第四个参数名为 selectionArgs,这

个参数需要配合第三个参数使用,如果在第三个参数里面有问号,那么在 selectionArgs 写入的数据就会替换掉"?",如果第三个参数为 null,则此参数也应该为 null;第五个参数名为 sortOrder,表示按照什么要求进行排序。此外,查询手机短信需要在 AndroidManifest.xml 文件中添加< uses-permission android:name = " android. permission. READ_SMS"/> 权限。此实例的完整项目在 MyCode\MySample228 文件夹中。

234 使用 ContentResolver 获取通话记录

此实例主要通过在 getContentResolver()的 query()方法中使用 CallLog. Calls. CONTENT_URI 参数,实现查询当前手机所有的通话记录。当实例运行之后,将在 ListView 控件中显示当前手机所有的通话记录,效果分别如图 234.1 的左图和右图所示。

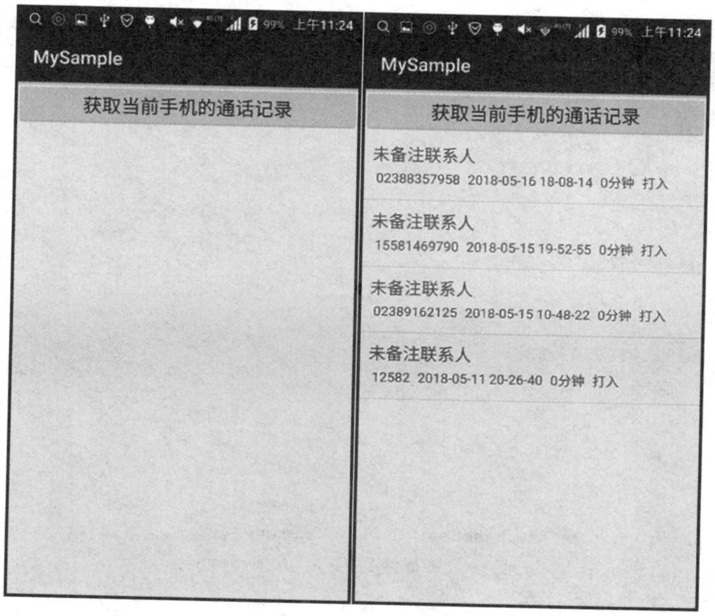

图 234.1

主要代码如下:

```
public void onClickButton1(View v) {                    //响应单击"获取当前手机的通话记录"按钮
  List<Map<String, String>> myData = getDataList();
  SimpleAdapter mySimpleAdapter = new SimpleAdapter(this, myData,
      R.layout.myitem, new String[]{"name", "number", "date",
      "duration", "type"}, new int[]{R.id.myName, R.id.myNumber,
      R.id.myDate, R.id.myDuration, R.id.myType});
  myListView.setAdapter(mySimpleAdapter);
}
private List<Map<String, String>> getDataList() {
  try {
    ContentResolver myContentResolver = getContentResolver();
    myCursor = myContentResolver.query(
        CallLog.Calls.CONTENT_URI,                      //通话记录的 URI
        new String[]{CallLog.Calls.CACHED_NAME,         //通话记录的联系人
        CallLog.Calls.NUMBER,                           //通话记录的电话号码
```

```java
            CallLog.Calls.DATE,                              //通话记录的日期
            CallLog.Calls.DURATION,                          //通话时长
            CallLog.Calls.TYPE},                             //通话类型
            null, null, CallLog.Calls.DEFAULT_SORT_ORDER);   //按照时间逆序排列
} catch (SecurityException e) { }
List<Map<String, String>> myArrayList = new ArrayList<Map<String, String>>();
while (myCursor.moveToNext()) {
  String myName =
        myCursor.getString(myCursor.getColumnIndex(CallLog.Calls.CACHED_NAME));
  String myNumber =
            myCursor.getString(myCursor.getColumnIndex(CallLog.Calls.NUMBER));
  long myDate = myCursor.getLong(myCursor.getColumnIndex(CallLog.Calls.DATE));
  String newDate =
        new SimpleDateFormat("yyyy-MM-dd HH-mm-ss").format(new Date(myDate));
  int myDuration =
            myCursor.getInt(myCursor.getColumnIndex(CallLog.Calls.DURATION));
  int myType = myCursor.getInt(myCursor.getColumnIndex(CallLog.Calls.TYPE));
  String myCatagory = "";
  switch (myType) {
    case CallLog.Calls.INCOMING_TYPE:
      myCatagory = "打入";
      break;
    case CallLog.Calls.OUTGOING_TYPE:
      myCatagory = "打出";
      break;
    case CallLog.Calls.MISSED_TYPE:
      myCatagory = "未接";
      break;
    default:
      break;
  }
  Map<String, String> map = new HashMap<String, String>();
  map.put("name", (myName == null) ? "未备注联系人" : myName);
  map.put("number", myNumber);
  map.put("date", newDate);
  map.put("duration", (myDuration / 60) + "分钟");
  map.put("type", myCatagory);
  myArrayList.add(map);
}
return myArrayList;
}
```

上面这段代码在 MyCode\MySample671\app\src\main\java\com\bin\luo\mysample\MainActivity.java 文件中。在这段代码中，mySimpleAdapter = new SimpleAdapter(this, myData, R.layout.myitem, new String[]{"name", "number", "date", "duration", "type"}, new int[]{R.id.myName, R.id.myNumber, R.id.myDate, R.id.myDuration, R.id.myType})用于在自定义适配器中根据 myitem 布局的样式设置列表项的对应字段。关于 myitem 布局的详细内容请参考源代码中的 MyCode\MySample671\app\src\main\res\layout\myitem.xml 文件。此外，查询通话记录需要在 AndroidManifest.xml 文件中添加<uses-permission android:name="android.permission.READ_CALL_LOG"/>权限。此实例的完整项目在 MyCode\MySample671 文件夹中。

235　使用 ContentResolver 获取 SD 卡的文件

此实例主要通过在 getContentResolver() 的 query() 方法中指定查询参数为某种文件类型，从而获取手机外部存储介质上的指定类型的所有文件。当实例运行之后，单击"获取 SD 卡上的所有 MP3 文件"按钮，则在 ListView 控件中显示手机外部存储上的所有 MP3 文件，单击在 ListView 控件中的任意列表项，则在弹出的 Toast 中显示该文件的路径信息，如图 235.1 的左图所示。单击"获取 SD 卡上的所有 PNG 文件"按钮，则在 ListView 控件中显示手机外部存储上的所有 PNG 文件，单击在 ListView 控件中的任意列表选项，则在弹出的 Toast 中显示该文件的路径信息，如图 235.1 的右图所示。

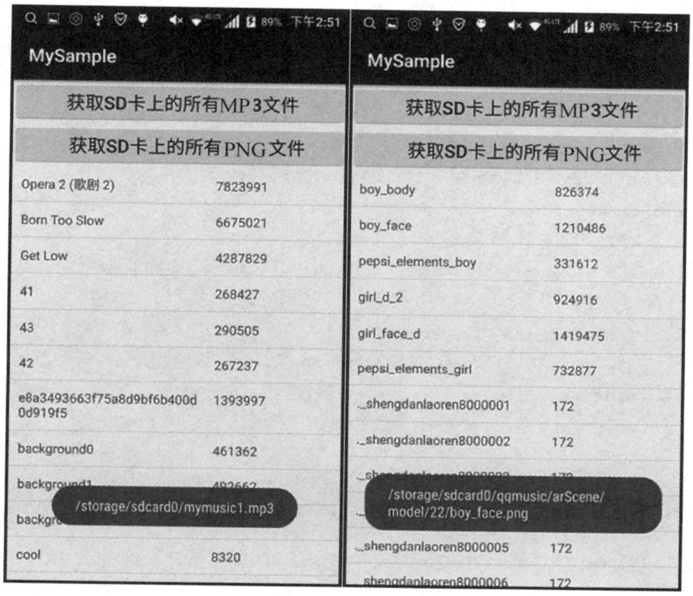

图　235.1

主要代码如下：

```java
//响应单击"获取 SD 卡上的所有 MP3 文件"按钮
void OnClickBtn1(View view){ GetFiles("audio/mpeg"); }
//响应单击"获取 SD 卡上的所有 PNG 文件"按钮
void OnClickBtn2(View view){ GetFiles("image/png"); }
private void GetFiles(String myType) {
    myFiles.clear();
    mySizes.clear();
    myItems.clear();
    myPathData.clear();
    //通过 ContentResolver 查询指定外部存储文件信息
    ContentResolver myContentResolver = getContentResolver();
    Cursor myCursor = myContentResolver.query(
                    MediaStore.Files.getContentUri ("external"), null,
                    "mime_type = \"" + myType + "\"", null, null);
    while (myCursor.moveToNext()) {
        String myName = myCursor.getString(myCursor.getColumnIndex(
            MediaStore.Files.FileColumns.TITLE));            //显示名称
```

```
    String mySize = myCursor.getString(myCursor.getColumnIndex(
      MediaStore.Files.FileColumns.SIZE));                //显示文件大小
    byte[] data = myCursor.getBlob(myCursor.getColumnIndex(
        MediaStore.Files.FileColumns.DATA));              //文件的位置数据
    myFiles.add(myName);
    mySizes.add(mySize);
    //将位置数据转换成字符串形式的路径
    String myPath = new String(data, 0, data.length - 1);
    myPathData.add(myPath);
  }
  for (int i = 0; i < myFiles.size(); i++) {
    HashMap<String, Object> myListItem = new HashMap<>();
    myListItem.put("name", myFiles.get(i));
    myListItem.put("size", mySizes.get(i));
    myItems.add(myListItem);
  }
  SimpleAdapter myAdapter = new SimpleAdapter(MainActivity.this, myItems,
            R.layout.myitem, new String[]{"name", "size"},
            new int[]{R.id.myName, R.id.mySize});
  myListView.setAdapter(myAdapter);
  //单击文件名称(列表项)显示文件路径信息
  myListView.setOnItemClickListener(new AdapterView.OnItemClickListener() {
    @Override
  public void onItemClick(AdapterView<?> parent,
                          View view, int position, long id){
    Toast.makeText(MainActivity.this,
                myPathData.get(position),Toast.LENGTH_SHORT).show();
  } });
}
```

上面这段代码在 MyCode\MySample672\app\src\main\java\com\bin\luo\mysample\MainActivity.java 文件中。在上面这段代码中，myAdapter = new SimpleAdapter（MainActivity.this，myItems，R.layout.myitem，new String[]{"name"，"size"}，new int[]{R.id.myName，R.id.mySize}）表示根据 myitem 布局的样式创建 ListView 的列表项，关于 myitem 布局的详细内容请参考源代码中的 MyCode\MySample672\app\src\main\res\layout\myitem.xml 文件。myCursor = myContentResolver.query(MediaStore.Files.getContentUri("external")，null,"mime_type=\""+myType+"\""，null，null)表示在手机外部存储介质上查找 mime_type 为指定文件类型的文件。文件类型和文件后缀名有相应的映射关系，常用的文件后缀名的映射关系如下。

（1）MimeTypeMap.loadEntry("application/pdf"，"pdf")。

（2）MimeTypeMap.loadEntry("application/zip"，"zip")。

（3）MimeTypeMap.loadEntry("application/msword"，"doc")。

（4）MimeTypeMap.loadEntry("application/vnd.ms-excel"，"xls")。

（5）MimeTypeMap.loadEntry("application/vnd.ms-powerpoint"，"ppt")。

（6）MimeTypeMap.loadEntry("audio/midi"，"mid")。

（7）MimeTypeMap.loadEntry("audio/mpeg"，"mp3")。

（8）MimeTypeMap.loadEntry("audio/x-pn-realaudio"，"rm")。

（9）MimeTypeMap.loadEntry("audio/x-wav"，"wav")。

（10）MimeTypeMap.loadEntry("image/bmp"，"bmp")。

（11）MimeTypeMap.loadEntry("image/gif", "gif")。
（12）MimeTypeMap.loadEntry("image/ico", "ico")。
（13）MimeTypeMap.loadEntry("image/jpeg", "jpg")。
（14）MimeTypeMap.loadEntry("image/png", "png")。
（15）MimeTypeMap.loadEntry("image/svg+xml", "svg")。
（16）MimeTypeMap.loadEntry("image/tiff", "tif")。
（17）MimeTypeMap.loadEntry("image/x-photoshop", "psd")。
（18）MimeTypeMap.loadEntry("text/plain", "text")。
（19）MimeTypeMap.loadEntry("video/mpeg", "mpeg")。
（20）MimeTypeMap.loadEntry("video/mp4", "mp4")。
（21）MimeTypeMap.loadEntry("video/x-ms-wmv", "wmv")。
（22）MimeTypeMap.loadEntry("video/x-msvideo", "avi")。

此外，读取存储卡上的文件需要在 AndroidManifest.xml 文件中添加< uses-permission android：name= "android.permission.READ_EXTERNAL_STORAGE"/>权限。需要说明的是，使用此实例的相关类需要在 gradle 中引入 compile 'com.android.support:design：25.0.1'依赖项。此实例的完整项目在 MyCode\MySample672 文件夹中。

236 使用 ContentResolver 改变屏幕亮度值

此实例主要实现了使用手指在屏幕上进行上下滑动，增大或减小屏幕的亮度。当实例运行后，如果手指向下滑动，则亮度减小；如果手指向上滑动，则亮度增大，效果分别如图 236.1 的左图和右图所示。注意：此实例在部分模拟器中测试无效，因此应在手机上直接测试；虽然实例屏幕截图看不出明显的亮度变化，但在手机上直接观察时变化非常明显。

图 236.1

主要代码如下：

```java
public void setBrightness(int myBrightness) {
    WindowManager.LayoutParams myParams = getWindow().getAttributes();
    myParams.screenBrightness = Float.valueOf(myBrightness) * (1f / 255f);
    getWindow().setAttributes(myParams);
    Uri myUri = android.provider.Settings.System.getUriFor("screen_brightness");
    ContentResolver myResolver = getContentResolver();
    android.provider.Settings.System.putInt(myResolver,
            "screen_brightness", myBrightness);                        //改变亮度值
    getContentResolver().notifyChange(myUri, null);
}
public int getScreenBrightness() {
    int currentBrightness = 0;
    ContentResolver myResolver = getContentResolver();
    try {
     currentBrightness = android.provider.Settings.System.getInt(
            myResolver, Settings.System.SCREEN_BRIGHTNESS);            //获取亮度值
    } catch (Exception e) { }
    return currentBrightness;
}
```

上面这段代码在 MyCode\MySample718\app\src\main\java\com\bin\luo\mysample\MainActivity.java 文件中。需要说明的是，操作手机设置需要在 AndroidManifest.xml 文件中添加 <uses-permission android:name="android.permission.WRITE_SETTINGS"/> 权限。此实例的完整项目在 MyCode\MySample718 文件夹中。

237 使用 ContentResolver 设置屏幕亮度值

此实例主要通过 ContentResolver 修改数据库的 screen_brightness 字段，实现改变手机的屏幕亮度。当实例运行之后，如果向右拖动滑块，则屏幕亮度增大；如果向左拖动滑块，则屏幕亮度减小，效果分别如图 237.1 的左图和右图所示。注意：此实例应在 Android 5.0 以上的手机中测试，在模拟器中测试可能无效。

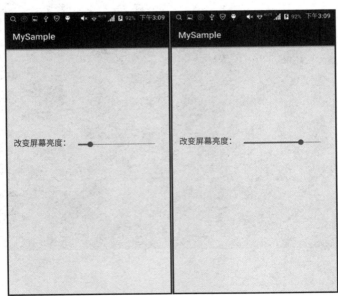

图　237.1

主要代码如下:

```
//根据指定值设置亮度
public void setBrightness(Activity activity, int brightness) {
    WindowManager.LayoutParams myParams = 
            activity.getWindow().getAttributes();                  //获取当前窗口属性参数对象
    myParams.screenBrightness = 
                    Float.valueOf(brightness) * (1f/255f);         //设置亮度值
    activity.getWindow().setAttributes(myParams);
    //如果不更新系统,则退出应用后,亮度改变立即失效
    Uri myUri = android.provider.Settings.System.getUriFor("screen_brightness");
    ContentResolver myContentResolver = getContentResolver();
    //根据新亮度值修改数据库对应字段
    android.provider.Settings.System.putInt(myContentResolver,
                            "screen_brightness",brightness);
    myContentResolver.notifyChange(myUri,null);                    //通知更新手机
}
```

上面这段代码在 MyCode\MySample817\app\src\main\java\com\bin\luo\mysample\MainActivity.java 文件中。此外,修改系统设置需要在 AndroidManifest.xml 文件中添加 < uses-permission android:name="android.permission.WRITE_SETTINGS"/>权限。此实例的完整项目在 MyCode\MySample817 文件夹中。

238 使用 ContentResolver 检测旋转屏幕功能

此实例主要通过在 Settings.System.getInt()方法的参数中使用 ContentResolver 参数和 android.provider.Settings.System.ACCELEROMETER_ROTATION 参数,实现检测当前手机的自动旋转屏幕功能的设置情况。当实例运行之后,单击"检测是否启用自动旋转屏幕功能"按钮,如果在手机设置中已经启用自动旋转屏幕功能按钮(注意:在实现横屏和竖屏自动切换时,特别需要检测此设置),如图 238.1 的左图所示,则在弹出的 Toast 中提示"已启用自动旋转屏幕!",如图 238.1 的右图所示,否则在弹出的 Toast 中提示"已禁用自动旋转屏幕!"。

图 238.1

主要代码如下：

```
//响应单击"检测是否启用自动旋转屏幕功能"按钮
public void onClickBtn1(View v) {
  ContentResolver myContentResolver = getContentResolver();
  //通过 ContentResolver 获取自动旋转屏幕功能标志
  int myFlag = Settings.System.getInt(myContentResolver,
              android.provider.Settings.System.ACCELEROMETER_ROTATION,0);
  if (myFlag == 1)                  //通过标志判断自动旋转屏幕功能是否启用
    Toast.makeText(MainActivity.this,
                   "已启用自动旋转屏幕!", Toast.LENGTH_SHORT).show();
  else if (myFlag == 0)
    Toast.makeText(MainActivity.this,
                   "已禁用自动旋转屏幕!", Toast.LENGTH_SHORT).show();
}
```

上面这段代码在 MyCode\MySample900\app\src\main\java\com\bin\luo\mysample\MainActivity.java 文件中。此实例的完整项目在 MyCode\MySample900 文件夹中。

239　使用 BroadcastReceiver 监听来电电话号码

此实例主要通过使用广播接收者 BroadcastReceiver，实现监听手机来电电话号码。当实例运行之后，从其他手机向当前手机拨打电话，则在弹出的 Toast 中显示来电方的电话号码，如图 239.1 所示。

图　239.1

主要代码如下：

```
public class CallReceiver extends BroadcastReceiver {
  @Override
  public void onReceive(Context context, Intent myIntent) {
    String myPhone =
      myIntent.getExtras().getString(TelephonyManager.EXTRA_INCOMING_NUMBER);
    Toast.makeText(context,"新来电号码:" + myPhone, Toast.LENGTH_SHORT).show();
} }
```

上面这段代码在 MyCode\MySample285\app\src\main\java\com\bin\luo\mysample\CallReceiver.java 文件中。在这段代码中，myPhone = myIntent.getExtras().getString(TelephonyManager.EXTRA_INCOMING_NUMBER)用于获取手机来电电话号码，但它主要是通过广播接收者 BroadcastReceiver 的 onReceive()方法以参数的形式传递的，因此通常需要以 BroadcastReceiver 为基类创建一个新的广播接收者，并且需要在 AndroidManifest.xml 文件中进行注册和添加权限，注册的主要代码如下：

```
<receiver android:name=".CallReceiver">
  <intent-filter>
    <action android:name="android.intent.action.PHONE_STATE" />
    <action android:name="android.intent.action.NEW_OUTGOING_CALL" />
  </intent-filter>
</receiver>
```

添加权限的主要代码如下：

```
<uses-permission android:name="android.permission.READ_PHONE_STATE" />
<uses-permission android:name="android.permission.PROCESS_OUTGOING_CALLS"/>
```

上面这段代码在 MyCode\MySample285\app\src\main\AndroidManifest.xml 文件中。此实例的完整项目在 MyCode\MySample285 文件夹中。

240 使用 BroadcastReceiver 判断手机电池是否正在充电

此实例主要通过使用 registerReceiver()方法注册电池服务请求，实现以广播接收的形式判断手机是否在充电。当实例运行之后，如果手机正在充电，则将每隔一定时间在弹出的 Toast 中显示"手机正在充电..."，如图 240.1 的左图所示。如果手机没有充电，则将每隔一定时间在弹出的 Toast 中显示"手机未充电！"，如图 240.1 的右图所示。

主要代码如下：

```
public class MainActivity extends Activity {
  @Override
  protected void onCreate(Bundle savedInstanceState) {
    super.onCreate(savedInstanceState);
    setContentView(R.layout.activity_main);
    IntentFilter myFilter = new IntentFilter(Intent.ACTION_BATTERY_CHANGED);
    registerReceiver(MyPowerConnectionReceiver, myFilter);
  }
```

```java
private BroadcastReceiver MyPowerConnectionReceiver = new BroadcastReceiver() {
    @Override
    public void onReceive(Context context, Intent intent) {
        int myStatus = intent.getIntExtra(BatteryManager.EXTRA_STATUS, -1);
        boolean isCharging = myStatus == BatteryManager.BATTERY_STATUS_CHARGING ||
                myStatus == BatteryManager.BATTERY_STATUS_FULL;
        if (isCharging) {
            Toast.makeText(getApplicationContext(),
                    "手机正在充电...", Toast.LENGTH_SHORT).show();
        } else {
            Toast.makeText(getApplicationContext(),
                    "手机未充电!", Toast.LENGTH_SHORT).show();
        }
    }
};
```

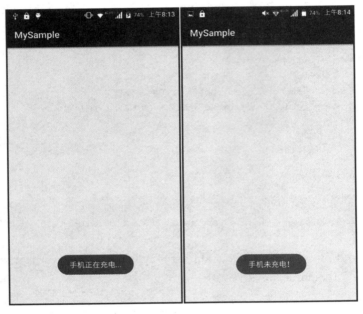

图 240.1

上面这段代码在 MyCode\MySample292\app\src\main\java\com\bin\luo\mysample\MainActivity.java 文件中。在这段代码中，int myStatus = intent.getIntExtra（BatteryManager.EXTRA_STATUS，-1）用于接收系统以广播形式发送的手机电池状态信息。boolean isCharging = myStatus == BatteryManager.BATTERY_STATUS_CHARGING||myStatus == BatteryManager.BATTERY_STATUS_FULL 用于判断当前手机电池是否正在充电。如果需要持续不断地接收系统以广播形式发送的手机电池状态信息，则应该在 AndroidManifest.xml 文件中添加 ACTION_POWER_CONNECTED 和 ACTION_POWER_DISCONNECTED，主要代码如下：

```xml
<receiver android:name=".MyPowerConnectionReceiver">
    <intent-filter>
        <action android:name="android.intent.action.ACTION_POWER_CONNECTED"/>
        <action android:name="android.intent.action.ACTION_POWER_DISCONNECTED"/>
    </intent-filter>
</receiver>
```

上面这段代码在 MyCode\MySample292\app\src\main\AndroidManifest.xml 文件中。此实例的完整项目在 MyCode\MySample292 文件夹中。

241 使用 BroadcastReceiver 监听屏幕开启或关闭

此实例主要通过使用广播接收者监听 Intent.ACTION_SCREEN_ON 和 Intent.ACTION_SCREEN_OFF,实现监视手机屏幕打开或关闭。当实例运行后,在"音乐文件:"输入框中输入在 SD 卡上存储的音乐文件名称,然后单击"开始播放"按钮,将播放指定的音乐文件,如图 241.1 所示。此时如果关闭屏幕,将停止播放音乐;如果打开屏幕,则将继续播放音乐。

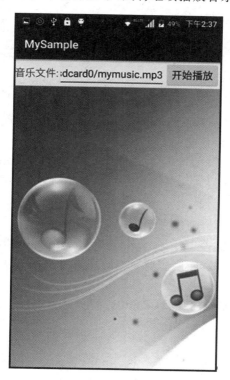

图 241.1

主要代码如下:

```java
public class MainActivity extends Activity {
    EditText myFile;
    MediaPlayer myPlayer;
    @Override
    protected void onCreate(Bundle savedInstanceState) {
        super.onCreate(savedInstanceState);
        setContentView(R.layout.activity_main);
        myFile = (EditText) findViewById(R.id.myFile);
        myFile.setText(Environment.getExternalStorageDirectory()
            + "/mymusic.mp3");                    //存储在 SD 卡上的音乐文件 mymusic.mp3
        myPlayer = new MediaPlayer();
        IntentFilter myFilter = new IntentFilter();
        myFilter.addAction(Intent.ACTION_SCREEN_OFF);
        myFilter.addAction(Intent.ACTION_SCREEN_ON);
        registerReceiver(myReceiver, myFilter);
```

```
    }
 public void onClickmyBtn1(View v) {                        //响应单击"开始播放"按钮
  try {
   myPlayer.setDataSource(myFile.getText().toString());
   myPlayer.prepare();
   myPlayer.start();
  } catch (Exception e) { e.printStackTrace(); }
 }
 private final BroadcastReceiver myReceiver = new BroadcastReceiver() {
  @Override
  public void onReceive(final Context myContext, final Intent myIntent) {
   final String myAction = myIntent.getAction();
   if (Intent.ACTION_SCREEN_ON.equals(myAction)) {
    myPlayer.start();                                       //如果打开屏幕,就继续播放
   } else if (Intent.ACTION_SCREEN_OFF.equals(myAction)) {
    myPlayer.pause();                                       //如果关闭屏幕,就停止播放
   } } };
}
```

上面这段代码在 MyCode\MySample278\app\src\main\java\com\bin\luo\mysample\MainActivity.java 文件中。在这段代码中,myFilter.addAction(Intent.ACTION_SCREEN_OFF) 和 myFilter.addAction(Intent.ACTION_SCREEN_ON) 用于在 IntentFilter 中添加打开屏幕和关闭屏幕这两个动作,然后通过 registerReceiver(myReceiver, myFilter) 向系统提交接收这两个动作的请求。当发生关闭或打开屏幕的动作时,系统就把通知发给 myReceiver,Intent.ACTION_SCREEN_ON.equals(myAction) 则根据这两个动作决定当前应用应该做什么。此外,读取在 SD 卡上的音乐文件需要在 AndroidManifest.xml 文件中添加 < uses-permission android:name="android.permission.READ_EXTERNAL_STORAGE"/> 权限。此实例的完整项目在 MyCode\MySample278 文件夹中。

242 自定义 BroadcastReceiver 实现短信拦截

此实例主要通过自定义 BroadcastReceiver,实现短信拦截功能。当实例运行后,单击"开启短信拦截功能"按钮,则当前应用处于短信拦截监听状态;单击"向 10086 发送查询短信"按钮,则跳转到短信应用界面,然后单击右侧的发送按钮(右箭头),如图 242.1 的左图所示;当 10086 收到查询请求之后,则立即向当前手机发送短信,此时当前应用就会将 10086 发送的短信拦截下来,如图 242.1 的右图所示。

主要代码如下:

```
public void onClickBtn1(View v) {                           //响应单击"开启短信拦截功能"按钮
  IntentFilter myFilter = new IntentFilter();               //动态设置过滤信息
  myFilter.addAction("android.provider.Telephony.SMS_RECEIVED");
  registerReceiver(new SmsReceiver(), myFilter);            //动态注册广播接收者
  Toast.makeText(MainActivity.this,
               "短信拦截功能已开启!", Toast.LENGTH_SHORT).show();
}
public void onClickBtn2(View v) {                           //响应单击"向 10086 发送查询短信"按钮
  Intent myIntent = new Intent(Intent.ACTION_SENDTO, Uri.parse("smsto:10086"));
  myIntent.putExtra("sms_body", "101");
  startActivity(myIntent);                                  //跳转到短信应用界面进行短信发送操作
}
```

```
public class SmsReceiver extends BroadcastReceiver {
    @Override
    public void onReceive(Context context, Intent intent){
        if(intent.getAction() == "android.provider.Telephony.SMS_RECEIVED"){
            Object[] myPdus = (Object[])intent.getExtras().get("pdus");
            ArrayList myArrayList = new ArrayList();
            for (Object myPdu:myPdus){
                SmsMessage myMessage = SmsMessage.createFromPdu((byte[])myPdu);
                String myAddress = myMessage.getOriginatingAddress();      //获取发送人号码
                String myBody = myMessage.getMessageBody();                //获取短信内容
                String myDate = new SimpleDateFormat("yyyy-MM-dd HH:mm:ss").
                        format(new Date(myMessage.getTimestampMillis()));  //获取短信发送时间
                myArrayList.add("短信发件人:" + myAddress + ",短信内容:"
                        + myBody + ",短信发送时间:" + myDate);              //拼接短信字符串,并封装至列表
            }
            //输出所拦截的短信内容
            MainActivity.myTextView.setText(myArrayList.toString());
}}}
```

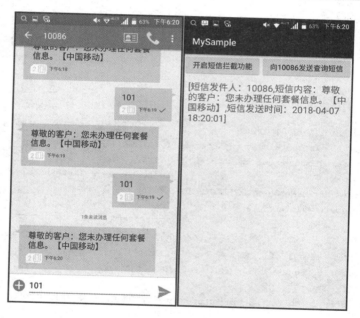

图 242.1

上面这段代码在 MyCode\MySample880\app\src\main\java\com\bin\luo\mysample\MainActivity.java 文件中。此外,操作短信需要在 AndroidManifest.xml 文件中添加权限< uses-permission android:name = "android.permission.RECEIVE_SMS"/>。此实例的完整项目在 MyCode\MySample880 文件夹中。

243 使用 RingtoneManager 设置手机闹钟铃声

此实例主要通过使用 RingtoneManager 的 setActualDefaultRingtoneUri()方法,实现设置手机的闹钟铃声。当实例运行后,单击"设置闹钟铃声"按钮,将从下向上滑出"使用文件管理完成操作"窗口,如图 243.1 的左图所示;单击"仅此一次"按钮,则将显示"音乐"窗口,在其中任意选择一个音乐文

件,该文件即成为手机的闹钟铃声,如图 243.1 的右图所示。当在手机"设置"窗口中调整"闹钟音量"时,即可听到刚才设置的闹钟铃声。

图 243.1

主要代码如下:

```java
public void onClickmyBtn1(View v) {                              //响应单击"设置闹钟铃声"按钮
    Intent myIntent = new Intent(RingtoneManager.ACTION_RINGTONE_PICKER);
    myIntent.putExtra(RingtoneManager.EXTRA_RINGTONE_TYPE,
                      RingtoneManager.TYPE_ALARM);
    myIntent.putExtra(RingtoneManager.EXTRA_RINGTONE_TITLE,"设置闹钟铃声");
    startActivityForResult(myIntent, 1);
}
@Override                                                        //设置铃声之后的回调函数
protected void onActivityResult(int requestCode, int resultCode, Intent data) {
    super.onActivityResult(requestCode, resultCode, data);
    try {
        Uri myUri = data.getParcelableExtra(RingtoneManager.EXTRA_RINGTONE_PICKED_URI);
        if (myUri != null) {
            switch (requestCode) {
                case 1:
                    RingtoneManager.setActualDefaultRingtoneUri(this,
                                     RingtoneManager.TYPE_ALARM, myUri);
                    Toast.makeText(MainActivity.this,
                        "铃声设置成功,请设置闹钟测试!", Toast.LENGTH_SHORT).show();
                    break;
                default:
                    break;
            } } } catch (Exception e) {
        Toast.makeText(MainActivity.this,
                       e.getMessage().toString(), Toast.LENGTH_SHORT).show();
    } }
```

上面这段代码在 MyCode\MySample423\app\src\main\java\com\bin\luo\mysample\MainActivity.java 文件中。在这段代码中，myUri = data.getParcelableExtra（RingtoneManager.EXTRA_RINGTONE_PICKED_URI）用于在选择音乐文件之后返回该音乐文件的 Uri。RingtoneManager.setActualDefaultRingtoneUri(this, RingtoneManager.TYPE_ALARM，myUri)用于将 myUri 代表的音乐文件设置为闹钟铃声。此外，设置铃声需要在 AndroidManifest.xml 文件中添加< uses-permission android:name = "android.permission.WRITE_SETTINGS"/>权限。此实例的完整项目在 MyCode\MySample423 文件夹中。

244　使用 RingtoneManager 设置手机通知铃声

此实例主要通过使用 RingtoneManager 的 setActualDefaultRingtoneUri()方法，实现设置手机的通知提示铃声。当实例运行后，单击"设置通知铃声"按钮，将从下向上滑出"使用文件管理完成操作"窗口，如图 244.1 的左图所示；单击"仅此一次"按钮，则将显示"音乐"窗口，在其中任意选择一个音乐文件，该文件即成为手机的通知提示铃声，如图 244.1 的右图所示。当在手机"设置"窗口中调整"铃声音量"时，即可听到刚才设置的通知铃声。

图　244.1

主要代码如下：

```java
public void onClickmyBtn1(View v) {                         //响应单击"设置通知铃声"按钮
    Intent myIntent = new Intent(RingtoneManager.ACTION_RINGTONE_PICKER);
    myIntent.putExtra(RingtoneManager.EXTRA_RINGTONE_TYPE,
        RingtoneManager.TYPE_NOTIFICATION);
    myIntent.putExtra(RingtoneManager.EXTRA_RINGTONE_TITLE,"设置通知铃声");
    startActivityForResult(myIntent, 2);
}
@Override                                                   //设置铃声之后的回调函数
protected void onActivityResult(int requestCode, int resultCode, Intent data) {
    super.onActivityResult(requestCode, resultCode, data);
```

```
try {
    Uri myUri = data.getParcelableExtra(RingtoneManager.EXTRA_RINGTONE_PICKED_URI);
    if (myUri != null) {
        switch (requestCode) {
        case 2:
            RingtoneManager.setActualDefaultRingtoneUri(this,
                        RingtoneManager.TYPE_NOTIFICATION, myUri);
            Toast.makeText(MainActivity.this,
                    "铃声设置成功,请设置铃声音量测试!", Toast.LENGTH_SHORT).show();
            break;
        default:
            break;
        } } } catch (Exception e) {
    Toast.makeText(MainActivity.this,
                e.getMessage().toString(), Toast.LENGTH_SHORT).show();
} }
```

上面这段代码在 MyCode\MySample424\app\src\main\java\com\bin\luo\mysample\MainActivity.java 文件中。在这段代码中，myUri = data.getParcelableExtra（RingtoneManager.EXTRA_RINGTONE_PICKED_URI）用于在选择音乐文件之后返回该音乐文件的 Uri。RingtoneManager.setActualDefaultRingtoneUri(this, RingtoneManager.TYPE_NOTIFICATION, myUri)用于将 myUri 代表的音乐文件设置为通知提示铃声。此外，设置铃声需要在 AndroidManifest.xml 文件中添加< uses-permission android:name="android.permission.WRITE_SETTINGS"/>权限。此实例的完整项目在 MyCode\MySample424 文件夹中。

245　使用 AlarmManager 以指定时间执行操作

此实例主要实现了使用 AlarmManager 配合 Intent，定时通过 Broadcast 发送提示。当实例运行之后，在弹出的 Toast 中提示"10 秒后 alarm 开启"，如图 245.1 的左图所示；然后退出当前实例，等待 10 秒，将收到广播服务，并在弹出的 Toast 提示"响应 alarm"，如图 245.1 的右图所示。

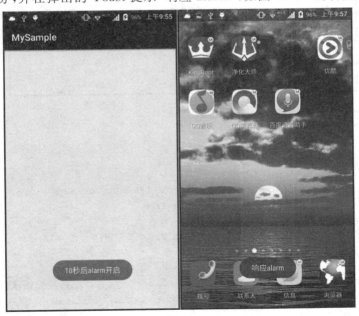

图　245.1

主要代码如下：

```java
public class MainActivity extends Activity {
  @Override
  protected void onCreate(Bundle savedInstanceState) {
    super.onCreate(savedInstanceState);
    setContentView(R.layout.activity_main);
    Intent myIntent = new Intent(MainActivity.this,
            AlarmReceiver.class);                    //发送一个广播,广播接收后 Toast 提示定时操作
    myIntent.setAction("myFlag");
    PendingIntent myPendingIntent =
            PendingIntent.getBroadcast(MainActivity.this, 0, myIntent, 0);
    AlarmManager myAlarmManager = (AlarmManager)getSystemService(ALARM_SERVICE);
    myAlarmManager.set(AlarmManager.RTC_WAKEUP,
            System.currentTimeMillis() + 10 * 1000, myPendingIntent);
    Toast.makeText(MainActivity.this,
            "10 秒后 alarm 开启", Toast.LENGTH_LONG).show();
} }
```

上面这段代码在 MyCode\MySample662\app\src\main\java\com\bin\luo\mysample\MainActivity.java 文件中。在这段代码中，myAlarmManager.set(AlarmManager.RTC_WAKEUP，System.currentTimeMillis() + 10 * 1000，myPendingIntent)表示 10 秒后唤醒 AlarmManager。myIntent = new Intent(MainActivity.this，AlarmReceiver.class)表示使用 AlarmReceiver 广播接收者响应 AlarmManager 的唤醒服务。AlarmReceiver 广播接收者的主要代码如下：

```java
public class AlarmReceiver extends BroadcastReceiver {
  @Override
  public void onReceive(Context context, Intent intent) {
    if(intent.getAction().equals("myFlag")){
      Toast.makeText(context, "响应 alarm", Toast.LENGTH_LONG).show();
} } }
```

上面这段代码在 MyCode\MySample662\app\src\main\java\com\bin\luo\mysample\AlarmReceiver.java 文件中。此外，使用 AlarmReceiver 广播接收者需要在 AndroidManifest.xml 文件中注册，主要代码如下：

```xml
<receiver android:name=".AlarmReceiver"/>
```

此实例的完整项目在 MyCode\MySample662 文件夹中。

246 使用 AudioManager 获取和设置音量

此实例主要通过在 SeekBar 控件的 onProgressChanged 事件响应方法中调用 AudioManager 的 setStreamVolume()方法，实现以拖动 SeekBar 滑块的方式改变手机音量的大小。在实例运行之前，先使用任一音乐播放器播放歌曲；当实例运行之后，SeekBar 控件的滑块位置即决定当前音量的大小，向右拖动滑块，音量变大；向左拖动滑块，音量变小，如图 246.1 的左图和右图所示。模拟器和手机实测效果均相同。

主要代码如下：

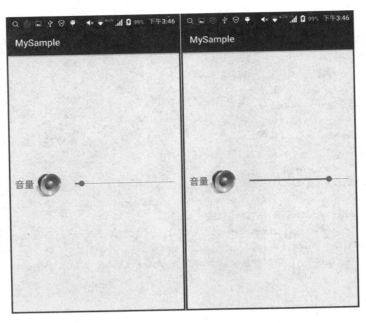

图 246.1

```java
public class MainActivity extends Activity {
 public AudioManager myAudioManager;
 private int maxVolume, currentVolume;
 @Override
 protected void onCreate(Bundle savedInstanceState) {
  super.onCreate(savedInstanceState);
  setContentView(R.layout.activity_main);
  myAudioManager = (AudioManager)getSystemService(Context.AUDIO_SERVICE);
  //获取系统最大音量
  maxVolume = myAudioManager.getStreamMaxVolume(AudioManager.STREAM_MUSIC);
  SeekBar mySeekBar = (SeekBar) findViewById(R.id.mySeekBar);
  mySeekBar.setMax(maxVolume);                    //设置拖动条最高值与系统最大音量匹配
  //获取音量当前值
  currentVolume = myAudioManager.getStreamVolume(AudioManager.STREAM_MUSIC);
  mySeekBar.setProgress(currentVolume);
  mySeekBar.setOnSeekBarChangeListener(new SeekBar.OnSeekBarChangeListener() {
   //音量大小改变时调用
   public void onProgressChanged(SeekBar arg0, int progress, boolean fromUser) {
    //根据当前滑块值设置音量大小
    myAudioManager.setStreamVolume(AudioManager.STREAM_MUSIC, progress, 0);
   }
   @Override
   public void onStartTrackingTouch(SeekBar seekBar) { }
   @Override
   public void onStopTrackingTouch(SeekBar seekBar) { }
  });} }
```

上面这段代码在 MyCode \ MySample063 \ app \ src \ main \ java \ com \ bin \ luo \ mysample \ MainActivity.java 文件中。在这段代码中，myAudioManager.setStreamVolume（AudioManager. STREAM_MUSIC，progress，0）用于根据 SeekBar 的当前值改变音量的大小。此外，在使用 AudioManager 调节音量大小时，必须获取系统的最大音量，即 maxVolume = myAudioManager.

getStreamMaxVolume（AudioManager. STREAM_MUSIC）。此实例的完整项目在 MyCode\MySample063 文件夹中。

247　使用 PowerManager 实现屏幕一直亮着

此实例主要通过创建 PowerManager.WakeLock 的实例，实现在应用运行时保持屏幕一直亮着。当实例运行后，即使在手机上设置了"屏幕延时"时间，屏幕也会一直亮着，如图 247.1 所示。

图　247.1

主要代码如下：

```
public class MainActivity extends Activity {
  PowerManager myPowerManager = null;
  PowerManager.WakeLock myWakeLock = null;
  @Override
  protected void onCreate(Bundle savedInstanceState) {
   super.onCreate(savedInstanceState);
   setContentView(R.layout.activity_main);
   this.myPowerManager =
              (PowerManager)this.getSystemService(Context.POWER_SERVICE);
   this.myWakeLock =
          this.myPowerManager.newWakeLock(PowerManager.FULL_WAKE_LOCK, "");
  }
  @Override
  protected void onResume() {
   super.onResume();
   this.myWakeLock.acquire();
```

```
  }
  @Override
  protected void onPause() {
    super.onPause();
    this.myWakeLock.release();
  }
}
```

上面这段代码在 MyCode\MySample273\app\src\main\java\com\bin\luo\mysample\MainActivity.java 文件中。在这段代码中,this.myWakeLock = this.myPowerManager.newWakeLock(PowerManager.FULL_WAKE_LOCK,"")用于创建 PowerManager.WakeLock 实例,this.myWakeLock.acquire()用于保持一直亮着的状态。此外,使用 PowerManager.WakeLock 需要在 AndroidManifest.xml 文件中添加< uses-permission android:name = "android.permission.WAKE_LOCK"/>权限。此实例的完整项目在 MyCode\MySample273 文件夹中。

248　使用 WallpaperManager 设置壁纸

此实例主要通过使用壁纸管理器 WallpaperManager,实现将选中的图像设置为壁纸。当实例运行之后,单击图像即可显示下一幅图像,当显示到图库的最后一幅图像时,自动跳转显示第一幅图像,效果如图 248.1 的左图所示。单击"设置手机壁纸"按钮,则当前显示的图像自动设置为手机壁纸,效果如图 248.1 的右图所示。

图　248.1

主要代码如下:

```
public void onClickButton1(View v) {                          //响应单击"设置手机壁纸"按钮
    WallpaperManager myWallpaperManager =
                WallpaperManager.getInstance(this);           //获取壁纸管理器
    BitmapDrawable myBitmapDrawable =
            (BitmapDrawable) myImageView.getDrawable();       //获取图像资源
```

```
            try {                                         //根据图像资源设置壁纸
                myWallpaperManager.setBitmap(myBitmapDrawable.getBitmap());
                Toast.makeText(getApplicationContext(),
                    "成功设置手机壁纸!", Toast.LENGTH_SHORT).show();
            } catch (Exception e) { e.printStackTrace(); }
        }
```

上面这段代码在 MyCode\MySample060\app\src\main\java\com\bin\luo\mysample\MainActivity.java 文件中。在这段代码中，myWallpaperManager.setBitmap（myBitmapDrawable.getBitmap()）用于将 ImageView 控件的当前图像设置为壁纸。除了可以使用 setBitmap()方法设置壁纸外，常用的设置壁纸方法还有 setResource()、setWallpaper()等。此外，设置手机壁纸需要在 AndroidManifest.xml 文件中添加< uses-permission android: name = " android. permission. SET_WALLPAPER"/>权限。此实例的完整项目在 MyCode\MySample060 文件夹中。

249 使用 PackageManager 获取支持分享的应用

此实例主要通过使用 PackageManager 类的 queryIntentActivities()方法，获取当前手机安装的支持分享功能的所有应用信息。当实例运行之后，单击"获取手机中支持分享功能的应用"按钮，则在弹出的 Toast 中显示当前手机安装的支持分享功能的应用包名信息，如图 249.1 的左图和右图所示。

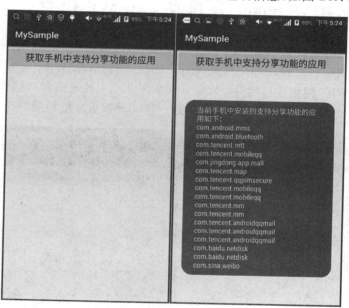

图 249.1

主要代码如下：

```
//响应单击"获取手机中支持分享功能的应用"按钮
public void onClickmyBtn1(View v) {
    String myInfo = "当前手机中安装的支持分享功能的应用如下:";
    Intent myIntent = new Intent(Intent.ACTION_SEND, null);
    myIntent.addCategory(Intent.CATEGORY_DEFAULT);
    myIntent.setType("text/plain");
```

```
PackageManager myPackageManager = this.getPackageManager();
List < ResolveInfo > myResolveInfos =
                myPackageManager.queryIntentActivities(myIntent,
                        PackageManager.COMPONENT_ENABLED_STATE_DEFAULT);
for (int i = 0; i < myResolveInfos.size(); i++) {
 ResolveInfo myResolveInfo = myResolveInfos.get(i);
 myInfo += "\n" + myResolveInfo.activityInfo.packageName;
}
Toast.makeText(getApplicationContext(),
                        myInfo, Toast.LENGTH_SHORT).show();
}
```

上面这段代码在 MyCode\MySample209\app\src\main\java\com\bin\luo\mysample\MainActivity.java 文件中。在这段代码中，Intent myIntent = new Intent(Intent.ACTION_SEND,null)用于创建支持分享功能的 Intent。List < ResolveInfo > myResolveInfos = myPackageManager.queryIntentActivities(myIntent,PackageManager.COMPONENT_ ENABLED_STATE _DEFAULT)用于根据 myIntent 查找所有符合分享功能的应用包。此实例的完整项目在 MyCode\MySample209 文件夹中。

250 使用 WifiManager 开启或关闭 WiFi 信号

此实例主要通过使用 WifiManager 的 setWifiEnabled()方法，实现启用 WiFi 网络信号或关闭 WiFi 网络信号。当实例运行之后，单击"打开 WiFi"按钮，则将启用 WiFi 网络连接信号，单击"关闭 WiFi"按钮，则将关闭 WiFi 网络连接信号，效果分别如图 250.1 的左图和右图所示。注意：由于启用或关闭 WiFi 信号有一个时间过程，因此需要在延迟一定的时间后才能检测到正确的执行结果。

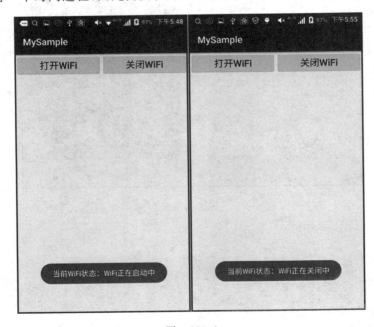

图 250.1

主要代码如下：

```java
public class MainActivity extends Activity {
  WifiManager myWifiManager;
  Handler myHandler = new Handler();
  @Override
  protected void onCreate(Bundle savedInstanceState) {
    super.onCreate(savedInstanceState);
    setContentView(R.layout.activity_main);
    myWifiManager = (WifiManager) this.getSystemService(Context.WIFI_SERVICE);
    myHandler.postDelayed(new Runnable() {
      @Override
      public void run() {
        checkWifiState();
      } },1500);
  }
  //响应单击"打开 WiFi"按钮
  public void onClickmyBtn1(View v) {
    myWifiManager.setWifiEnabled(true);
    //打开 WiFi 有个时间过程，因此需要延迟一定的时间进行检测
    myHandler.postDelayed(new Runnable() {
      @Override
      public void run() {
        checkWifiState();
      } },1500);
  }
  //响应单击"关闭 WiFi"按钮
  public void onClickmyBtn2(View v) {
    myWifiManager.setWifiEnabled(false);
    //关闭 WiFi 有个时间过程，因此需要延迟一定的时间进行检测
    myHandler.postDelayed(new Runnable() {
      @Override
      public void run() {
          checkWifiState();
      } },1500);
  }
  public void checkWifiState() {
    String myInfo = "当前 WiFi 状态:";
    switch (myWifiManager.getWifiState()) {
    case WifiManager.WIFI_STATE_DISABLED:
      myInfo += "WiFi 已经关闭";
      break;
    case WifiManager.WIFI_STATE_DISABLING:
      myInfo += "WiFi 正在关闭中";
      break;
    case WifiManager.WIFI_STATE_ENABLED:
      myInfo += "WiFi 已经启用";
      break;
    case WifiManager.WIFI_STATE_ENABLING:
      myInfo += "WiFi 正在启动中";
      break;
    case WifiManager.WIFI_STATE_UNKNOWN:
      myInfo += "未知 WiFi 状态";
      break;
    }
    Toast.makeText(MainActivity.this,myInfo, Toast.LENGTH_SHORT).show();
  }
}
```

上面这段代码在 MyCode\MySample263\app\src\main\java\com\bin\luo\mysample\MainActivity.java 文件中。在这段代码中，setWifiEnabled()用于启用或禁用 WiFi，当此方法的参数值为 true 时，表示启用 WiFi；当此方法的参数值为 false 时，表示禁用 WiFi。此外，操作 WiFi 需要相应的权限，主要代码如下：

```
<uses-permission android:name = "android.permission.ACCESS_WIFI_STATE"/>
<uses-permission android:name = "android.permission.ACCESS_NETWORK_STATE"/>
<uses-permission android:name = "android.permission.CHANGE_NETWORK_STATE"/>
<uses-permission android:name = "android.permission.CHANGE_WIFI_STATE"/>
```

上面这段代码在 MyCode\MySample263\app\src\main\AndroidManifest.xml 文件中。此实例的完整项目在 MyCode\MySample263 文件夹中。

251　使用 WifiManager 获取 IP 地址

此实例主要通过使用 WifiInfo 的 getIpAddress()方法，实现以 WiFi 管理服务的方式来获取 IP 地址。当实例运行之后，单击"通过 WiFi 管理服务获取 IP 地址"按钮，则在弹出的 Toast 中显示 IP 地址，如图 251.1 的左图和右图所示。注意：由于启用或关闭 WiFi 信号有一个时间过程，因此需要在延迟一定的时间后才能检测到正确的执行结果。

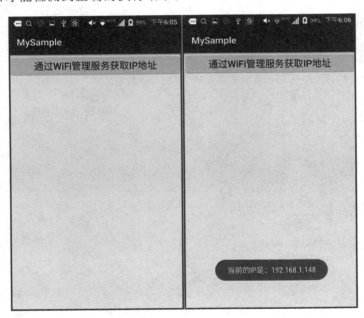

图　251.1

主要代码如下：

```
private String ConvertIP(int myIntIP) {
    return ((myIntIP >> 0) & 0xFF) + "." + ((myIntIP >> 8) & 0xFF) + "."
            + ((myIntIP >> 16) & 0xFF) + "." + ((myIntIP >> 24) & 0xFF);
}
//响应单击"通过 WiFi 管理服务获取 IP 地址"按钮
public void onClickmyBtn1(View v){
```

```
WifiManager myWifiManager =
    (WifiManager) getSystemService(Context.WIFI_SERVICE);    //获取 WiFi 管理服务
if (!myWifiManager.isWifiEnabled()) {                        //判断 WiFi 是否开启
    //开启 WiFi,需要设置权限:android.permission.CHANGE_WIFI_STATE
    myWifiManager.setWifiEnabled(true);
    Toast.makeText(MainActivity.this,
            "正在开启 WiFi,请稍后重新获取 IP 地址", Toast.LENGTH_SHORT).show();
}
//获取 WiFi 信息,需要设置权限:android.permission.ACCESS_WIFI_STATE
WifiInfo myWifiInfo = myWifiManager.getConnectionInfo();
int myIntIP = myWifiInfo.getIpAddress();                     //获得整型 IP 地址
String myStringIP = ConvertIP(myIntIP);                      //将整型 IP 地址转换为字符串
Toast.makeText(MainActivity.this,
        "当前的 IP 是:" + myStringIP, Toast.LENGTH_SHORT).show();
}
```

上面这段代码在 MyCode \ MySample350 \ app \ src \ main \ java \ com \ bin \ luo \ mysample \ MainActivity.java 文件中。在这段代码中,myIntIP = myWifiInfo.getIpAddress()用于获取 IP 地址。此外,操作 WiFi 需要在 AndroidManifest.xml 文件中添加< uses-permission android:name = "android.permission.ACCESS_WIFI_STATE"/>权限 和 < uses-permission android:name = "android.permission.CHANGE_WIFI_STATE "/>权限。此实例的完整项目在 MyCode\MySample350 文件夹中。

252　使用 ConnectivityManager 判断网络状态

此实例主要通过使用 ConnectivityManager 类的 getNetworkInfo()方法,从而判断当前手机的网络连接状态。当实例运行之后,单击"判断当前手机的网络连接状态"按钮,如果当前手机网络使用的是数据连接,则在弹出的 Toast 中显示"正在使用移动数据连接网络!",如图 252.1 的左图所示;如果当前手机网络使用的是 WiFi 连接,则在弹出的 Toast 中显示"正在使用 WiFi 连接网络!",如图 252.1 的右图所示。

图　252.1

主要代码如下：

```java
//响应单击"判断当前手机的网络连接状态"按钮
public void onClickmyBtn1(View v) {
    Context myContext = this.getApplicationContext();
    ConnectivityManager myConnectivityManager = (ConnectivityManager)
            myContext.getSystemService(Context.CONNECTIVITY_SERVICE);
    if (myConnectivityManager == null){ return; }
    else {
    NetworkInfo myMobileNet = myConnectivityManager.
            getNetworkInfo(ConnectivityManager.TYPE_MOBILE);
    NetworkInfo myWifiNet = myConnectivityManager.
            getNetworkInfo(ConnectivityManager.TYPE_WIFI);
    if(myWifiNet.isConnected()){
     Toast.makeText(MainActivity.this,
            "正在使用WiFi连接网络!",Toast.LENGTH_SHORT).show();
    }else if(myMobileNet.isConnected()){
     Toast.makeText(MainActivity.this,
            "正在使用移动数据连接网络!",Toast.LENGTH_SHORT).show();
    }else if(!myMobileNet.isConnected()&!myWifiNet.isConnected()){
     Toast.makeText(MainActivity.this,
            "当前没有网络连接!",Toast.LENGTH_SHORT).show();
} } }
```

上面这段代码在 MyCode\MySample215\app\src\main\java\com\bin\luo\mysample\MainActivity.java 文件中。在这段代码中，myMobileNet = myConnectivityManager.getNetworkInfo(ConnectivityManager.TYPE_MOBILE)用于获取移动数据连接信息，该方法的参数支持的类型包括：TYPE_WIFI、TYPE_BLUETOOTH、TYPE_DUMMY、TYPE_ETHERNET、TYPE_MOBILE、TYPE_MOBILE_DUN、TYPE_VPN、TYPE_WIMAX，因此可以创建多个不同类型的网络连接进行测试。myMobileNet.isConnected()用于判断移动数据网络连接是否成功，如果连接成功，则返回true，否则返回false。此外，检测网络连接需要在AndroidManifest.xml文件中添加<uses-permission android:name="android.permission.ACCESS_NETWORK_STATE"/>权限。此实例的完整项目在 MyCode\MySample215 文件夹中。

253　使用 BluetoothAdapter 打开或关闭蓝牙

此实例主要通过使用getSystemService()方法获取系统蓝牙管理器BluetoothManager的适配器BluetoothAdapter，然后调用BluetoothAdapter的enable()和disable()方法，实现动态启用或禁用手机的蓝牙功能。当实例运行之后，单击"打开蓝牙"按钮，则将启动蓝牙，并跳转到系统的蓝牙配对设置界面，如图253.1的左图所示。单击"关闭蓝牙"按钮，则将关闭蓝牙，如图253.1的右图所示。

主要代码如下：

```java
public class MainActivity extends Activity {
 BluetoothAdapter myBluetoothAdapter;
 @Override
 protected void onCreate(Bundle savedInstanceState) {
  super.onCreate(savedInstanceState);
  setContentView(R.layout.activity_main);
```

```
        //通过系统服务获取蓝牙管理器
        BluetoothManager myBluetoothManager = (BluetoothManager)
                            getSystemService(Context.BLUETOOTH_SERVICE);
        //通过蓝牙管理器获取蓝牙适配器
        myBluetoothAdapter = myBluetoothManager.getAdapter();
    }
    //响应单击"打开蓝牙"按钮
    public void onClickBtn1(View v) {
        myBluetoothAdapter.enable();                           //启用蓝牙功能
        Intent myIntent = new Intent(Settings.ACTION_BLUETOOTH_SETTINGS);
        startActivity(myIntent);                               //跳转至蓝牙设置页面,以配对蓝牙
        Toast.makeText(this,"蓝牙已打开!",Toast.LENGTH_SHORT).show();
    }
    //响应单击"关闭蓝牙"按钮
    public void onClickBtn2(View v) {
        myBluetoothAdapter.disable();                          //禁用蓝牙功能
        Toast.makeText(this,"蓝牙已关闭!",Toast.LENGTH_SHORT).show();
    } }
```

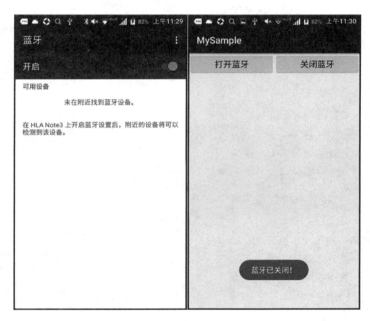

图 253.1

上面这段代码在 MyCode\MySample874\app\src\main\java\com\bin\luo\mysample\MainActivity.java 文件中。需要注意的是,启用或禁用蓝牙需要在 AndroidManifest.xml 文件中添加<uses-permission android:name="android.permission.BLUETOOTH"/>权限和<uses-permission android:name="android.permission.BLUETOOTH_ADMIN"/>权限。此实例的完整项目在 MyCode\MySample874 文件夹中。

254 使用 LocationListener 获取当前经纬度值

此实例主要通过在 LocationListener 类的 onLocationChanged()方法中监听位置信息,获取当前手机的经度和纬度值。当实例运行之后,将显示当前手机的经度值和纬度值,如图 254.1 所示。注

意：在测试时，一定要先开启 GPS，并且尽可能在无遮挡的空旷地方，否则可能不能显示数据。

图 254.1

主要代码如下：

```java
public class MainActivity extends Activity {
 private LocationManager myLocationManager;
 private LocationListener myLocationlistener;
 String myLocationprovider;
 private TextView myTextView;
 @Override
 protected void onCreate(Bundle savedInstanceState) {
  super.onCreate(savedInstanceState);
  setContentView(R.layout.activity_main);
  myTextView = (TextView) this.findViewById(R.id.myTextView);
  try {
   Criteria myCriteria = new Criteria();
   myCriteria.setAccuracy(Criteria.ACCURACY_FINE);              //设置精度
   myCriteria.setAltitudeRequired(false);                       //不提供海拔高度信息
   myCriteria.setBearingRequired(false);                        //不提供方向信息
   myCriteria.setCostAllowed(true);                             //允许运营商计费
   myCriteria.setPowerRequirement(Criteria.POWER_LOW);          //设置电池消耗为低耗费
   myLocationManager =
           (LocationManager) getSystemService(Context.LOCATION_SERVICE);
   if (checkgps()) {                                            //检查 GPS 功能开启
    myLocationprovider = myLocationManager.getBestProvider(myCriteria, true);
    myLocationlistener = new MyLocationListener();              //注册位置监听器
    if (Build.VERSION.SDK_INT >= Build.VERSION_CODES.M) {
```

```java
            if (checkSelfPermission(Manifest.permission.ACCESS_FINE_LOCATION)
                    != PackageManager.PERMISSION_GRANTED & checkSelfPermission(
                    Manifest.permission.ACCESS_COARSE_LOCATION)
                    != PackageManager.PERMISSION_GRANTED) { return; }
        }
        myLocationManager.requestLocationUpdates(
                        myLocationprovider, 1000, 0, myLocationlistener);
    }
  }catch(Exception e){
    Toast.makeText(MainActivity.this,
                    "异常错误:" + e.toString(),Toast.LENGTH_LONG).show();
} }
  private class MyLocationListener implements LocationListener{
    //若位置发生变化,onLocationChanged 方法被调用
    @Override
    public void onLocationChanged(Location myLocation) {
      if(myLocation != null){
        String myLatitude = Double.toString(myLocation.getLatitude());
        String myLongitude = Double.toString(myLocation.getLongitude());
        myTextView = (TextView)MainActivity.this.findViewById(R.id.myTextView);
        myTextView.setText("当前手机的位置如下:\n 经度:"
                                       + myLongitude + "\n 纬度:" + myLatitude);
    } }
    @Override
    public void onProviderDisabled(String provider){ }        //若屏蔽提供商,该方法被调用
    @Override
    public void onProviderEnabled(String provider){ }         //若激活提供商,该方法被调用
    //若状态发生变化,该方法被调用
    @Override
    public void onStatusChanged(String provider, int status, Bundle extras) { }
  }
  private boolean checkgps(){
    boolean providerEnabled
            = myLocationManager.isProviderEnabled(LocationManager.GPS_PROVIDER);
    if(providerEnabled == true){                              //若被激活,则返回真值
      Toast.makeText(this, "GPS 模块活动正常", Toast.LENGTH_SHORT).show();
      return true;
    }
    else{
      Toast.makeText(this, "请开启 GPS", Toast.LENGTH_SHORT).show();
      return false;
} } }
```

上面这段代码在 MyCode\MySample318\app\src\main\java\com\bin\luo\mysample\MainActivity.java 文件中。在这段代码中,onLocationChanged(Location myLocation)用于响应手机位置的改变,并通过 myLocation 参数传递当前的经度和纬度值。另外,监听位置变化在 AndroidManifest.xml 文件中添加< uses-permission android:name="android.permission.ACCESS_FINE_LOCATION"/>权限和< uses-permission android:name="android.permission.ACCESS_LOCATION_EXTRA_COMMANDS"/>权限。此实例的完整项目在 MyCode\MySample318 文件夹中。

255 使用 SensorManager 获取传感器信息

此实例主要通过使用 getSystemService()和 getSensorList()方法,实现获取手机的传感器信息。当实例运行之后,单击"获取手机的传感器信息"按钮,则在下面显示当前手机的所有传感器信息,如图 255.1 的左图和右图所示。

图 255.1

主要代码如下:

```java
//响应单击"获取手机的传感器信息"按钮
public void onClickmyBtn1(View v) {
    TextView myTextView = (TextView) findViewById(R.id.myTextView);
    myTextView.setMovementMethod(ScrollingMovementMethod.getInstance());
    SensorManager mySensorManager = (SensorManager)
                        getSystemService(Context.SENSOR_SERVICE);
    //从传感器管理器中获得全部的传感器列表
    List<Sensor> mySensors = mySensorManager.getSensorList(Sensor.TYPE_ALL);
    myTextView.setText("经检测该手机有" +
            mySensors.size() + "个传感器,他们分别是:\n");      //显示有多少个传感器
    for (Sensor mySensor : mySensors) {                        //显示每个传感器的具体信息
    String myInfo = "\n" + " 设备名称:" + mySensor.getName() + "\n"
            + " 设备版本:" + mySensor.getVersion() + "\n" + " 供应商:"
            + mySensor.getVendor() + "\n";
    switch (mySensor.getType()) {
     case Sensor.TYPE_ACCELEROMETER:
      myTextView.setText(myTextView.getText().toString() +
                mySensor.getType() + " 加速度传感器 accelerometer" + myInfo);
      break;
     case Sensor.TYPE_GYROSCOPE:
      myTextView.setText(myTextView.getText().toString() +
```

```
                    mySensor.getType() + "陀螺仪传感器 gyroscope" + myInfo);
            break;
        case Sensor.TYPE_LIGHT:
            myTextView.setText(myTextView.getText().toString() +
                    mySensor.getType() + "环境光线传感器 light" + myInfo);
            break;
        case Sensor.TYPE_MAGNETIC_FIELD:
            myTextView.setText(myTextView.getText().toString() +
                    mySensor.getType() + "电磁场传感器 magnetic field" + myInfo);
            break;
        case Sensor.TYPE_ORIENTATION:
            myTextView.setText(myTextView.getText().toString() +
                    mySensor.getType() + "方向传感器 orientation" + myInfo);
            break;
        case Sensor.TYPE_PRESSURE:
            myTextView.setText(myTextView.getText().toString() +
                    mySensor.getType() + "压力传感器 pressure" + myInfo);
            break;
        case Sensor.TYPE_PROXIMITY:
            myTextView.setText(myTextView.getText().toString() +
                    mySensor.getType() + "距离传感器 proximity" + myInfo);
            break;
        case Sensor.TYPE_TEMPERATURE:
            myTextView.setText(myTextView.getText().toString() +
                    mySensor.getType() + "温度传感器 temperature" + myInfo);
            break;
        default:
            myTextView.setText(myTextView.getText().toString() +
                    mySensor.getType() + "未知传感器" + myInfo);
            break;
    } }
}
```

上面这段代码在 MyCode \ MySample267 \ app \ src \ main \ java \ com \ bin \ luo \ mysample \ MainActivity.java 文件中。在这段代码中，mySensors = mySensorManager.getSensorList（Sensor.TYPE_ALL）用于获取手机的所有传感器。对于某个具体的传感器，它的具体信息可以使用下列方法进行获取。

(1) getMaximumRange()方法，该方法用于获取最大取值范围。
(2) getName()方法，该方法用于获取设备名称。
(3) getPower()方法，该方法用于获取功率。
(4) getResolution()方法，该方法用于获取精度。
(5) getType()方法，该方法用于获取传感器类型。
(6) getVentor()方法，该方法用于获取设备供应商。
(7) getVersion()方法，该方法用于获取设备版本号。
此实例的完整项目在 MyCode\MySample267 文件夹中。

256 使用传感器监测耳朵与手机听筒的距离

此实例主要通过在 SensorEventListener 的 onSensorChanged()事件响应方法中监听 Proximity 传感器的值，实现监测耳朵与手机听筒的距离。当实例运行之后，使用手指（耳朵或其他物体也一样）

在手机听筒附近任意移动,当手指(在三维)距离手机听筒大约1厘米左右时,将提示距离较近信息,如图256.1的左图所示。当手指超出手机听筒1厘米范围时,将提示距离较远信息,如图256.1的右图所示。这主要是因为Proximity传感器检测物体与手机的距离单位是厘米,一些Proximity传感器只能返回远和近两个状态,即大于或等于最大距离返回远状态,小于最大距离返回近状态。Proximity传感器可用于近距离接听电话时自动关闭LCD屏幕以节省电量。

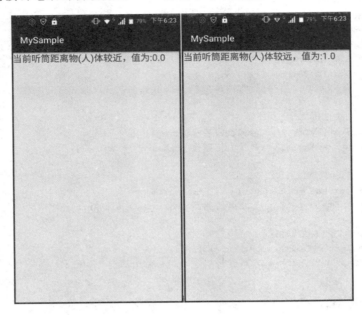

图 256.1

主要代码如下:

```java
public class MainActivity extends Activity implements SensorEventListener {
  TextView myText;
  SensorManager mySensorManager = null;              // 传感器管理器
  Sensor mySensor = null;                            // 传感器实例
  @Override
  protected void onCreate(Bundle savedInstanceState) {
    super.onCreate(savedInstanceState);
    setContentView(R.layout.activity_main);
    myText = (TextView) findViewById(R.id.myText);
    mySensorManager = (SensorManager) getSystemService(Context.SENSOR_SERVICE);
    mySensor = mySensorManager.getDefaultSensor(Sensor.TYPE_PROXIMITY);
  }
  @Override
  protected void onResume() {
    super.onResume();
    mySensorManager.registerListener(this, mySensor,
                SensorManager.SENSOR_DELAY_NORMAL);   //注册传感器
  }
  @Override
  protected void onPause() {
    super.onPause();
    mySensorManager.unregisterListener(this);         // 取消注册传感器
  }
```

```
@Override
public void onSensorChanged(SensorEvent event) {
  float myRange = event.values[0];
  if (myRange == mySensor.getMaximumRange()) {
    myText.setText("当前听筒距离物(人)体较远,值为:" + myRange);
  } else {
    myText.setText("当前听筒距离物(人)体较近,值为:" + myRange);
  }
  myText.setTextSize(20);
}
@Override
public void onAccuracyChanged(Sensor sensor, int accuracy) { }
}
```

上面这段代码在 MyCode \ MySample271 \ app \ src \ main \ java \ com \ bin \ luo \ mysample \ MainActivity.java 文件中。其中,mySensor = mySensorManager.getDefaultSensor(Sensor.TYPE_PROXIMITY)表示当前获取的传感器是 Proximity 传感器,该传感器通过 onSensorChanged (SensorEvent event)的 event.values[0]获取手机听筒与物体之间的距离。此外,监听传感器值发生变化的监听器必须经过注册才有效,即: mySensorManager.registerListener(this, mySensor, SensorManager.SENSOR_DELAY_NORMAL)。此实例的完整项目在 MyCode\MySample271 文件夹中。

257　使用加速度传感器监听手机的三维变化

此实例主要通过解析 SensorEventListener 的 onSensorChanged()事件参数值 SensorEvent,实现在晃动手机时监听加速度传感器的变化。当实例运行之后,不停地任意晃动手机,则加速度传感器的 x、y、z 值将不断变化,效果分别如图 257.1 的左图和右图所示。

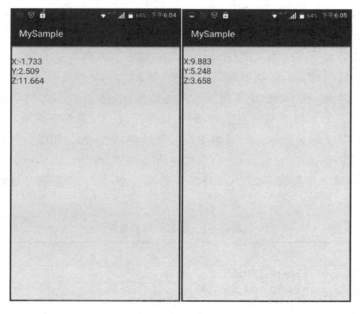

图　257.1

主要代码如下：

```java
public class MainActivity extends Activity {
    private long lastTime = 0,curTime = 0;
    TextView myTextView;
    @Override
    protected void onCreate(Bundle savedInstanceState) {
        super.onCreate(savedInstanceState);
        setContentView(R.layout.activity_main);
        myTextView = (TextView) findViewById(R.id.myTextView);
        SensorManager mySensorManager =
                (SensorManager) getSystemService(Context.SENSOR_SERVICE);
        Sensor myAccelomererSensor =                              // 获取加速度传感器
                mySensorManager.getDefaultSensor(Sensor.TYPE_ACCELEROMETER);
        SensorEventListener mySensorEventListener = new SensorEventListener() {
            @Override
            public void onAccuracyChanged(Sensor sensor, int accuracy) { }
            public void onSensorChanged(SensorEvent event) {      //传感器数据变动事件
                curTime = System.currentTimeMillis();             //获取当前时刻的 ms 数
                if ((curTime - lastTime) > 100) {                 //100ms 检测一次
                    //获取加速度传感器的三个参数
                    float x = event.values[SensorManager.DATA_X];
                    float y = event.values[SensorManager.DATA_Y];
                    float z = event.values[SensorManager.DATA_Z];
                    myTextView.setText("\nX:" + x + "\nY:" + y + "\nZ:" + z);
                    lastTime = curTime;
                } } };
        //在传感器管理器中注册监听器
        mySensorManager.registerListener(mySensorEventListener,
                    myAccelomererSensor, SensorManager.SENSOR_DELAY_NORMAL);
} }
```

上面这段代码在 MyCode\MySample269\app\src\main\java\com\bin\luo\mysample\MainActivity.java 文件中。在这段代码中，x＝event.values[SensorManager.DATA_X]、y＝event.values[SensorManager.DATA_Y]和 z＝event.values[SensorManager.DATA_Z]分别代表三维坐标的三个方向，它们的值是某个方向加速度（Acceleration）减去该方向的重力值（Gx、Gy、Gz），所以手机静止时其范围是［－10,10］，例如，手机屏幕朝上平放，则 $x=0$、$y=0$、$z=10$；这是因为手机静止不动时，所有方向都没有加速度，平放则产生了向下的重力加速度，即 $z=-10$，因为重力方向与 z 轴正向相反，所以相减后是 10。如果将手机向左倾斜，则 x 轴为正值。如果将手机向右倾斜，则 x 轴为负值。如果将手机向上倾斜，则 y 轴为负值。如果将手机向下倾斜，则 y 轴为正值。

此外，监听器必须使用 registerListener()方法注册才有效，registerListener()方法的第一个参数表示监听 Sensor 事件，第二个参数是 Sensor 目标传感器类型，第三个参数是延迟时间的精度。延迟时间参数如表 257-1 所示。

表 257-1　延迟时间参数

参　　数	延迟时间（ms）
SensorManager.SENSOR_DELAY_FASTEST	0
SensorManager.SENSOR_DELAY_GAME	20
SensorManager.SENSOR_DELAY_UI	60
SensorManager.SENSOR_DELAY_NORMAL	200

由于传感器 Sensor 服务是否频繁和快慢都与电量的消耗有关,同时也会影响处理的效率,所以需要兼顾到消耗电池和处理效率的平衡。此实例的完整项目在 MyCode\MySample269 文件夹中。

258　通过传感器实现自动进行横屏和竖屏切换

此实例主要通过使用自定义类 MyUtils 监听传感器的变化,实现横屏和竖屏自动切换。当实例运行之后,在"网址:"输入框中输入一个网址,如"http://www.sohu.com/",单击"浏览"按钮,则在 WebView 控件中显示该网页,如果竖屏放置手机,则此应用的竖屏显示效果如图 258.1 的左图所示;如果横屏放置手机,则此应用的横屏显示效果如图 258.1 的右图所示;即此应用能够通过传感器的变化自动决定是以横屏显示或是以竖屏显示。

图　258.1

主要代码如下:

```
@Override
protected void onPause() {                          //停止监听
    super.onPause();
    MyUtils.getInstance(this).stop();
}
@Override
protected void onResume() {                         //开始监听
    super.onResume();
    MyUtils.getInstance(this).start(this);
    onClickmyBtn1(null);
}
public void onClickmyBtn1(View v) {                 //响应单击"浏览"按钮
    myWebView.loadUrl(myEditText.getText().toString());
}
```

上面这段代码在 MyCode \ MySample762 \ app \ src \ main \ java \ com \ bin \ luo \ mysample \ MainActivity.java 文件中。在这段代码中,MyUtils.getInstance(this).start(this)表示启用手机传感

器的监听功能。MyUtils是一个自定义类，该类的主要代码如下：

```java
public class MyUtils {
 private static MyUtils myInstance;
 private Activity myActivity;
 private boolean isLandscape = false;                              //默认是竖屏
 private SensorManager mySensorManager;
 private OrientationSensorListener myOrientationSensorListener;
 private Sensor mySensor;
 //根据重力感应监听结果,改变屏幕朝向
 private Handler myHandler = new Handler(Looper.getMainLooper()) {
  public void handleMessage(Message msg) {
   if (msg.what == 888) {
    int orientation = msg.arg1;
    //根据手机屏幕的朝向角度,设置内容的横竖屏,并且记录状态
    if (orientation > 45 & orientation < 135) {
     myActivity.setRequestedOrientation(
                    ActivityInfo.SCREEN_ORIENTATION_REVERSE_LANDSCAPE);
     isLandscape = true;
    } else if (orientation > 135 & orientation < 225) {
     myActivity.setRequestedOrientation(
                    ActivityInfo.SCREEN_ORIENTATION_REVERSE_PORTRAIT);
     isLandscape = false;
    } else if (orientation > 225 & orientation < 315) {
      myActivity.setRequestedOrientation(
                    ActivityInfo.SCREEN_ORIENTATION_LANDSCAPE);
     isLandscape = true;
    } else if ((orientation > 315 & orientation < 360)
                         || (orientation > 0 & orientation < 45)) {
     myActivity.setRequestedOrientation(
                        ActivityInfo.SCREEN_ORIENTATION_PORTRAIT);
     isLandscape = false;
 } } } };
public static MyUtils getInstance(Context context) {              //获取实例对象
  if (myInstance == null) {
   synchronized (MyUtils.class) {
    if (myInstance == null) { myInstance = new MyUtils(context); }
 } }
  return myInstance;
}
private MyUtils(Context context) {                                //初始化重力感应器
  mySensorManager =
        (SensorManager) context.getSystemService(Context.SENSOR_SERVICE);
  mySensor = mySensorManager.getDefaultSensor(Sensor.TYPE_GRAVITY);
  myOrientationSensorListener = new OrientationSensorListener(myHandler);
}
public class OrientationSensorListener implements SensorEventListener {
   private static final int _DATA_X = 0;
   private static final int _DATA_Y = 1;
   private static final int _DATA_Z = 2;
   public static final int ORIENTATION_UNKNOWN = -1;
   private Handler rotateHandler;
   public OrientationSensorListener(Handler handler){ rotateHandler = handler;}
```

```
public void onAccuracyChanged(Sensor arg0, int arg1) { }
public void onSensorChanged(SensorEvent event) {
  float[] values = event.values;
  int orientation = ORIENTATION_UNKNOWN;
  float X = - values[_DATA_X];
  float Y = - values[_DATA_Y];
  float Z = - values[_DATA_Z];
  float magnitude = X * X + Y * Y;
  if (magnitude * 4 >= Z * Z) {                              //在屏幕旋转时
    float OneEightyOverPi = 57.29577957855f;
    float angle = (float) Math.atan2(-Y, X) * OneEightyOverPi;
    orientation = 90 - Math.round(angle);
    while (orientation >= 360) { orientation -= 360; }
    while (orientation < 0) { orientation += 360; }
  }
  if (rotateHandler != null) {
    rotateHandler.obtainMessage(888, orientation, 0).sendToTarget();
  } } }
public void start(Activity activity) {                       //开始监听
  myActivity = activity;
  mySensorManager.registerListener(myOrientationSensorListener,
                        mySensor, SensorManager.SENSOR_DELAY_UI);
}
public void stop() {                                         //停止监听
  mySensorManager.unregisterListener(myOrientationSensorListener);
  myActivity = null;                                         //防止内存泄漏
} }
```

上面这段代码在 MyCode\MySample762\app\src\main\java\com\bin\luo\mysample\ MyUtils. java 文件中。需要说明的是,访问网络需要在 AndroidManifest. xml 文件中添加< uses-permission android:name = " android. permission. INTERNET "/> 权限。此实例的完整项目在 MyCode\ MySample762 文件夹中。

259 使用 setRequestedOrientation()实现横屏

此实例主要通过在 setRequestedOrientation () 方法中设置 ActivityInfo. SCREEN _ ORIENTATION_LANDSCAPE 参数或 ActivityInfo. SCREEN_ORIENTATION_PORTRAIT 参数,实现限制应用屏幕始终以横屏或竖屏显示。当实例运行之后,单击"竖屏显示"按钮,则应用屏幕将以竖屏显示,如图 259.1 的左图所示。单击"横屏显示"按钮,则应用屏幕将以横屏显示,如图 259.1 的右图所示。

主要代码如下:

```
public void onClickmyBtn1(View v) {                          //响应单击"横屏显示"按钮
  setRequestedOrientation(ActivityInfo.SCREEN_ORIENTATION_LANDSCAPE);
}
public void onClickmyBtn2(View v) {                          //响应单击"竖屏显示"按钮
  setRequestedOrientation(ActivityInfo.SCREEN_ORIENTATION_PORTRAIT);
}
```

上面这段代码在 MyCode \ MySample235 \ app \ src \ main \ java \ com \ bin \ luo \ mysample \

图 259.1

MainActivity.java 文件中。在这段代码中，setRequestedOrientation（ActivityInfo.SCREEN_ORIENTATION_LANDSCAPE）用于横屏显示，setRequestedOrientation（ActivityInfo.SCREEN_ORIENTATION_PORTRAIT）用于竖屏显示。实际上，在许多 Android 应用中，为了避免横竖屏切换引发的不必要麻烦，通常禁止默认的横竖屏切换，即在 AndroidManifest.xml 中通过设置 android：screenOrientation 属性值，android：screenOrientation 属性有以下几个属性值。

（1）unspecified，默认值，由系统来判断显示方向，判定的策略是和设备相关的，所以不同的设备会有不同的显示方向。

（2）landscape，横屏显示，宽比高长。

（3）portrait，竖屏显示，高比宽长。

（4）user，用户当前首选的方向。

（5）behind，和该 Activity 下面的那个 Activity 的方向一致（在 Activity 堆栈中）。

（6）sensor，由物理感应器决定，如果用户旋转设备，则屏幕会进行横竖屏切换。

（7）nosensor，忽略物理感应器，这样就不会随着用户旋转设备而更改了（unspecified 设置除外）。

比如，如果设置 android：screenOrientation＝"portrait"，则无论手机如何变动，拥有这个属性的 Activity 都竖屏显示；如果设置 android：screenOrientation＝"landscape"，则横屏显示。此实例的完整项目在 MyCode\MySample235 文件夹中。

260 根据手机是横屏或是竖屏进行控件布局

此实例主要通过监听 onConfigurationChanged()事件响应方法，实现根据手机是横屏或是竖屏进行控件布局。当实例运行之后，如果当前手机处于竖屏状态，则将同时显示两幅图像，如图 260.1 的左图所示。如果当前手机处于横屏状态，则仅显示下面的图像，如图 260.1 的右图所示。注意：如果测试不成功，请将手机设置中的"竖屏"改为"自动旋转"。

第8章 系统和设备

图 260.1

主要代码如下：

```
public void onConfigurationChanged(Configuration myConfig) {
    super.onConfigurationChanged(myConfig);
    ViewGroup.LayoutParams myParams = myImageView2.getLayoutParams();
    if(myConfig.orientation == Configuration.ORIENTATION_PORTRAIT){
        myImageView1.setVisibility(View.VISIBLE);              //当竖屏时,显示两幅图像
    }else if(myConfig.orientation == Configuration.ORIENTATION_LANDSCAPE){
        myImageView1.setVisibility(View.GONE);                 //当横屏时,仅显示下面的图像
    }
    myImageView2.setLayoutParams(myParams);
}
```

上面这段代码在 MyCode\MySample316\app\src\main\java\com\bin\luo\mysample\MainActivity.java 文件中。在这段代码中,if(myConfig.orientation == Configuration.ORIENTATION_PORTRAIT)表示当前手机是竖屏,if(myConfig.orientation == Configuration.ORIENTATION_LANDSCAPE)表示当前手机是横屏,myImageView1.setVisibility(View.GONE)表示暂时不显示 myImageView1 控件。另外,监听 onConfigurationChanged()需要首先在 AndroidManifest.xml 文件中修改配置<activity android:name=".MainActivity" android:configChanges="orientation|screenSize">和添加权限<uses-permission android:name="android.permission.CHANGE_CONFIGURATION"/>。此实例的完整项目在 MyCode\MySample316 文件夹中。

261 使用 FLAG_FULLSCREEN 标志实现全屏显示

此实例主要通过使用 getWindow()的 setFlags()方法,实现控制应用的界面是否以全屏模式显示。当实例运行之后,单击"全屏显示"按钮,则应用界面的标题栏将隐藏,如图 261.1 的左图所示。

单击"退出全屏"按钮,则将恢复默认界面,显示标题栏,如图 261.1 的右图所示。

图 261.1

主要代码如下:

```
public class MainActivity extends Activity {
    @Override
    protected void onCreate(Bundle savedInstanceState) {
        super.onCreate(savedInstanceState);
        boolean isNoTitle = getIntent().getBooleanExtra("isNoTitle",false);
        requestWindowFeature(isNoTitle?
                Window.FEATURE_NO_TITLE:Window.FEATURE_ACTION_BAR);
        getWindow().setFlags(WindowManager.LayoutParams.FLAG_FULLSCREEN,
            WindowManager.LayoutParams.FLAG_FULLSCREEN);            //设置全屏
        setContentView(R.layout.activity_main);
    }
    //响应单击"全屏显示"按钮
    public void onClickmyBtn1(View v) {
        Intent myIntent = getIntent();
        myIntent.putExtra("isNoTitle",true);
        finish();
        startActivity(myIntent);
    }
    //响应单击"退出全屏"按钮
    public void onClickmyBtn2(View v) {
        Intent myIntent = getIntent();
        myIntent.putExtra("isNoTitle",false);
        finish();
        startActivity(myIntent);
    } }
```

上面这段代码在 MyCode\MySample247\app\src\main\java\com\bin\luo\mysample\MainActivity.java 文件中。在这段代码中,getWindow().setFlags(WindowManager.

LayoutParams.FLAG_FULLSCREEN,WindowManager.LayoutParams.FLAG_FULLSCREEN)用于全屏显示应用。此实例的完整项目在 MyCode\MySample247 文件夹中。

262　使用 Display 获取屏幕宽度和高度

此实例主要通过使用 Display 的 getWidth()方法和 getHeight()方法，获取当前手机的屏幕宽度和屏幕高度。当实例运行之后，单击"获取屏幕的宽度和高度"按钮，则在弹出的 Toast 中显示当前手机的屏幕宽度和屏幕高度，如图 262.1 所示。

图　262.1

主要代码如下：

```
//响应单击"获取屏幕的宽度和高度"按钮
public void onClickmyBtnDiaplay(View v) {
    Display myDisplay = getWindowManager().getDefaultDisplay();
    int myWidth  = myDisplay.getWidth();                     //获取屏幕宽度
    int myHeight = myDisplay.getHeight();                    //获取屏幕高度
    Toast.makeText(getApplicationContext(),"屏幕宽度:" + myWidth
            +",屏幕高度:" + myHeight, Toast.LENGTH_SHORT).show();
}
```

上面这段代码在 MyCode \ MySample078 \ app \ src \ main \ java \ com \ bin \ luo \ mysample \ MainActivity.java 文件中。此实例的完整项目在 MyCode\MySample078 文件夹中。

263　使用 StatFs 获取内部总空间和可用空间大小

此实例主要通过使用 StatFs 的 getBlockCount()方法、getAvailableBlocks()方法、getBlockSize()方法和 Environment 的 getDataDirectory()方法，从而获取手机内部的总空间和可用空间大小。当实

例运行之后,单击"获取手机内部总空间大小"按钮,则在弹出的 Toast 中显示手机内部的总空间大小,如图 263.1 的左图所示;单击"获取手机内部可用空间大小"按钮,则在弹出的 Toast 中显示手机内部的可用空间大小,如图 263.1 的右图所示。

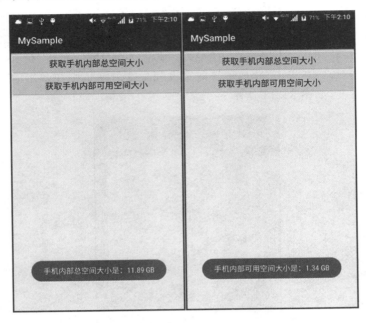

图 263.1

主要代码如下:

```java
//响应单击"获取手机内部总空间大小"按钮
public void onClickmyBtn1(View v) {
    File myPath = Environment.getDataDirectory();
    StatFs myStatFs = new StatFs(myPath.getPath());
    long myBlockSize = myStatFs.getBlockSize();
    long myBlockCount = myStatFs.getBlockCount();
    long myTotalSize = myBlockCount * myBlockSize;
    String mySize = Formatter.formatFileSize(this, myTotalSize);
    Toast.makeText(this, "手机内部总空间大小是:"
            + mySize, Toast.LENGTH_SHORT).show();
}
//响应单击"获取手机内部可用空间大小"按钮
public void onClickmyBtn2(View v) {
    File myPath = Environment.getDataDirectory();
    StatFs myStatFs = new StatFs(myPath.getPath());
    long myBlockSize = myStatFs.getBlockSize();
    long myAvailableBlocks = myStatFs.getAvailableBlocks();
    long myAvailableSize = myAvailableBlocks * myBlockSize;
    String mySize = Formatter.formatFileSize(this, myAvailableSize);
    Toast.makeText(this, "手机内部可用空间大小是:"
            + mySize, Toast.LENGTH_SHORT).show();
}
```

上面这段代码在 MyCode \ MySample667 \ app \ src \ main \ java \ com \ bin \ luo \ mysample \ MainActivity.java 文件中。在上面这段代码中,myBlockSize = myStatFs.getBlockSize()用于获取

单个扇区的大小。myBlockCount = myStatFs.getBlockCount()用于获取全部扇区的数量。myAvailableBlocks = myStatFs.getAvailableBlocks()用于获取可用扇区的数量。然后扇区大小 * 扇区数量即为空间大小。myPath = Environment.getDataDirectory()用于获取根目录/data 内部存储路径。Environment 获取其他目录的方法还有。

（1）Environment.getDownloadCacheDirectory()，该方法用于获取缓存目录/cache。

（2）Environment.getExternalStorageDirectory()，该方法用于获取 SD 卡目录/mnt/ sdcard，即手机外置 sd 卡的路径。

（3）Environment.getRootDirectory()，该方法用于获取系统目录/system。

此实例的完整项目在 MyCode\MySample667 文件夹中。

264　使用 GestureDetector 实现纵向滑动切换

此实例主要通过使用 GestureDetector 动态监听控件操作，并重写 onFling()方法监听滑动事件，从而响应在 ViewFlipper 控件中上下滑动时切换图像。当实例运行之后，向上滑动手指，则 ViewFlipper 控件滑入下一幅图像，向下滑动手指，则 ViewFlipper 控件滑入上一幅图像，效果分别如图 264.1 的左图和右图所示。

图　264.1

主要代码如下：

```
public class MainActivity extends Activity {
    int[] myImages = {R.mipmap.myimage1, R.mipmap.myimage2,
                R.mipmap.myimage3, R.mipmap.myimage4, R.mipmap.myimage5};
    ViewFlipper myViewFlipper;
    @Override
    protected void onCreate(Bundle savedInstanceState) {
     super.onCreate(savedInstanceState);
     setContentView(R.layout.activity_main);
```

```java
myViewFlipper = (ViewFlipper) findViewById(R.id.myViewFlipper);
for (int i = 0; i < myImages.length; i++) {
    ImageView myImageView = new ImageView(MainActivity.this);
    myImageView.setImageResource(myImages[i]);
    myImageView.setScaleType(ImageView.ScaleType.FIT_XY);
    myViewFlipper.addView(myImageView);
}
myViewFlipper.setClickable(true);                          //设置 ViewFlipper 控件可点击
final GestureDetector myGestureDetector =
        new GestureDetector(new GestureDetector.SimpleOnGestureListener() {
    @Override
    //检测手势滑动事件
    public boolean onFling(MotionEvent e1, MotionEvent e2,
                           float velocityX, float velocityY) {
        if (e2.getY() > e1.getY()) showNextView();          //向下滑动
        else if (e1.getY() > e2.getY()) showPreviousView(); //向上滑动
        return false;
    } });
myViewFlipper.setOnTouchListener(new View.OnTouchListener() {
    @Override
    public boolean onTouch(View v, MotionEvent event) {
        //传递事件至手势识别器对象
        return myGestureDetector.onTouchEvent(event);
    } });
}
public void showPreviousView() {
    //设置下滑时,ViewFlipper 子视图进场和退场动画
    myViewFlipper.setInAnimation(AnimationUtils.loadAnimation(this,
                                        R.anim.slide_bottom_in));
    myViewFlipper.setOutAnimation(AnimationUtils.loadAnimation(this,
                                        R.anim.slide_top_out));
    myViewFlipper.showPrevious();                           //显示上一幅图像
}
public void showNextView() {
    //设置上滑时,ViewFlipper 子视图进场和退场动画
    myViewFlipper.setInAnimation(AnimationUtils.loadAnimation(this,
                                        R.anim.slide_top_in));
    myViewFlipper.setOutAnimation(AnimationUtils.loadAnimation(this,
                                        R.anim.slide_bottom_out));
    myViewFlipper.showNext();                               //显示下一幅图像
}
```

上面这段代码在 MyCode\MySample835\app\src\main\java\com\bin\luo\mysample\MainActivity.java 文件中。在这段代码中,myViewFlipper.setInAnimation(AnimationUtils.loadAnimation(this, R.anim.slide_bottom_in))用于设置在手指上滑时图像的进场动画,关于 R.anim.slide_bottom_in 进场动画的详细内容请参考源代码中的 MyCode\MySample835\app\src\main\res\anim\slide_bottom_in.xml 文件。此实例的完整项目在 MyCode\MySample835 文件夹中。

265 自定义手机振动器(Vibrator)的振动模式

此实例主要通过设置 Vibrator 的 vibrate() 方法的模式参数,实现自定义手机振动器(Vibrator)的振动模式。当实例运行之后,单击"自定义振动模式 1"按钮,则手机将按照指定的模式 1 进行振动;单击"自定义振动模式 2"按钮,则手机将按照指定的模式 2 进行振动,效果分别如图 265.1 的左图和右图所示。

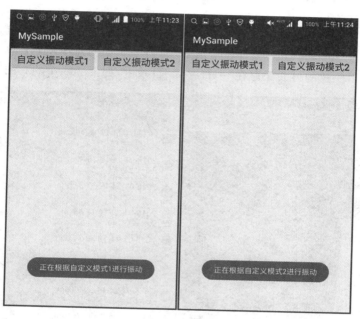

图 265.1

主要代码如下:

```
//响应单击"自定义振动模式 1"按钮
public void onClickmyBtn1(View v) {
    Vibrator myVibrator = (Vibrator) getSystemService(VIBRATOR_SERVICE);
    long[] myPattern = {3000, 1000, 2000, 5000, 3000, 1000};
    myVibrator.vibrate(myPattern, -1);
    Toast.makeText(this,
            "正在根据自定义模式 1 进行振动", Toast.LENGTH_SHORT).show();
}
//响应单击按钮"自定义振动模式 2"
public void onClickmyBtn2(View v) {
    Vibrator myVibrator = (Vibrator) getSystemService(VIBRATOR_SERVICE);
    long[] myPattern = {300, 100, 200, 500, 300, 100};
    myVibrator.vibrate(myPattern, -1);
    Toast.makeText(this,
            "正在根据自定义模式 2 进行振动", Toast.LENGTH_SHORT).show();
}
```

上面这段代码在 MyCode \ MySample370 \ app \ src \ main \ java \ com \ bin \ luo \ mysample \ MainActivity.java 文件中。在这段代码中,vibrate() 方法如果只有 1 个参数,则该参数用来指定振动的 ms 数。如果该方法有两个参数,则第 1 个参数用来指定振动的模式,第 2 个参数用来指定是否需要循环,如 vibrate(myPattern,-1)中的 myPattern 即是预置的振动模式,myPattern = {3000,1000,2000,5000,3000,1000}表示等待 3 秒后,振动 1 秒,再等待 2 秒后,振动 5 秒,再等待 3 秒后,振动 1 秒。此外,操作振动器需要在 AndroidManifest.xml 文件中添加< uses-permission android:name="android.permission.VIBRATE"/>权限。此实例的完整项目在 MyCode\MySample370 文件夹中。

266 使用 SurfaceView 实现照相机的预览功能

此实例主要实现了使用 SurfaceView 控件实现照相机的预览功能。当实例运行之后,摄像头获取的预览图像直接通过 SurfaceView 控件显示出来,如图 266.1 的左图所示;单击"开始照相"按钮,则当前显示的图像将以照片的形式保存在"手机存储"目录中,如图 266.1 的右图所示。

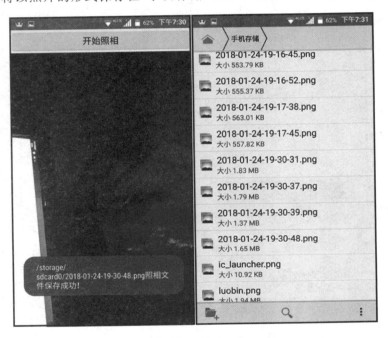

图 266.1

主要代码如下:

```
public class MainActivity extends Activity {
  Camera myCamera;
  SurfaceView mySurfaceView;
  SurfaceHolder myHolder;
  @Override
  protected void onCreate(Bundle savedInstanceState) {
    super.onCreate(savedInstanceState);
    setContentView(R.layout.activity_main);
    mySurfaceView = (SurfaceView) findViewById(R.id.mySurfaceView);
    myHolder = mySurfaceView.getHolder();
    myHolder.setFixedSize(176, 155);
    myHolder.setKeepScreenOn(true);
    myHolder.setType(SurfaceHolder.SURFACE_TYPE_PUSH_BUFFERS);
    myHolder.addCallback(new SurfaceHolder.Callback() {
```

```java
@Override
public void surfaceCreated(SurfaceHolder holder) {
    try {
        myCamera = Camera.open();
        if (myCamera == null) {
            int myCameraCount = Camera.getNumberOfCameras();
            myCamera = Camera.open(myCameraCount - 1);
        }
        Camera.Parameters myParams = myCamera.getParameters();
        myParams.setJpegQuality(80);                           //设置照片质量
        myParams.setPictureSize(1024, 768);                    //设置照片大小
        myParams.setPreviewFrameRate(5);                       //预览帧率
        myCamera.setDisplayOrientation(90);                    //默认为横屏,旋转90度至竖屏
        myCamera.setPreviewDisplay(holder);
        myCamera.setPreviewCallback(new Camera.PreviewCallback() {
            @Override
            public void onPreviewFrame(byte[] data, Camera camera) { }
        });
        myCamera.startPreview();
    } catch (Exception e) { e.printStackTrace(); }
}
@Override
public void surfaceChanged(SurfaceHolder holder,
                           int format, int width, int height) { }
@Override
public void surfaceDestroyed(SurfaceHolder holder) {
    holder.removeCallback(this);
    myCamera.setPreviewCallback(null);
    myCamera.stopPreview();
    myCamera.lock();
    myCamera.release();
    myCamera = null;
} });}
public void onClickButton(View v) {                            //响应单击"开始照相"按钮
    myCamera.takePicture(null, null, new Camera.PictureCallback() {
        @Override
        public void onPictureTaken(byte[] data, Camera camera) {
            String myFileName =
                    new SimpleDateFormat("yyyy-MM-dd-HH-mm-ss").format(new Date());
            File myFile = new File(Environment.getExternalStorageDirectory()
                    + "/" + myFileName + ".png");
            try {
                FileOutputStream myStream = new FileOutputStream(myFile);
                myStream.write(data);
                myStream.close();
                Toast.makeText(MainActivity.this,
                    myFile.getPath() + "照相文件保存成功!", Toast.LENGTH_LONG).show();
                camera.startPreview();                         //重新开始预览
            } catch (Exception e) { e.printStackTrace(); } });
}
}
```

上面这段代码在 MyCode\MySample678\app\src\main\java\com\bin\luo\mysample\MainActivity.java 文件中。需要说明的是，照相需要在 AndroidManifest.xml 文件中添加< uses-permission android：name = " android.permission.CAMERA"/> 和 < uses-permission android：name = "android.permission.WRITE_EXTERNAL_STORAGE"/>权限。此实例的完整项目在 MyCode\MySample678 文件夹中。

267　使用 Camera 实现缩小和放大预览画面

此实例主要通过使用 setZoom()方法控制照相机参数 Camera.Parameters，实现在摄像头预览时缩小和放大预览画面。当实例运行之后，将自动打开摄像头的预览画面，单击"放大当前视角"按钮，则预览画面放大，如图 267.1 的左图所示；单击"缩小当前视角"按钮，则预览画面缩小，如图 267.1 的右图所示。

图　267.1

主要代码如下：

```java
public class MainActivity extends Activity implements SurfaceHolder.Callback {
    Camera myCamera;
    int myZoom = 5;
    private SurfaceHolder myHolder;
    SurfaceView mySurfaceView;                          //用于显示预览画面
    @Override
    protected void onCreate(Bundle savedInstanceState) {
        super.onCreate(savedInstanceState);
        setContentView(R.layout.activity_main);
        mySurfaceView = (SurfaceView) findViewById(R.id.mySufaceView);
        myHolder = mySurfaceView.getHolder();
        myHolder.addCallback(this);
    }
```

```java
@Override
public void surfaceCreated(SurfaceHolder holder) {
  myCamera = Camera.open();
  try {
   myCamera.setDisplayOrientation(90);
   myCamera.setPreviewDisplay(holder);           //通过 surfaceview 显示预览画面
   myCamera.startPreview();                       //开始预览
   myCamera.getParameters().setZoom(myZoom);
  } catch (IOException e) { e.printStackTrace(); }
}
@Override
 public void surfaceChanged(SurfaceHolder holder,
                            int format, int width, int height){ }
@Override
public void surfaceDestroyed(SurfaceHolder holder) {
 myCamera.stopPreview();
 myCamera.release();
 myCamera = null;
}
public void onClickBtn1(View v) {                //响应单击"放大当前视角"按钮
 if (myCamera.getParameters().isZoomSupported()) {
  Camera.Parameters myParams = myCamera.getParameters();
  myParams.setZoom(myZoom += 1);
  myCamera.setParameters(myParams);
} }
public void onClickBtn2(View v) {                //响应单击"缩小当前视角"按钮
 if (myCamera.getParameters().isZoomSupported()) {
  Camera.Parameters myParams = myCamera.getParameters();
  myParams.setZoom(myZoom -= 1);
  myCamera.setParameters(myParams);
} } }
```

上面这段代码在 MyCode\MySample764\app\src\main\java\com\bin\luo\mysample\MainActivity.java 文件中。在这段代码中，myParams = myCamera.getParameters()用于获取照相机参数，myParams.setZoom(myZoom -= 1)用于设置照相机的缩放系数，myCamera.setParameters(myParams)用于设置照相机的参数。此外，操作照相机需要在 AndroidManifest.xml 文件中添加< uses-permission android:name="android.permission.CAMERA"/>权限。此实例的完整项目在 MyCode\MySample764 文件夹中。

268 使用 Camera 实现预览时摄像头手动对焦

此实例主要通过使用 SurfaceView 控件预览摄像头画面，并通过 Camera 对象的方法 onAutoFocus()获取对焦结果，从而实现为摄像头预览添加手动对焦功能。当实例运行之后，将自动打开摄像头的预览画面；如果某处画面模糊不清，点击模糊不清区域则重新以该处为焦点进行对焦操作（出现红色的方框）；当对焦操作完成时，红色对焦方框自动消失，该区域画面会逐渐变得清晰，且在屏幕下方以 Toast 形式提示对焦结果，如图 268.1 所示。测试时，单击预览画面的非中心位置，当这些地方从非焦点变成焦点之后，画面立刻变得清晰起来。

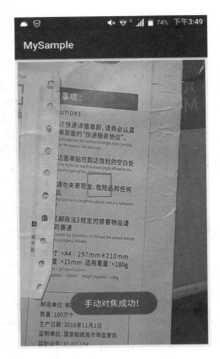

图 268.1

主要代码如下：

```java
public class MainActivity extends Activity implements SurfaceHolder.Callback, View.OnClickListener, View.OnTouchListener {
 Camera myCamera;
 private SurfaceView mySurfaceView;
 private SurfaceHolder myHolder;
 private ImageView myRectView;                               //对焦方框
 @Override
 protected void onCreate(Bundle savedInstanceState) {
  super.onCreate(savedInstanceState);
  setContentView(R.layout.activity_main);
  mySurfaceView = (SurfaceView) findViewById(R.id.MySurfaceView);
  myHolder = mySurfaceView.getHolder();
  myHolder.addCallback(this);
  mySurfaceView.setOnClickListener(this);
  mySurfaceView.setOnTouchListener(this);
  myRectView = (ImageView) findViewById(R.id.MyRectView);
  myRectView.setVisibility(View.GONE);
 }
 @Override
 public void surfaceCreated(SurfaceHolder holder) {
  if (myCamera == null) {
   myCamera = Camera.open();                                 //开启相机
   try {
    myCamera.setDisplayOrientation(90);
    myCamera.setPreviewDisplay(holder);                      //通过 SurfaceView 预览画面
    myCamera.startPreview();                                 //开始预览
   } catch (IOException e) { e.printStackTrace(); }
```

```java
} }
  @Override
    public void surfaceChanged(SurfaceHolder holder,
                               int format, int width, int height){ }
  @Override
  public void surfaceDestroyed(SurfaceHolder holder) {
   myCamera.stopPreview();
   myCamera.release();
   myCamera = null;
  }
  @Override
  public void onClick(View v) {
   if (v.getId() == R.id.MySurfaceView) {
    if (myCamera != null) {
     myCamera.autoFocus(new Camera.AutoFocusCallback() {
      @Override
      public void onAutoFocus(boolean success, Camera camera) {
       if (success) {
        Toast.makeText(MainActivity.this,
                "手动对焦成功!", Toast.LENGTH_LONG).show();
       } else {
        camera.autoFocus(this);                        //如果失败,自动对焦
        Toast.makeText(MainActivity.this,
            "手动对焦失败,已调整为自动对焦!", Toast.LENGTH_LONG).show();
} } }); } } }
  @Override
  public boolean onTouch(View v, MotionEvent event){
   if (event.getAction() == MotionEvent.ACTION_DOWN) {     //绘制对焦方框
    float x = event.getX();
    float y = event.getY();
    Paint myPaint = new Paint();
    myPaint.setStrokeWidth(5);
    myPaint.setStyle(Paint.Style.STROKE);
    myPaint.setColor(Color.RED);
    Bitmap myBitmap = Bitmap.createBitmap(
       getWindowManager().getDefaultDisplay().getWidth(),
       getWindowManager().getDefaultDisplay().getHeight(),
       Bitmap.Config.ARGB_8888);
    Canvas myCanvas = new Canvas(myBitmap);
    myCanvas.drawRect(x - 75, y - 75, x + 75, y + 75, myPaint);
    myRectView.setImageBitmap(myBitmap);
    myRectView.setVisibility(View.VISIBLE);
   } else if (event.getAction() == MotionEvent.ACTION_UP) {
    myRectView.setVisibility(View.GONE);
   }
   return false;
} }
```

上面这段代码在 MyCode\MySample765\app\src\main\java\com\bin\luo\mysample\MainActivity.java 文件中。此外,操作照相机需要在 AndroidManifest.xml 文件中添加< uses-permission android:name="android.permission.CAMERA"/>权限。此实例的完整项目在 MyCode\MySample765 文件夹中。

269　从相册中选择图像并设置为手机壁纸

此实例主要通过使用 setWallpaper() 方法，实现在相册中选择图像，并设置为手机壁纸。当实例运行之后，单击"从相册中选择图像"按钮，则将打开相册目录，然后在其中任意选择一幅图像，则该图像将显示在 ImageView 控件中；单击"设置为桌面壁纸"按钮，则显示的图像将成为桌面壁纸，效果分别如图 269.1 的左图和右图所示。

图　269.1

主要代码如下：

```java
void onClickBtn1(View view){                                      //响应单击"从相册中选择图像"按钮
    myImageView.setDrawingCacheEnabled(true);
    Intent myIntent = new Intent(Intent.ACTION_PICK);
    myIntent.setType("image/*");
    startActivityForResult(myIntent,1);
}
@Override
protected void onActivityResult(int requestCode,
                    int resultCode,Intent data){                  //获取在相册中选择的图像
    Uri myUri = data.getData();                                    //获取图像数据
    myImageView.setImageURI(myUri);                                //预览所选图像
}
void onClickBtn2(View view){                                       //响应单击"设置为桌面壁纸"按钮
    try{
        Bitmap myBitmap = Bitmap.createBitmap(
    myImageView.getDrawingCache());                                //获取缓存的图像数据
        myImageView.setDrawingCacheEnabled(false);                 //清空图像缓存数据
        setWallpaper(myBitmap);                                    //设置图像为手机壁纸
        Toast.makeText(MainActivity.this,"设置壁纸成功!",Toast.LENGTH_SHORT).show();
    }catch(Exception e){ e.printStackTrace(); }
}
```

上面这段代码在 MyCode\MySample657\app\src\main\java\com\bin\luo\mysample\MainActivity.java 文件中。此外,设置壁纸需要在 AndroidManifest.xml 文件中添加 <uses-permission android:name="android.permission.SET_WALLPAPER"/>权限。此实例的完整项目在 MyCode\MySample657 文件夹中。

270 使用 Runnable 间隔执行重复的任务

此实例主要通过使用 Runnable,实现间隔执行重复的任务。当实例运行之后,两幅电影海报图像每间隔 5 秒相互切换一次,并且永不停歇,效果分别如图 270.1 的左图和右图所示。

图 270.1

主要代码如下:

```
public class MainActivity extends Activity {
  public TransitionDrawable myTransition;
  private ImageView myImageView;
  boolean bChecked = true;
  final Handler myHandler = new Handler();
  @Override
  protected void onCreate(Bundle savedInstanceState) {
    super.onCreate(savedInstanceState);
    setContentView(R.layout.activity_main);
    myImageView = (ImageView) findViewById(R.id.myImageView);
    myHandler.postDelayed(myRunnable, 5000);                    //每间隔 5 秒执行一次任务
  }
  //设置要执行的任务
Runnable myRunnable = new Runnable() {
  @Override
  public void run() {
    try {
```

```
            myHandler.postDelayed(this, 5000);
            if (bChecked) {                                  //从 myimage1 切换到 myimage2
              myTransition = new TransitionDrawable(new Drawable[]{
                  getDrawable(R.mipmap.myimage1),
                  getDrawable(R.mipmap.myimage2)});
            } else {                                         //从 myimage2 切换到 myimage1
              myTransition = new TransitionDrawable(new Drawable[]{
                  getDrawable(R.mipmap.myimage2),
                  getDrawable(R.mipmap.myimage1)});
            }
            myImageView.setImageDrawable(myTransition);
            myTransition.startTransition(5000);
            bChecked = !bChecked;
          } catch (Exception e) { e.printStackTrace();}
}};}
```

上面这段代码在 MyCode\MySample655\app\src\main\java\com\bin\luo\mysample\MainActivity.java 文件中。此实例的完整项目在 MyCode\MySample655 文件夹中。

271 使用 Timer 实现促销活动的倒计时功能

此实例主要通过使用 Timer 实现促销活动的倒计时功能。当实例运行之后，立即开始进行倒计时，效果分别如图 271.1 的左图和右图所示。

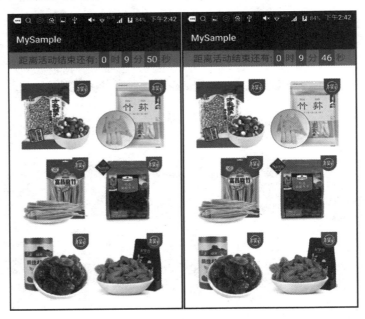

图 271.1

主要代码如下：

```
public class MainActivity extends Activity {
  long myCount = 600;                                        //总时间是 600 秒
  @Override
```

```java
protected void onCreate(Bundle savedInstanceState) {
  super.onCreate(savedInstanceState);
  setContentView(R.layout.activity_main);
  final TextView myHour = (TextView) findViewById(R.id.myHour);
  final TextView myMinute = (TextView) findViewById(R.id.myMinute);
  final TextView mySecond = (TextView) findViewById(R.id.mySecond);
  final Timer myTimer = new Timer();
  myTimer.schedule(new TimerTask() {
   @Override
   public void run() {
    runOnUiThread(new Runnable() {
     @Override
     public void run() {
      myCount --;
      String myString = formatLongToTimeStr(myCount);
      String[] mySpilt = myString.split(":");
      for (int i = 0; i < mySpilt.length; i++) {
       myHour.setText(mySpilt[0]);
       myMinute.setText(mySpilt[1]);
       mySecond.setText(mySpilt[2]);
      }
      if (myCount <= 0) { myTimer.cancel(); }
    } }); } },1000,1000);                               //每秒刷新一次,并递减 1 秒
}
public String formatLongToTimeStr(Long l) {             //格式化时间
 int hour = 0;
 int minute = 0;
 int second = l.intValue();
 if (second > 60) {
  minute = second / 60;
  second = second % 60;
 }
 if (minute > 60) {
  hour = minute / 60;
  minute = minute % 60;
 }
 String myTime = hour + ":" + minute + ":" + second;
 return myTime;
} }
```

上面这段代码在 MyCode\MySample707\app\src\main\java\com\bin\luo\mysample\MainActivity.java 文件中。此实例的完整项目在 MyCode\MySample707 文件夹中。

272 使用 Runtime 执行系统命令静默安装应用包

此实例主要通过使用 Runtime.getRuntime().exec()方法执行系统指令"pm install－r",实现静默安装指定的应用安装包。当实例运行之后,单击"选择安装文件"按钮,然后在弹出的"内部存储空

间"窗口中选择安装文件包,即可实现静默安装,效果分别如图 272.1 的左图和右图所示。

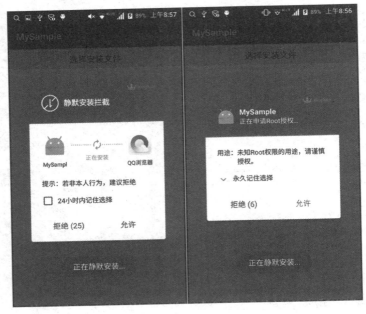

图 272.1

主要代码如下:

```
public class MainActivity extends Activity {
 Handler myHandler = new Handler(){
  @Override
  public void handleMessage(Message msg){
   if(msg.what == 0){
    Toast.makeText(MainActivity.this,
                    "正在静默安装...",Toast.LENGTH_SHORT).show();
   }else if(msg.what == 1){
    Toast.makeText(MainActivity.this,"安装失败!",Toast.LENGTH_SHORT).show();
    finish();
   }else if(msg.what == 2){
    Toast.makeText(MainActivity.this,"安装成功!",Toast.LENGTH_SHORT).show();
    finish();
 } } };
 @Override
 protected void onCreate(Bundle savedInstanceState) {
  super.onCreate(savedInstanceState);
  setContentView(R.layout.activity_main);
 }
 public void onClickBtn1(View v) {                    //响应单击"选择安装文件"按钮
  Intent myIntent = new Intent(Intent.ACTION_GET_CONTENT);
   //筛选文件后缀名是.apk 的安装文件
  myIntent.setType("application/vnd.android.package-archive");
  myIntent.addCategory(Intent.CATEGORY_OPENABLE);
  startActivityForResult(myIntent,1);                 //跳转至文件选择界面
```

```java
    }
    @Override
    protected void onActivityResult(int requestCode, int resultCode, Intent data){
      if(resultCode == RESULT_OK){                    //返回用户所选 APK 文件信息
        myHandler.sendEmptyMessage(0);                //开始静默安装操作
        Uri myUri = data.getData();
        final String myPath = getRealPathFromURI(myUri);    //将 Uri 转换为绝对路径
        myHandler.postDelayed(new Runnable(){
          @Override
          public void run(){                          //进行静默安装并返回安装结果
            boolean myInstallStatus = silenceInstall(myPath);
            if(!myInstallStatus){                     //表示安装失败
              myHandler.sendEmptyMessage(1);
            }else{                                    //表示安装成功
              myHandler.sendEmptyMessage(2);
            } } },1000);                              //使用延时操作是为了等待应用返回主界面,否则会黑屏
      } }
  public String getRealPathFromURI(Uri uri){
    String myPath = null;
    String[] myProjection = {MediaStore.Images.Media.DATA};
    Cursor myCursor = getContentResolver().query(uri,myProjection,null,null,null);
    //在数据库中查询指定 Uri,并返回符合条件的文件路径
    if(myCursor.moveToFirst()){
      int myIndex = myCursor.getColumnIndexOrThrow(MediaStore.Images.Media.DATA);
      myPath = myCursor.getString(myIndex);
    }
    myCursor.close();
    return myPath;
  }
  public boolean silenceInstall(String apkPath){
    boolean myResult = false;
    DataOutputStream myOutputStream = null;
    BufferedReader myErrorReader = null;
    try{
      Process myProcess = Runtime.getRuntime().exec("su");
      myOutputStream = new DataOutputStream(myProcess.getOutputStream());
      String myCommand = "pm install - r " + apkPath + "\n";
      myOutputStream.write(myCommand.getBytes(Charset.forName("utf - 8")));
      myOutputStream.flush();
      myOutputStream.writeBytes("exit\n");
      myOutputStream.flush();
      myProcess.waitFor();
      myErrorReader = new BufferedReader(
                      new InputStreamReader(myProcess.getErrorStream()));
      String myInstallMessage = "",myMessageLine;
      //读取命令的执行结果
      while((myMessageLine = myErrorReader.readLine())!= null){
        myInstallMessage += myMessageLine;
      }
      if (!myInstallMessage.contains("Failure")) myResult = true;
    }catch(Exception e){ }
    finally{
      try{
```

```
        myOutputStream.close();
        myErrorReader.close();
      }catch(Exception e){ }
    }
    return myResult;
} }
```

上面这段代码在 MyCode\MySample873\app\src\main\java\com\bin\luo\mysample\MainActivity.java 文件中。此外,读取 SD 卡文件需要在 AndroidManifest.xml 文件中添加< uses-permission android:name="android.permission.READ_EXTERNAL_STORAGE"/>权限。此实例的完整项目在 MyCode\MySample873 文件夹中。

第三方SDK开发

273 使用腾讯 SDK 获取授权 QQ 账户的简介

此实例主要通过使用腾讯 SDK，实现在登录 QQ 账户成功之后，从当前应用获取 QQ 个人账户的简介信息。当实例运行之后，单击"登录 QQ 账户"按钮，则自动跳转到 QQ 登录界面执行登录 QQ 账户的操作，如果登录成功，则在弹出的 Toast 中显示"登录成功！"，如图 273.1 的左图所示。只有在登录 QQ 账户成功之后，单击"获取账户信息"按钮，才能在弹出的 Toast 中显示该 QQ 账户的简介信息，如图 273.1 的右图所示。

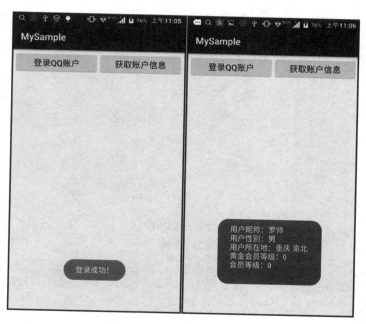

图 273.1

主要代码如下：

```
public class MainActivity extends Activity implements IUiListener {
    Tencent myTencent;
    @Override
    protected void onCreate(Bundle savedInstanceState) {
```

```java
    super.onCreate(savedInstanceState);
    myTencent = Tencent.createInstance("1106416651", getApplicationContext());
    setContentView(R.layout.activity_main);
}
public void onClickBtn1(View v) {                           //响应单击"登录QQ账户"按钮
    myTencent.login(MainActivity.this, "all", this);        //使用QQ账号进行授权登录
}
public void onClickBtn2(View v) {                           //响应单击"获取账户信息"按钮
 UserInfo myUserInfo = new UserInfo(this, myTencent.getQQToken());
    myUserInfo.getUserInfo(new IUiListener() {              //通过回调函数解析用户信息
     @Override
     public void onComplete(Object o) {
      try {
        JSONObject myJsonObject = new JSONObject(o.toString());
        String myNickName = myJsonObject.getString("nickname");
        String myGender = myJsonObject.getString("gender");
        String myProvince = myJsonObject.getString("province");
        String myCity = myJsonObject.getString("city");
        String myYellowVipLevel = myJsonObject.getString("yellow_vip_level");
        String myVipLevel = myJsonObject.getString("level");
        //使用StringBuilder拼接字符串内容
        StringBuilder myBuilder = new StringBuilder();
        myBuilder.append("用户昵称:" + myNickName + "\n");
        myBuilder.append("用户性别:" + myGender + "\n");
        myBuilder.append("用户所在地:" + myProvince + " " + myCity + "\n");
        myBuilder.append("黄金会员等级:" + myYellowVipLevel + "\n");
        myBuilder.append("会员等级:" + myVipLevel + "\n");
        Toast.makeText(MainActivity.this,
                       myBuilder.toString(), Toast.LENGTH_SHORT).show();
      } catch (Exception e) { e.printStackTrace(); }
     }
     @Override
     public void onError(UiError uiError) { }
     @Override
     public void onCancel() { }
    }); }
@Override
public void onComplete(Object o) {
 try {
    //在进行授权登录操作时,获取其access_token、openid和expire_in值,
    //并通过这些字段值来获取授权用户信息
    JSONObject myJsonObject = new JSONObject(o.toString());
    myTencent.setAccessToken(myJsonObject.getString("access_token"),
                             myJsonObject.getString("expires_in"));
    myTencent.setOpenId(myJsonObject.getString("openid"));
    Toast.makeText(MainActivity.this,"登录成功!", Toast.LENGTH_SHORT).show();
 } catch (Exception e) { e.printStackTrace(); }
}
@Override
public void onError(UiError uiError) { }
@Override
public void onCancel() { }
@Override
protected void onActivityResult(int requestCode,
                      int resultCode, Intent data) {    //处理登录回调操作
    Tencent.onActivityResultData(requestCode, resultCode, data, this);
} }
```

上面这段代码在 MyCode\MySample939\app\src\main\java\com\bin\luo\mysample\MainActivity.java 文件中。需要说明的是，此实例需要在 MyCode\MySample939\app\libs 目录下添加库文件 open_sdk_r5886_lite.jar，该库文件的下载地址是 http://wiki.connect.qq.com/sdk%E4%B8%8B%E8%BD%BD。然后在 MyCode\MySample939\app 目录下的 build.gradle 文件的根节点下添加如下代码：

```
repositories { flatDir { dirs 'libs' } }
```

并在该文件的 dependencies 节点下添加依赖项：

```
compile files('libs/open_sdk_r5886_lite.jar')
compile 'com.android.support:support-v4:26.0.0-alpha1'
```

同时还要在 MyCode\MySample939\app\src\main\AndroidManifest.xml 文件中按照下列粗体字所示的内容进行修改：

```xml
<?xml version="1.0" encoding="utf-8"?>
<manifest xmlns:android="http://schemas.android.com/apk/res/android"
    package="com.bin.luo.mysample">
<application
    android:allowBackup="true"
    android:icon="@mipmap/ic_launcher"
    android:label="@string/app_name"
    android:supportsRtl="true"
    android:theme="@style/AppTheme">
<activity android:name=".MainActivity">
 <intent-filter>
   <action android:name="android.intent.action.MAIN"/>
   <category android:name="android.intent.category.LAUNCHER"/>
 </intent-filter>
</activity>
<activity
    android:name="com.tencent.tauth.AuthActivity"
    android:launchMode="singleTask"
    android:noHistory="true">
 <intent-filter>
   <action android:name="android.intent.action.VIEW"/>
   <category android:name="android.intent.category.DEFAULT"/>
   <category android:name="android.intent.category.BROWSABLE"/>
   <data android:scheme="tencent1106416651"/><!-- 填入你的 APP ID -->
 </intent-filter>
</activity>
<activity
    android:name="com.tencent.connect.common.AssistActivity"
    android:configChanges="orientation|keyboardHidden|screenSize"/>
</application>
<uses-permission android:name="android.permission.INTERNET"/>
<uses-permission android:name="android.permission.ACCESS_NETWORK_STATE"/>
</manifest>
```

此外，使用腾讯 SDK 服务需要在腾讯开放平台上创建应用并获取 AppID，步骤如下。

(1) 登录腾讯开放平台(http://open.qq.com)并进入管理中心页面。

(2) 单击右侧的"创建应用"按钮,然后选择"移动应用 安卓"选项,再单击"创建应用"按钮进行下一步操作。在弹出的对话框中选择"软件"选项,并单击"确定"按钮进行下一步操作。

(3) 在表单中填写应用的相关信息,填写完成后单击底部的"提交审核"按钮即可完成应用创建操作。应用审核一般会在 24 小时内完成并反馈 ID 号码,即 AndroidManifest.xml 文件中的 ID 号。

此实例的完整项目在 MyCode\MySample939 文件夹中。

274 使用腾讯 SDK 实现以第三方登录 QQ 账户

此实例主要通过使用腾讯 SDK,实现以第三方形式登录 QQ 账户。当实例运行之后,单击"登录 QQ 账户"按钮,则将显示 QQ 登录窗口,如果登录成功,将在弹出的 Toast 中显示登录结果,效果分别如图 274.1 的左图和右图所示。

图 274.1

主要代码如下:

```
public class MainActivity extends Activity implements IUiListener {
  Tencent myTencent;
  @Override
  protected void onCreate(Bundle savedInstanceState) {
    super.onCreate(savedInstanceState);
    //1106416651 是腾讯反馈给申请人的 ID,各个应用都不相同
    myTencent = Tencent.createInstance("1106416651", getApplicationContext());
    setContentView(R.layout.activity_main);
  }
  public void onClickBtn1(View v) {                       //响应单击"登录 QQ 账户"按钮
    myTencent.login(MainActivity.this, "all", this);
  }
  @Override
  protected void onActivityResult(int requestCode,
```

```
                                int resultCode, Intent data) {        //处理登录回调操作
        Tencent.onActivityResultData(requestCode, resultCode, data, this);
    }
    @Override
    public void onComplete(Object o) {
        Toast.makeText(MainActivity.this, "登录成功!", Toast.LENGTH_SHORT).show();  //登录成功时的回调操作
    }
    @Override
    public void onError(UiError uiError) {
        Toast.makeText(MainActivity.this, "登录失败!", Toast.LENGTH_SHORT).show();  //登录失败时的回调操作
    }
    @Override
    public void onCancel() {
        Toast.makeText(MainActivity.this,                                          //登录取消时的回调操作
                "取消登录操作!", Toast.LENGTH_SHORT).show();
    }
}
```

上面这段代码在 MyCode\MySample942\app\src\main\java\com\bin\luo\mysample\MainActivity.java 文件中。需要说明的是，此实例需要参考实例273（或直接看源代码）的内容在 MyCode\MySample942\app\libs 目录下添加库文件 open_sdk_r5886_lite.jar，及修改 MyCode\MySample942\app\build.gradle 文件和 MyCode\MySample942\app\src\main\AndroidManifest.xml 文件。此实例的完整项目在 MyCode\MySample942 文件夹中。

275　使用腾讯 SDK 将指定文本分享给 QQ 好友

此实例主要通过使用腾讯 SDK，实现将指定的文本内容分享给 QQ 好友。当实例运行之后，在各个输入框中输入对应的内容，如图 275.1 的左图所示；然后单击"分享给 QQ 好友"按钮，则将启用 QQ 分享功能，如图 275.1 的右图所示。

图　275.1

主要代码如下:

```
public void onClickBtn1(View v) {                              //响应单击"分享给QQ好友"按钮
    Bundle myBundle = new Bundle();
    myBundle.putInt(QQShare.SHARE_TO_QQ_KEY_TYPE,
            QQShare.SHARE_TO_QQ_TYPE_APP);                     //设置分享类型为应用分享类型
    myBundle.putString(QQShare.SHARE_TO_QQ_TITLE,
            myEditTitle.getText().toString());                 //设置分享标题
    myBundle.putString(QQShare.SHARE_TO_QQ_SUMMARY,
            myEditSummary.getText().toString());               //设置分享摘要
    myBundle.putString(QQShare.SHARE_TO_QQ_APP_NAME,
            myEditAppname.getText().toString());               //设置分享应用的显示名称
    myTencent.shareToQQ(MainActivity.this, myBundle, this);    //分享应用至QQ好友
}
```

上面这段代码在 MyCode\MySample940\app\src\main\java\com\bin\luo\mysample\MainActivity.java 文件中。需要说明的是,此实例需要参考实例273(或直接看源代码)的内容在 MyCode\MySample940\app\libs 目录下添加库文件 open_sdk_r5886_lite.jar,及修改 MyCode\MySample940\app\build.gradle 文件和 MyCode\MySample940\app\src\main\AndroidManifest.xml 文件。此实例的完整项目在 MyCode\MySample940 文件夹中。

276 使用腾讯 SDK 将本地图像发表到 QQ 空间

此实例主要通过使用腾讯 SDK,实现将本地图像发表到 QQ 空间。当实例运行之后,单击"选择图像文件"按钮,然后在弹出的窗口中选择图像文件,则对应的图像将显示在 ImageView 控件中,如图 276.1 的左图所示;然后单击"发表到 QQ 空间"按钮,则将跳转到发送窗口;在发送窗口中单击"发送"按钮,则该图像将会出现在 QQ 空间中,如图 276.1 的右图所示。

图 276.1

主要代码如下:

```java
public class MainActivity extends Activity implements IUiListener {
  Tencent myTencent;
  String myPath = Environment.getExternalStorageDirectory() + "/myimg2.jpg";
  ImageView myImageView;
  @Override
  protected void onCreate(Bundle savedInstanceState) {
    super.onCreate(savedInstanceState);
    //1106416651 是腾讯反馈给申请人的 ID,各个应用都不相同
    myTencent = Tencent.createInstance("1106416651", getApplicationContext());
    setContentView(R.layout.activity_main);
    myImageView = (ImageView) findViewById(R.id.myImageView);
  }
  public void onClickBtn1(View v) {                    //响应单击"选择图像文件"按钮
    Intent myIntent = new Intent(Intent.ACTION_PICK);
    myIntent.setType("image/*");
    //打开相册窗口,并返回所选图像文件路径
    startActivityForResult(myIntent, 1);
  }
  public void onClickBtn2(View v) {                    //响应单击"发表到 QQ 空间"按钮
    final Bundle myBundle = new Bundle();
    myBundle.putInt(QQShare.SHARE_TO_QQ_KEY_TYPE,
                    QQShare.SHARE_TO_QQ_TYPE_IMAGE);
    myBundle.putString(QQShare.SHARE_TO_QQ_IMAGE_LOCAL_URL, myPath);
    myBundle.putString(QQShare.SHARE_TO_QQ_APP_NAME, "测试应用");
    myBundle.putInt(QQShare.SHARE_TO_QQ_EXT_INT,
                    QQShare.SHARE_TO_QQ_FLAG_QZONE_AUTO_OPEN);
    myTencent.shareToQQ(MainActivity.this, myBundle, this);
  }
  public static String getRealFilePath(final Context context,
          final Uri uri) {                             //获取图像文件的路径
    if (null == uri) return null;
    final String scheme = uri.getScheme();
    String data = null;
    if (scheme == null) data = uri.getPath();
    else if (ContentResolver.SCHEME_FILE.equals(scheme)){data = uri.getPath();}
    else if (ContentResolver.SCHEME_CONTENT.equals(scheme)) {
      Cursor cursor = context.getContentResolver().query(uri,
          new String[]{MediaStore.Images.ImageColumns.DATA}, null, null, null);
      if (null != cursor) {
        if (cursor.moveToFirst()) {
          int index = cursor.getColumnIndex(MediaStore.Images.ImageColumns.DATA);
          if (index > -1) { data = cursor.getString(index); }
        }
        cursor.close();
    } }
    return data;
  }
  @Override
  protected void onActivityResult(int requestCode,
                  int resultCode, Intent data) {
    if (resultCode == RESULT_OK) {
      Uri myUri = data.getData();
      myImageView.setImageURI(myUri);                  //显示所选图像文件
      myPath = getRealFilePath(MainActivity.this, myUri);
    } else {Tencent.onActivityResultData(requestCode, resultCode, data, this);}
```

```
    }
    @Override
    public void onComplete(Object o) {              //发表成功时的回调函数
      Toast.makeText(MainActivity.this, "发表成功!", Toast.LENGTH_SHORT).show();
    }
    @Override
    public void onError(UiError uiError) {          //发表失败时的回调函数
      Toast.makeText(MainActivity.this, "发表失败!"
                        + uiError.errorMessage, Toast.LENGTH_SHORT).show();
    }
    @Override
    public void onCancel() {                        //取消发表时的回调函数
      Toast.makeText(MainActivity.this,"取消发表!",Toast.LENGTH_SHORT).show();
    }
} }
```

上面这段代码在 MyCode\MySample948\app\src\main\java\com\bin\luo\mysample\MainActivity.java 文件中。需要说明的是，此实例需要参考实例 273（或直接看源代码）的内容在 MyCode\MySample948\app\libs 目录下添加库文件 open_sdk_r5886_lite.jar，及修改 MyCode\MySample948\app\build.gradle 文件和 MyCode\MySample948\app\src\main\AndroidManifest.xml 文件。此实例的完整项目在 MyCode\MySample948 文件夹中。

277　使用微信 SDK 将视频链接分享给微信好友

此实例主要通过使用微信 SDK 的 WXMediaMessage 和 SendMessageToWX，实现将指定的视频链接分享到指定的微信好友对话中。当实例运行之后，在"视频文件链接:"输入框中输入视频链接，如"http://v.youku.com/v_show/id_XMzA5MDk5OTcxMg==.html? spm= a2hww.20027244.m_250166.5~5! 2~5~5~5~5~A"，如图 277.1 的左图所示；然后单击"将下面的视频链接分享至微信好友"按钮，则该视频链接内容将出现在指定的微信好友对话中，如图 277.1 的右图所示。

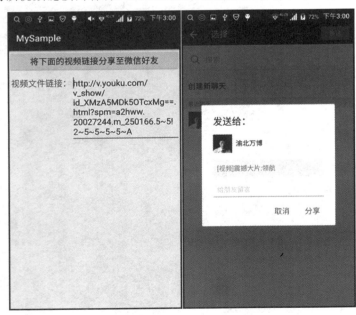

图　277.1

主要代码如下：

```java
public class MainActivity extends Activity {
 IWXAPI myWeChat;
 EditText myEditText;
 @Override
 protected void onCreate(Bundle savedInstanceState) {
  super.onCreate(savedInstanceState);
   myWeChat = WXAPIFactory.createWXAPI(this,
     "wx9fbf8c966226923a", true);                    //创建微信接口对象
   myWeChat.registerApp("wx9fbf8c966226923a");       //将应用的 AppID 注册至微信
   setContentView(R.layout.activity_main);
   myEditText = (EditText) findViewById(R.id.myEditText);
 }
 //响应单击"将下面的视频链接分享至微信好友"按钮
 public void onClickBtn1(View v) {
  WXVideoObject myVideoObject = new WXVideoObject();
  //设置分享视频 url 链接
  myVideoObject.videoUrl = myEditText.getText().toString();
  WXMediaMessage myWXMediaMessage = new WXMediaMessage(myVideoObject);
  myWXMediaMessage.title = "震撼大片:领航";           //设置分享视频标题
  SendMessageToWX.Req myRequest = new SendMessageToWX.Req();
  //设置请求唯一标识符
  myRequest.transaction = String.valueOf(System.currentTimeMillis());
  myRequest.message = myWXMediaMessage;
  myRequest.scene = SendMessageToWX.Req.WXSceneSession;  //微信好友对话
  myWeChat.sendReq(myRequest);                       //发起分享请求
 } }
```

上面这段代码在 MyCode\MySampleX88\app\src\main\java\com\bin\luo\mysample\MainActivity.java 文件中。需要说明的是，此实例需要在 MyCode\MySampleX88\app\build.gradle 文件中添加 compile 'com.tencent.mm.opensdk:wechat-sdk-android-without-mta:+'依赖项，同时还要在 MyCode\MySampleX88\app\src\main\AndroidManifest.xml 文件中添加下列权限：

```
<uses-permission android:name = "android.permission.INTERNET"/>
<uses-permission android:name = "android.permission.ACCESS_NETWORK_STATE"/>
<uses-permission android:name = "android.permission.ACCESS_WIFI_STATE"/>
<uses-permission android:name = "android.permission.READ_PHONE_STATE"/>
```

此外，使用微信 SDK 需要在微信开放平台上创建应用并获取 AppID,具体步骤如下。

（1）登录微信开放平台账号（https://open.weixin.qq.com），进入"管理中心"，单击"创建移动应用"按钮创建新应用。

（2）在表单内填写应用的相关信息，完成后单击"下一步"按钮。

（3）在应用平台选项中选择"Android 应用"，并在其下方表单内填入相关信息，完成后单击"提交审核"按钮即可完成应用创建操作。一般应用审核需要 7 个工作日左右时间，其中应用签名需要通过签名生成工具来获取，该工具可在微信开放平台的资源中心下载。

（4）应用创建审核通过后，会自动生成 AppID 值。用户需要将该 AppID 值传入微信接口对象构造器中以调用微信 SDK 相关功能。详细内容请参考官方网站（https://open.weixin.qq.com）。

此实例的完整项目在 MyCode\MySampleX88 文件夹中。

278　使用微信 SDK 将音乐链接分享到朋友圈

此实例主要通过使用微信 SDK 的 WXMediaMessage 和 SendMessageToWX，实现将指定的音乐链接分享到微信的朋友圈。当实例运行之后，在"音乐文件链接："输入框中输入音乐链接，如"http://staff2.ustc.edu.cn/～wdw/softdown/index.asp/0042515_05.ANDY.mp3"，如图 278.1 的左图所示；然后单击"将下面的音乐链接分享至微信朋友圈"按钮，则该音乐链接将出现在微信的朋友圈中，如图 278.1 的右图所示。

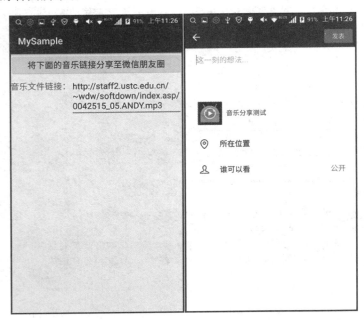

图　278.1

主要代码如下：

```
//响应单击"将下面的音乐链接分享至微信朋友圈"按钮
public void onClickBtn1(View v) {
    WXMusicObject myMusicObject = new WXMusicObject();
    //设置分享音乐的url链接
    myMusicObject.musicUrl = myEditText.getText().toString();
    WXMediaMessage myMessage = new WXMediaMessage(myMusicObject);
    myMessage.mediaObject = myMusicObject;
    myMessage.title = "音乐分享测试";                              //设置音乐链接分享标题
    //设置音乐链接分享缩略图样式
    myMessage.setThumbImage(BitmapFactory.decodeResource(getResources(),
    R.mipmap.ic_launcher));
    SendMessageToWX.Req myRequest = new SendMessageToWX.Req();
    //设置该请求对象所对应的唯一标识符
    myRequest.transaction = String.valueOf(System.currentTimeMillis());
    myRequest.message = myMessage;
    myRequest.scene = SendMessageToWX.Req.WXSceneTimeline;        //微信朋友圈
    myWeChat.sendReq(myRequest);                                  //发送分享请求
}
```

上面这段代码在 MyCode\MySample926\app\src\main\java\com\bin\luo\mysample\MainActivity.java 文件中。需要说明的是，此实例需要在 MyCode\MySample926\app\build.gradle 文件中添加 compile 'com.tencent.mm.opensdk:wechat-sdk-android-without-mta:+'依赖项，同时还要在 MyCode\MySample926\app\src\main\AndroidManifest.xml 文件中添加下列权限：

```
<uses-permission android:name="android.permission.INTERNET"/>
<uses-permission android:name="android.permission.ACCESS_NETWORK_STATE"/>
<uses-permission android:name="android.permission.ACCESS_WIFI_STATE"/>
<uses-permission android:name="android.permission.READ_PHONE_STATE"/>
```

此外，使用微信 SDK 需要在微信开放平台上创建应用并获取 AppID，具体步骤请参考官方网站（https://open.weixin.qq.com）。此实例的完整项目在 MyCode\MySample926 文件夹中。

279　使用百度 SDK 根据起点和终点规划步行线路

此实例主要通过使用百度地图 SDK，实现根据起点和终点规划步行线路。当实例运行之后，在"城市："输入框中输入城市名称，如"北京"，在"起点："输入框中输入起点名称，如"清华大学"，在"终点："输入框中输入终点名称，如"北京动物园"，然后单击"规划步行线路"按钮，则在地图中显示起点和终点之间的步行线路；单击该线路上的中转站图标，则在弹出的 Toast 中显示在该中转站点的步行建议，效果分别如图 279.1 的左图和右图所示。

图　279.1

主要代码如下：

```
public class MainActivity extends Activity {
    BaiduMap myBaiduMap;
    MapView myMapView;
    EditText myEditEnd;
    EditText myEditBegin;
```

```java
EditText myEditCity;
 @Override
 protected void onCreate(Bundle savedInstanceState) {
   super.onCreate(savedInstanceState);
   SDKInitializer.initialize(getApplicationContext());
   setContentView(R.layout.activity_main);
   myMapView = (MapView) findViewById(R.id.myMapView);
   myBaiduMap = myMapView.getMap();
   myEditBegin = (EditText) findViewById(R.id.myEditBegin);
   myEditEnd = (EditText) findViewById(R.id.myEditEnd);
   myEditCity = (EditText) findViewById(R.id.myEditCity);
 }
 public void onClickBtn1(View v) {                         //响应单击"规划步行线路"按钮
   RoutePlanSearch myRoutePlanSearch = RoutePlanSearch.newInstance();
   myRoutePlanSearch.setOnGetRoutePlanResultListener(
     new OnGetRoutePlanResultListener() {
     @Override
     public void onGetWalkingRouteResult(WalkingRouteResult result) {
       if (result.getRouteLines() == null) {
         Toast.makeText(MainActivity.this,
             "抱歉,未搜索到相关路线!", Toast.LENGTH_SHORT).show();
       } else {
         myBaiduMap.clear();                          //重置地图,防止路线重叠
         MyRouteOverlay myRouteOverlay = new MyRouteOverlay(myBaiduMap);
         //将路线数据传递至图层
         myRouteOverlay.setData(result.getRouteLines().get(0));
         myRouteOverlay.addToMap();                  //将路线图显示在地图上
         myRouteOverlay.zoomToSpan();                //自适应缩放
         //设置路线节点点击监听
         myBaiduMap.setOnMarkerClickListener(myRouteOverlay);
       }
     }
     @Override
     public void onGetTransitRouteResult(TransitRouteResult transitRouteResult){}
     @Override
     public void onGetMassTransitRouteResult(MassTransitRouteResult
                                                 massTransitRouteResult) { }
     @Override
     public void onGetDrivingRouteResult(DrivingRouteResult result) { }
     @Override
     public void onGetIndoorRouteResult(IndoorRouteResult indoorRouteResult){}
     @Override
     public void onGetBikingRouteResult(BikingRouteResult bikingRouteResult){}
   });
   //根据地址信息初始化起点对象
   PlanNode myStartPoint = PlanNode.withCityNameAndPlaceName(
           myEditCity.getText().toString(), myEditBegin.getText().toString());
   //根据地址信息初始化终点对象
   PlanNode myEndPoint = PlanNode.withCityNameAndPlaceName(
           myEditCity.getText().toString(), myEditEnd.getText().toString());
   //开始搜索最佳步行路线,并将路线显示在地图上
   myRoutePlanSearch.walkingSearch(
         (new WalkingRoutePlanOption()).from(myStartPoint).to(myEndPoint));
 }
```

```
class MyRouteOverlay extends WalkingRouteOverlay {
    public MyRouteOverlay(BaiduMap baiduMap) { super(baiduMap); }
    @Override
    public boolean onRouteNodeClick(int i) {    //通过 Toast 显示步行建议信息
        Toast.makeText(MainActivity.this, "此处建议:" + mRouteLine.getAllStep().
                get(i).getInstructions(), Toast.LENGTH_SHORT).show();
        return true;
} } }
```

上面这段代码在 MyCode\MySample922\app\src\main\java\com\bin\luo\mysample\MainActivity.java 文件中。注意：显示百度地图的 com.baidu.mapapi.map.MapView 控件在 MyCode\MySample922\app\src\main\res\layout\activity_main.xml 文件中添加。需要说明的是，此实例需要引入百度地图组件，即 MyCode\MySample922\app\libs 目录下的所有内容，有了这些文件之后，就不再需要安装百度地图 App，即使在纯净的模拟器环境中也能进行正常测试。然后在 MyCode\MySample922\app\build.gradle 文件中添加 compile files('libs/BaiduLBS_Android.jar') 和 compile 'com.google.android.gms:play-services-appindexing:8.4.0' 依赖项。此外，还要按照下面粗体字所示的内容修改 AndroidManifest.xml 文件：

```xml
<?xml version="1.0" encoding="utf-8"?>
<manifest xmlns:android="http://schemas.android.com/apk/res/android"
        package="com.bin.luo.mysample">
 <application
        android:allowBackup="true"
        android:icon="@mipmap/ic_launcher"
        android:label="@string/app_name"
        android:supportsRtl="true"
        android:theme="@style/AppTheme">
  <meta-data
        android:name="com.baidu.lbsapi.API_KEY"
        android:value="21xq8w6f8yS6kmvzLxM8MeB0c1k0TXIX" />
  <activity android:name=".MainActivity">
   <intent-filter>
    <action android:name="android.intent.action.MAIN" />
    <category android:name="android.intent.category.LAUNCHER" />
   </intent-filter>
  </activity>
 </application>
 <uses-permission android:name="android.permission.ACCESS_NETWORK_STATE" />
 <uses-permission android:name="android.permission.INTERNET" />
 <uses-permission
            android:name="com.android.launcher.permission.READ_SETTINGS" />
 <uses-permission android:name="android.permission.WAKE_LOCK" />
 <uses-permission android:name="android.permission.CHANGE_WIFI_STATE" />
 <uses-permission android:name="android.permission.ACCESS_WIFI_STATE" />
 <uses-permission android:name="android.permission.GET_TASKS" />
 <uses-permission android:name="android.permission.WRITE_EXTERNAL_STORAGE" />
 <uses-permission android:name="android.permission.WRITE_SETTINGS" />
</manifest>
```

百度地图开发说明请参考官方网址：http://lbsyun.baidu.com/sdk。此实例的完整项目在 MyCode\MySample922 文件夹中。

280　使用百度 SDK 实现将驾车线路分享给好友

此实例主要通过使用百度地图 SDK 的 ShareUrlSearch，实现将指定起点和终点的驾车线路分享给好友。当实例运行之后，在"起点纬度："输入框中输入起点纬度值，如"29.722571"，在"起点经度："输入框中输入起点经度值，如"106.590326"，在"终点纬度："输入框中输入终点纬度值，如"29.919309"，在"终点经度："输入框中输入终点经度值，如"107.248061"，然后单击"显示驾车线路"按钮，则在百度地图中显示指定起点和终点之间的驾车线路，如图 280.1 的左图所示。单击"分享驾车线路"按钮，则将把指定起点和终点之间的驾车线路（地址短串）分享到第三方应用，如 QQ 好友等，如图 280.1 的右图所示。

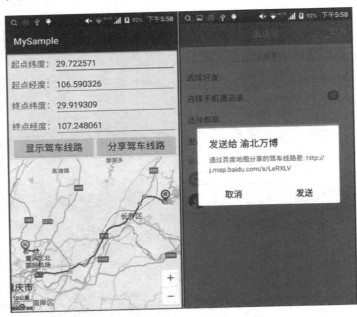

图　280.1

主要代码如下：

```java
public class MainActivity extends Activity {
    BaiduMap myBaiduMap;
    MapView myMapView;
    EditText myEditPoint1A;
    EditText myEditPoint1B;
    EditText myEditPoint2A;
    EditText myEditPoint2B;
    PlanNode myStartNode;
    PlanNode myEndNode;
    @Override
    protected void onCreate(Bundle savedInstanceState) {
        super.onCreate(savedInstanceState);
        SDKInitializer.initialize(getApplicationContext());
        setContentView(R.layout.activity_main);
        myMapView = (MapView) findViewById(R.id.myMapView);
        myBaiduMap = myMapView.getMap();
```

```java
        myEditPoint1A = (EditText) findViewById(R.id.myEditPoint1A);
        myEditPoint1B = (EditText) findViewById(R.id.myEditPoint1B);
        myEditPoint2A = (EditText) findViewById(R.id.myEditPoint2A);
        myEditPoint2B = (EditText) findViewById(R.id.myEditPoint2B);
    }
    public void onClickBtn1(View v) {                        //响应单击"显示驾车线路"按钮
        RoutePlanSearch myRoutePlan = RoutePlanSearch.newInstance();
        myRoutePlan.setOnGetRoutePlanResultListener(
                            new OnGetRoutePlanResultListener(){
            @Override
            public void onGetWalkingRouteResult(WalkingRouteResult walkingRouteResult){ }
            @Override
            public void onGetTransitRouteResult(TransitRouteResult transitRouteResult){ }
            @Override
            public void onGetMassTransitRouteResult(
                            MassTransitRouteResult massTransitRouteResult){ }
            @Override
            public void onGetDrivingRouteResult(DrivingRouteResult result){
                if(result.getRouteLines() == null){
                    Toast.makeText(MainActivity.this,
                            "抱歉,未搜索到驾车路线!",Toast.LENGTH_SHORT).show();
                }else{
                    myBaiduMap.clear();
                    MyRouteOverlay myRouteOverlay = new MyRouteOverlay(myBaiduMap);
                    //将路线数据传递至图层
                    myRouteOverlay.setData(result.getRouteLines().get(0));
                    myRouteOverlay.addToMap();                //将路线图显示在地图上
                    myRouteOverlay.zoomToSpan();              //自适应缩放
                    //设置路线节点点击监听
                    myBaiduMap.setOnMarkerClickListener(myRouteOverlay);
                } }
            @Override
            public void onGetIndoorRouteResult(IndoorRouteResult indoorRouteResult){ }
            @Override
            public void onGetBikingRouteResult(BikingRouteResult bikingRouteResult){ }
        });
        myStartNode = PlanNode.withLocation(new LatLng(
                Double.parseDouble(myEditPoint1A.getText().toString()),
                Double.parseDouble(myEditPoint1B.getText().toString())));
        myEndNode = PlanNode.withLocation(new LatLng(
                Double.parseDouble(myEditPoint2A.getText().toString()),
                Double.parseDouble(myEditPoint2B.getText().toString())));
        //开始搜索最佳驾车路线,并将路线显示在地图上
        myRoutePlan.drivingSearch((
                new DrivingRoutePlanOption()).from(myStartNode).to(myEndNode));
        myRoutePlan.destroy();                               //销毁搜索对象,释放内存
    }
    public void onClickBtn2(View v) {                        //响应单击"分享驾车线路"按钮
        //创建分享检索实例;
        ShareUrlSearch myShareUrlSearch = ShareUrlSearch.newInstance();
        //创建、设置分享检索监听者;
        myShareUrlSearch.setOnGetShareUrlResultListener(
                            new OnGetShareUrlResultListener() {
```

```java
    @Override
    public void onGetPoiDetailShareUrlResult(ShareUrlResult shareUrlResult){ }
    @Override
    public void onGetLocationShareUrlResult(ShareUrlResult shareUrlResult){ }
    @Override
    //使用 Intent 执行分享驾车线路动作
    public void onGetRouteShareUrlResult(ShareUrlResult shareUrlResult) {
      Intent myIntent = new Intent(Intent.ACTION_SEND);
      myIntent.putExtra(Intent.EXTRA_TEXT,
                "通过百度地图分享的驾车线路是:" + shareUrlResult.getUrl());
      myIntent.setType("text/plain");
      startActivity(Intent.createChooser(myIntent, "百度地图驾车线路分享"));
    }});
    RouteShareURLOption.RouteShareMode myRouteShareMode =
            RouteShareURLOption.RouteShareMode.CAR_ROUTE_SHARE_MODE;
    myShareUrlSearch.requestRouteShareUrl(new RouteShareURLOption()
            .from(myStartNode).to(myEndNode).routMode(myRouteShareMode));
    myShareUrlSearch.destroy();                  //销毁分享检索实例;
}
    class MyRouteOverlay extends DrivingRouteOverlay{
        public MyRouteOverlay(BaiduMap baiduMap){super(baiduMap);}
        @Override
        public boolean onRouteNodeClick(int i){   //通过 Toast 显示驾车路线建议信息
        Toast.makeText(MainActivity.this,"驾车建议:" + myRouteLine.getAllStep().get(i).getInstructions(),
Toast.LENGTH_SHORT).show();
            return true;
        }}}
```

上面这段代码在 MyCode\MySampleX65\app\src\main\java\com\bin\luo\mysample\MainActivity.java 文件中。注意：显示百度地图的 com.baidu.mapapi.map.MapView 控件在 MyCode\MySampleX65\app\src\main\res\layout\activity_main.xml 文件中添加。此外，此实例需要引入百度地图组件，即 MyCode\MySampleX65\app\libs 目录下的所有内容，然后在 MyCode\MySampleX65\app\build.gradle 文件中添加 compile files（'libs/BaiduLBS_Android.jar'）和 compile 'com.google.android.gms:play-services-appindexing：8.4.0'依赖项。此外，还要在 AndroidManifest.xml 文件中添加开发密钥和操作权限（参考实例 279 或直接看源代码）。百度地图开发说明请参考官方网址：http://lbsyun.baidu.com/sdk。此实例的完整项目在 MyCode\MySampleX65 文件夹中。

281 使用百度 SDK 调用百度地图 App 的驾车导航

此实例主要通过使用百度地图 SDK 的 BaiduMapNavigation，实现根据起点和终点对驾车线路进行导航。当实例运行之后，在"起点纬度："输入框中输入起点纬度值，如"29.733082"，在"起点经度："输入框中输入起点经度值，如"106.590041"，在"终点纬度："输入框中输入终点纬度值，如"29.919309"，在"终点经度："输入框中输入终点经度值，如"107.248061"，如图 281.1 的左图所示；然后单击"调用百度地图的驾车导航功能"按钮，则将启动百度地图的驾车导航功能，如图 281.1 的右图所示。注意：测试手机需要安装百度地图 App。

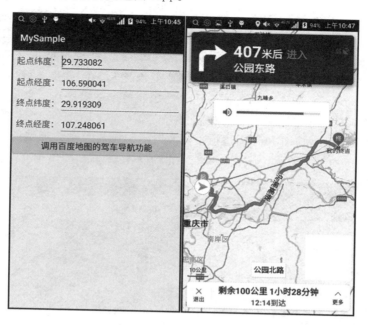

图　281.1

主要代码如下：

```
public void onClickBtn1(View v) {                    //响应单击"调用百度地图的驾车导航功能"按钮
    LatLng myPoint1 =
            new LatLng(Double.valueOf(myEditPoint1A.getText().toString()),
                    Double.valueOf(myEditPoint1B.getText().toString()));
    LatLng myPoint2 =
            new LatLng(Double.valueOf(myEditPoint2A.getText().toString()),
                    Double.valueOf(myEditPoint2B.getText().toString()));
    NaviParaOption myNaviParaOption = new NaviParaOption()
            .startPoint(myPoint1).endPoint(myPoint2)
            .startName("我的起点").endName("我的终点");
    try {//调用百度地图的驾车导航功能
        BaiduMapNavigation.openBaiduMapNavi(myNaviParaOption, this);
    } catch (Exception e) {
        Toast.makeText(MainActivity.this,
                e.getMessage().toString(), Toast.LENGTH_SHORT).show();
    } }
```

上面这段代码在 MyCode\MySampleX55\app\src\main\java\com\bin\luo\mysample\MainActivity.java 文件中。此外，此实例需要引入百度地图组件，即 MyCode\MySampleX55\app\libs 目录下的所有内容，然后在 MyCode\MySampleX55\app\build.gradle 文件中添加 compile files ('libs/BaiduLBS_Android.jar') 和 compile 'com.google.android.gms：play－services－appindexing：8.4.0' 依赖项。并且还要在 AndroidManifest.xml 文件中添加开发密钥和操作权限（参考实例279或直接看源代码）。百度地图开发说明请参考官方网址：http://lbsyun.baidu.com/sdk。此实例的完整项目在 MyCode\MySampleX55 文件夹中。

282　使用百度SDK调用百度地图App的POI检索

此实例主要通过使用百度地图SDK的BaiduMapPoiSearch,实现根据指定的经度、纬度、搜索半径及名称调用百度地图的POI周边检索页面。当实例运行之后,在"纬度:"输入框中输入重庆园博园的纬度,如"29.680962",在"经度:"输入框中输入重庆园博园的经度,如"106.570132",在"搜索半径(米):"输入框中输入搜索半径,如"2000",在"名称:"输入框中输入"重庆园博园",如图282.1的左图所示,然后单击"启动百度地图POI周边检索页面"按钮,则将启动百度地图POI周边检索,并显示搜索结果,如图282.1的右图所示。注意:测试手机需要安装百度地图App。

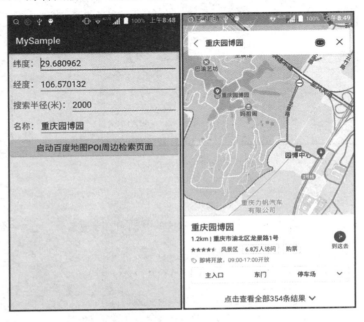

图　282.1

主要代码如下:

```java
public void onClickBtn1(View v) {                //响应单击"启动百度地图POI周边检索页面"按钮
    LatLng myPoint =
            new LatLng(Double.valueOf(myEditPointA.getText().toString()),
                    Double.valueOf(myEditPointB.getText().toString()));
    PoiParaOption myPoiParaOption = new PoiParaOption()
            .key(myEditName.getText().toString()).center(myPoint)
            .radius(Integer.parseInt(myEditRadius.getText().toString()));
    try {                                        //调启百度地图poi周边检索页面
        BaiduMapPoiSearch.openBaiduMapPoiNearbySearch(myPoiParaOption, this);
    } catch (Exception e) {
        Toast.makeText(MainActivity.this,
                e.getMessage().toString(), Toast.LENGTH_SHORT).show();
    }
}
```

上面这段代码在MyCode\MySampleX60\app\src\main\java\com\bin\luo\mysample\MainActivity.java文件中。需要说明的是,此实例需要引入百度地图组件,即MyCode\MySampleX60\app\libs目录下的所有内容,然后在MyCode\MySampleX60\app\build.gradle文件

中添加 compile files('libs/BaiduLBS_Android.jar')和 compile 'com.google.android.gms:play-services-appindexing:8.4.0'依赖项，并在 AndroidManifest.xml 文件中添加开发密钥和操作权限（参考实例 279 或直接看源代码）。百度地图开发说明请参考官方网址：http://lbsyun.baidu.com/sdk。此实例的完整项目在 MyCode\MySampleX60 文件夹中。

283　使用百度 SDK 实现在地图中定位手机位置

此实例主要通过使用百度地图 SDK，实现在百度地图中定位当前手机位置。当实例运行之后，单击"在百度地图中定位当前手机位置"按钮，则在百度地图中显示以当前手机位置为中心的地图，效果如图 283.1 所示。

图　283.1

主要代码如下：

```java
public class MainActivity extends Activity {
    MapView myMapView;
    BaiduMap myBaiduMap;
    @Override
    protected void onCreate(Bundle savedInstanceState) {
        super.onCreate(savedInstanceState);
        SDKInitializer.initialize(getApplicationContext());
        setContentView(R.layout.activity_main);
        myMapView = (MapView) findViewById(R.id.myMapView);
        myBaiduMap = myMapView.getMap();
        MapStatusUpdate myMapStatusUpdate =
                    MapStatusUpdateFactory.zoomTo(20);           //放大显示当前地图
        myBaiduMap.setMapStatus(myMapStatusUpdate);
    }
```

```java
public void onClickBtn1(View v){                              //响应单击"在百度地图中定位当前手
                                                              机位置"按钮
    myBaiduMap.setMyLocationEnabled(true);                    //启用定位图层
    LocationClient myLocationClient = new LocationClient(this);
    myLocationClient.registerLocationListener(new BDLocationListener(){
        @Override
        public void onReceiveLocation(BDLocation location){   //获取定位结果
            if (location == null || myMapView == null){
                Toast.makeText(MainActivity.this,
                        "未获取到当前位置,请稍后重试!",Toast.LENGTH_SHORT).show();
                return;
            }else{
                LatLng myLocation =
                        new LatLng(location.getLatitude(),location.getLongitude());
                MapStatusUpdate myMapStatusUpdate =
                        MapStatusUpdateFactory.newLatLng(myLocation);
                myBaiduMap.setMapStatus(myMapStatusUpdate);   //将地图中心设置为当前位置
                MyLocationData myLocationData = new MyLocationData.Builder()
                        .accuracy(location.getRadius()).latitude(location.getLatitude())
                        .longitude(location.getLongitude()).build();
                myBaiduMap.setMyLocationData(myLocationData); //设置定位数据

                //在地图上显示当前位置
                myBaiduMap.setMyLocationConfigeration(new MyLocationConfiguration(
                        MyLocationConfiguration.LocationMode.NORMAL, true, null));
                Toast.makeText(MainActivity.this,"定位成功!",Toast.LENGTH_SHORT).show();
            }}});
    LocationClientOption myLocationClientOption = new LocationClientOption();
    myLocationClientOption.setOpenGps(true);                  //打开 GPS 功能
    myLocationClientOption.setCoorType("bd09ll");             //设置坐标类型
    myLocationClient.setLocOption(myLocationClientOption);    //设置定位参数
    myLocationClient.start();                                 //开始定位
}}
```

上面这段代码在 MyCode\MySample921\app\src\main\java\com\bin\luo\mysample\MainActivity.java 文件中。注意：显示百度地图的 com.baidu.mapapi.map.MapView 控件在 MyCode\MySample921\app\src\main\res\layout\activity_main.xml 文件中添加。需要说明的是，此实例需要引入百度地图组件，即 MyCode\MySample921\app\libs 目录下的所有内容，然后在 MyCode\MySample921\app\build.gradle 文件中添加 compile files（'libs/BaiduLBS_Android.jar'）和 compile 'com.google.android.gms:play-services-appindexing：8.4.0' 依赖项，并在 AndroidManifest.xml 文件中添加开发密钥和操作权限（参考实例 279 或直接看源代码）。百度地图开发说明请参考官方网址：http://lbsyun.baidu.com/sdk。此实例的完整项目在 MyCode\MySample921 文件夹中。

284 使用百度 SDK 获取在地图上点击位置的地名

此实例主要通过使用百度地图 SDK，实现在单击百度地图时显示该点的位置信息。当实例运行之后，在百度地图中任意点击（主干线），则在弹出的 Toast 中显示该单击位置的具体地址信息，效果分别如图 284.1 的左图和右图所示。

图 284.1

主要代码如下：

```java
public class MainActivity extends Activity {
 @Override
 protected void onCreate(Bundle savedInstanceState) {
  super.onCreate(savedInstanceState);
  SDKInitializer.initialize(getApplicationContext());
  setContentView(R.layout.activity_main);
  MapView myMapView = (MapView) findViewById(R.id.MyMapView);
  BaiduMap myBaiduMap = myMapView.getMap();
  LatLng myLatLng = new LatLng(29.6876, 106.5979);
  MapStatus myMapStatus =
          new MapStatus.Builder().target(myLatLng).zoom(15).build();
  myBaiduMap.setMapStatus(MapStatusUpdateFactory.newMapStatus(myMapStatus));
  //点击地图监听
  myBaiduMap.setOnMapClickListener(new BaiduMap.OnMapClickListener() {
   @Override
   public void onMapClick(LatLng latLng) {
    GeoCoder myGeoCoder = GeoCoder.newInstance();
    myGeoCoder.setOnGetGeoCodeResultListener(
                         new OnGetGeoCoderResultListener() {
     @Override
     public void onGetGeoCodeResult(GeoCodeResult geoCodeResult) { }
     @Override
     //坐标转换为地址的回调函数
     public void onGetReverseGeoCodeResult(
                         ReverseGeoCodeResult reverseGeoCodeResult) {
      Toast.makeText(MainActivity.this, "刚才点击的位置是:" +
```

```
                        reverseGeoCodeResult.getAddress(), Toast.LENGTH_SHORT).show();
            }});
            //将点击地点坐标转换为地址信息
            myGeoCoder.reverseGeoCode(new ReverseGeoCodeOption().location(latLng));
            myGeoCoder.destroy();
        }
        @Override
        public boolean onMapPoiClick(MapPoi mapPoi) { return false; }
    });}}
```

上面这段代码在 MyCode\MySample915\app\src\main\java\com\bin\luo\mysample\MainActivity.java 文件中。注意：显示百度地图的 com.baidu.mapapi.map.MapView 控件在 MyCode\MySample915\app\src\main\res\layout\activity_main.xml 文件中添加。此外，此实例需要引入百度地图组件，即 MyCode\MySample915\app\libs 目录下的所有内容，然后在 MyCode\MySample915\app\build.gradle 文件中添加 compile files('libs/BaiduLBS_Android.jar') 和 compile 'com.google.android.gms:play-services-appindexing:8.4.0' 依赖项，并在 AndroidManifest.xml 文件中添加开发密钥和操作权限(参考实例 279 或直接看源代码)。百度地图开发说明请参考官方网址：http://lbsyun.baidu.com/sdk。此实例的完整项目在 MyCode\MySample915 文件夹中。

285 使用百度 SDK 在地图的城市之间绘制连线

此实例主要通过使用百度 SDK 在百度地图中根据两个城市的经纬度值实现连线的效果。当实例运行之后，单击"显示四点连线效果"按钮，则将用直线将西安、郑州、长沙、重庆连接起来，形成一个封闭的四边形，效果分别如图 285.1 的左图和右图所示。

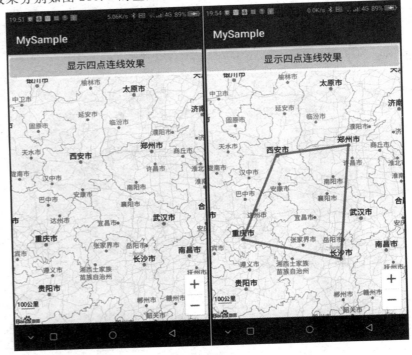

图 285.1

主要代码如下：

```java
public void onClickBtn1(View v) {                    //响应单击按钮"显示四点连线效果"
    LatLng myPoint1 = new LatLng(34.338739,108.896178);   //西安的纬度、经度
    LatLng myPoint2 = new LatLng(34.764961,113.605886);   //郑州的纬度、经度
    LatLng myPoint3 = new LatLng(28.245249,112.943583);   //长沙的纬度、经度
    LatLng myPoint4 = new LatLng(29.557184,106.552688);   //重庆的纬度、经度
    myPoints = new ArrayList<LatLng>();
    myPoints.add(myPoint1);
    myPoints.add(myPoint2);
    myPoints.add(myPoint3);
    myPoints.add(myPoint4);
    myPoints.add(myPoint1);
    OverlayOptions myPolyline = new PolylineOptions().width(10).
            color(0xAAFF0000).points(myPoints);           //绘制折线（连线）
    myBaiduMap.addOverlay(myPolyline);
}
```

上面这段代码在 MyCode\MySampleX10\app\src\main\java\com\bin\luo\mysample\MainActivity.java 文件中。注意：显示百度地图的 com.baidu.mapapi.map.MapView 控件在 MyCode\MySampleX10\app\src\main\res\layout\activity_main.xml 文件中添加。此外，此实例需要引入百度地图组件，即 MyCode\MySampleX10\app\libs 目录下的所有内容，然后在 MyCode\MySampleX10\app\build.gradle 文件中添加 compile files('libs/BaiduLBS_Android.jar')和 compile 'com.google.android.gms:play-services-appindexing:8.4.0'依赖项，并在 AndroidManifest.xml 文件中添加开发密钥和操作权限（参考实例 298 或直接看源代码）。百度地图开发说明请参考官方网址：http://lbsyun.baidu.com/sdk。此实例的完整项目在 MyCode\MySampleX10 文件夹中。

286 使用百度 SDK 在地图上添加图文悬浮框

此实例主要通过使用百度地图 SDK 中的 InfoWindow，实现在百度地图的指定位置添加图文悬浮框。当实例运行之后，单击"在指定的位置添加图文悬浮框"按钮，则在指定的位置（武汉的上方）添加一个图文悬浮框，效果分别如图 286.1 的左图和右图所示。

主要代码如下：

```java
public void onClickBtn1(View v) {                    //响应单击"在指定的位置添加图文悬浮框"按钮
    LatLng myPoint = new LatLng(30.53553,114.324234); //武汉的纬度、经度
    TextView myTextView = new TextView(getApplicationContext());
    NinePatchDrawable myNinePatchDrawable = (NinePatchDrawable)
        getDrawable(R.drawable.luobackground);        //使用指定图设置图文悬浮框的背景
    myTextView.setBackground(myNinePatchDrawable);
    Drawable myDrawable = getResources().getDrawable(R.mipmap.myimage1);
    //在调用 setCompoundDrawables()方法时,
    //必须首先调用 Drawable.setBounds()方法,否则图像不显示
    myDrawable.setBounds(-5, 0,
                myDrawable.getMinimumWidth(), myDrawable.getMinimumHeight());
    //设置图像顺序是：左、上、右、下,此代码仅在文字左边显示图像
    myTextView.setCompoundDrawables(myDrawable, null, null, null);
    myTextView.setWidth(780);                         //设置图文悬浮框宽度
```

```
    myTextView.setHeight(430);                          //设置图文悬浮框高度
    myTextView.setTextColor(Color.BLUE);                //设置文字颜色
    myTextView.setPadding(20,10,10,10);                 //设置内边距,顺序是:左、上、右、下
    myTextView.setText("武汉地处江汉平原东部,长江及其最大支流汉水横贯市境中央,将武汉城区一分为三,
    形成了武昌、汉口、汉阳三镇隔江鼎立的格局.");
    //在 InfoWindow 中加载图文悬浮框(在纵向偏移 -60,以露出武汉2字)
    InfoWindow myInfoWindow = new InfoWindow(myTextView, myPoint, -60);
    myBaiduMap.showInfoWindow(myInfoWindow);            //显示 InfoWindow
}
```

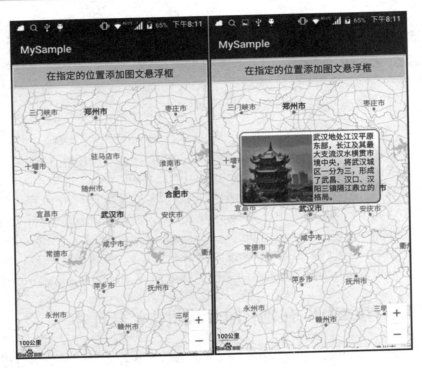

图 286.1

上面这段代码在 MyCode\MySampleX20\app\src\main\java\com\bin\luo\mysample\MainActivity.java 文件中。注意:显示百度地图的 com.baidu.mapapi.map.MapView 控件在 MyCode\MySampleX20\app\src\main\res\layout\activity_main.xml 文件中添加。此外,此实例需要引入百度地图组件,即 MyCode\MySampleX20\app\libs 目录下的所有内容,然后在 MyCode\MySampleX20\app\build.gradle 文件中添加 compile files('libs/BaiduLBS_Android.jar')和 compile 'com.google.android.gms:play-services-appindexing:8.4.0' 依赖项,并在 AndroidManifest.xml 文件中添加开发密钥和操作权限(参考实例 279 或直接看源代码)。百度地图开发说明请参考官方网址: http://lbsyun.baidu.com/sdk。此实例的完整项目在 MyCode\MySampleX20 文件夹中。

287 使用百度 SDK 在地图上添加淡入动画

此实例主要通过使用百度地图 SDK 中的 InfoWindow,实现在百度地图的指定位置添加淡入动画。当实例运行之后,单击"在指定的位置添加淡入动画"按钮,则在指定的位置(重庆江北嘴金融城)浮出一个由透明变为不透明的建筑物图像,效果分别如图 287.1 的左图和右图所示。

图 287.1

主要代码如下：

```java
public void onClickBtn1(View v) {                    //响应单击"在指定的位置添加淡入动画"按钮
    //重庆江北嘴金融城的纬度、经度
    LatLng myPoint = new LatLng(29.580759,106.576481);
    ImageView myImageView = new ImageView(getApplicationContext());
    myImageView.setImageResource(R.mipmap.myimage1);
    //创建淡入图像动画(0f 表示完全透明,1.0f 表示完全不透明)
    AlphaAnimation myAnimation = new AlphaAnimation(0f, 1.0f);
    //设置动画持续时间 5 秒
    myAnimation.setDuration(5000);
    //在 myImageView 上添加淡入动画 myAnimation
    myImageView.startAnimation(myAnimation);
    //创建线性布局管理器
    LinearLayout myLinearLayout = new LinearLayout(this);
    //以垂直方式布局界面控件
    myLinearLayout.setOrientation(LinearLayout.VERTICAL);
    //在线性布局管理器中添加 myImageView 控件
    myLinearLayout.addView(myImageView);
    //在 InfoWindow 中加载 myLinearLayout,实现淡入图像动画
    InfoWindow myInfoWindow = new InfoWindow(myLinearLayout, myPoint,160);
    //显示 InfoWindow
    myBaiduMap.showInfoWindow(myInfoWindow);
}
```

上面这段代码在 MyCode\MySampleX22\app\src\main\java\com\bin\luo\mysample\MainActivity.java 文件中。注意：显示百度地图的 com.baidu.mapapi.map.MapView 控件在 MyCode\MySampleX22\app\src\main\res\layout\activity_main.xml 文件中添加。此外,此实例需要引入百度地图组件,即 MyCode\MySampleX22\app\libs 目录下的所有内容,然后在 MyCode\MySampleX22\app\build.gradle 文件中添加 compile files('libs/BaiduLBS_Android.jar')和 compile 'com.google.android.gms:play-services-appindexing:8.4.0'依赖项,并在 AndroidManifest.xml 文件中添加开发密钥和操作权限(参考实例 279 或直接看源代码)。百度地图开发说明请参考官方网址:http://lbsyun.baidu.com/sdk。此实例的完整项目在 MyCode\MySampleX22 文件夹中。

288 使用百度 SDK 在地图上添加弹跳型动画

此实例主要通过使用百度地图 SDK 中的 MarkerOptions,实现在百度地图的指定位置添加弹跳型动画。当实例运行之后,单击"在指定的位置添加弹跳型动画"按钮,则篮球将在指定的位置(重庆奥林匹克体育中心)上下弹跳,效果分别如图 288.1 的左图和右图所示。

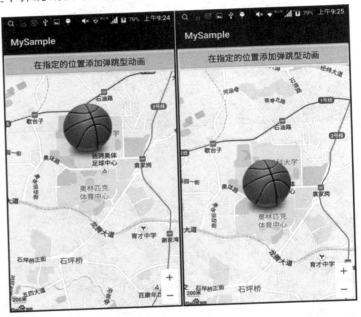

图 288.1

主要代码如下:

```java
public void onClickBtn1(View v) {                //响应单击"在指定的位置添加弹跳型动画"按钮
    myBaiduMap.clear();
    //定义动画图像的(重庆奥林匹克体育中心)坐标点(纬度、经度)
    LatLng myPoint = new LatLng(29.531138,106.512525);
    BitmapDescriptor myBitmap = BitmapDescriptorFactory.
        fromResource(R.mipmap.myimage1);              //获取图像资源
    MarkerOptions myMarkerOptions = new MarkerOptions().position(myPoint).
        icon(myBitmap).zIndex(0).period(10);          //根据图像创建 MarkerOptions
    //在 MarkerOptions 中设置弹跳型动画 MarkerAnimateType.jump
    myMarkerOptions.animateType(MarkerOptions.MarkerAnimateType.jump);
    myBaiduMap.addOverlay(myMarkerOptions);           //在百度地图中添加弹跳型动画
}
```

上面这段代码在 MyCode\MySampleX33\app\src\main\java\com\bin\luo\mysample\MainActivity.java 文件中。注意:显示百度地图的 com.baidu.mapapi.map.MapView 控件在 MyCode\MySampleX33\app\src\main\res\layout\activity_main.xml 文件中添加。此外,此实例需要引入百度地图组件,即 MyCode\MySampleX33\app\libs 目录下的所有内容,然后在 MyCode\

MySampleX33\app\build.gradle 文件中添加 compile files('libs/BaiduLBS_Android.jar')和 compile 'com.google.android.gms:play-services-appindexing：8.4.0'依赖项，并在 AndroidManifest.xml 文件中添加开发密钥和操作权限(参考实例 279 或直接看源代码)。百度地图开发说明请参考官方网址：http://lbsyun.baidu.com/sdk。此实例的完整项目在 MyCode\MySampleX33 文件夹中。

289　使用百度 SDK 在地图上查询指定城市兴趣点

此实例主要通过使用百度地图 SDK，实现在百度地图中查询指定城市的宾馆、美食等兴趣点（POI）信息。当实例运行之后，在"城市："输入框中输入城市名称，如"重庆"，在"兴趣点："输入框中输入"火锅"，然后单击"开始搜索"按钮，则在地图中显示重庆的火锅店分布图，单击其中悬浮的任意一个图标，则在弹出的 Toast 中显示该火锅店的名称，如图 289.1 的左图所示。如果在"城市："输入框中输入"渝北"，在"兴趣点："输入框中输入"楼盘"，然后单击"开始搜索"按钮，则在地图中显示渝北的楼盘分布图，单击其中悬浮的任意一个图标，则在弹出的 Toast 中显示该楼盘的名称，如图 289.1 的右图所示。

图　289.1

主要代码如下：

```java
public void onClickBtn1(View v) {                              //响应单击"开始搜索"按钮
    myPoiSearch = PoiSearch.newInstance();
    myPoiSearch.setOnGetPoiSearchResultListener(
                        new OnGetPoiSearchResultListener(){
        @Override
        public void onGetPoiResult(PoiResult result) {
            if (result.error == SearchResult.ERRORNO.NO_ERROR) {
                myBaiduMap.clear();                            //重置地图
                MarkerOverlay myOverlay = new MarkerOverlay(myBaiduMap);
                myBaiduMap.setOnMarkerClickListener(myOverlay);
```

```
            myOverlay.setData(result);                    //传入目标兴趣点相关数据
            myOverlay.addToMap();                         //在地图上显示
            myOverlay.zoomToSpan();                       //自适应缩放
        } else {
            Toast.makeText(MainActivity.this,
                    "暂时未搜索到相关" + myEditTarget, Toast.LENGTH_SHORT).show();
        }
    }
    @Override
    public void onGetPoiDetailResult(PoiDetailResult poiDetailResult) {}
    @Override
    public void onGetPoiIndoorResult(PoiIndoorResult poiIndoorResult) { } });
    myPoiSearch.searchInCity((new PoiCitySearchOption())
            .city(myEditCity.getText().toString())
            .keyword(myEditTarget.getText().toString())
            .pageNum(10));                                //在指定城市搜索目标,最多显示10个
}
class MarkerOverlay extends PoiOverlay {
    public MarkerOverlay(BaiduMap baiduMap) { super(baiduMap); }
    @Override
    //自定义标记点击监听事件
    public boolean onPoiClick(int i) {
        //获取点击位置对应的信息
        PoiInfo myPoiInfo = getPoiResult().getAllPoi().get(i);
        Toast.makeText(MainActivity.this, myPoiInfo.name, Toast.LENGTH_SHORT).show();
        return true;
    } }
```

上面这段代码在 MyCode\MySampleX41\app\src\main\java\com\bin\luo\mysample\MainActivity.java 文件中。注意:显示百度地图的 com.baidu.mapapi.map.MapView 控件在 MyCode\MySampleX41\app\src\main\res\layout\activity_main.xml 文件中添加。此外,此实例需要引入百度地图组件,即 MyCode\MySampleX41\app\libs 目录下的所有内容,然后在 MyCode\MySampleX41\app\build.gradle 文件中添加 compile files('libs/BaiduLBS_Android.jar')和 compile 'com.google.android.gms:play-services-appindexing:8.4.0'依赖项,并在 AndroidManifest.xml 文件中添加开发密钥和操作权限(参考实例279或直接看源代码)。百度地图开发说明请参考官方网址:http://lbsyun.baidu.com/sdk。此实例的完整项目在 MyCode\MySampleX41 文件夹中。

290 使用百度SDK在地图上为行政区添加边界线

此实例主要通过使用百度地图 SDK 的 DistrictSearch,实现为指定行政区添加自定义边界线。当实例运行之后,在"重庆市的区县名称:"输入框中输入区县名称,如"巴南区",然后单击"开始查询"按钮,则在百度地图中使用红色的虚线标注巴南区的行政区域,如图290.1的左图所示。如果在"重庆市的区县名称:"输入框中输入"北碚区",然后单击"开始查询"按钮,则在百度地图中使用红色的虚线标注北碚区的行政区域,如图290.1的右图所示。

图 290.1

主要代码如下：

```java
//响应单击"开始查询"按钮
public void onClickBtn1(View v) {
    myBaiduMap.clear();
    //建立查询条件
    DistrictSearchOption myDistrictSearchOption = new DistrictSearchOption().
            cityName("重庆").districtName(myEditText.getText().toString());
    //设置查询监听事件
    myDistrictSearch.setOnDistrictSearchListener(
                            new OnGetDistricSearchResultListener() {
        @Override
        public void onGetDistrictResult(DistrictResult districtResult) {
            if (districtResult.error == SearchResult.ERRORNO.NO_ERROR) {
                List<List<LatLng>> pointsList = districtResult.getPolylines();
                if (pointsList == null) return;
                for (List<LatLng> polyline : pointsList) {
                    OverlayOptions myPolylineOptions = new PolylineOptions().width(10)
                            .points(polyline).dottedLine(true).color(Color.RED);
                    //添加区域边界线 Overlay
                    myBaiduMap.addOverlay(myPolylineOptions);
                } } } });
    //执行行政区域的查询
    myDistrictSearch.searchDistrict(myDistrictSearchOption);
}
```

上面这段代码在 MyCode\MySampleX44\app\src\main\java\com\bin\luo\mysample\MainActivity.java 文件中。注意：显示百度地图的 com.baidu.mapapi.map.MapView 控件在 MyCode\MySampleX44\app\src\main\res\layout\activity_main.xml 文件中添加。此外，此实例需要引入百度地图组件，即 MyCode\MySampleX44\app\libs 目录下的所有内容，然后在 MyCode\

MySampleX44\app\build.gradle 文件中添加 compile files('libs/BaiduLBS_Android.jar')和 compile 'com.google.android.gms:play-services-appindexing:8.4.0'依赖项,并在 AndroidManifest.xml 文件中添加开发密钥和操作权限(参考实例 279 或直接看源代码)。百度地图开发说明请参考官方网址:http://lbsyun.baidu.com/sdk。此实例的完整项目在 MyCode\MySampleX44 文件夹中。

291 使用百度 SDK 在地图指定范围添加圆角矩形

此实例主要通过使用百度地图 SDK 中的 GroundOverlayOptions,实现在百度地图指定的地理范围内添加线性渐变的圆角矩形。当实例运行之后,单击"在指定的地理范围添加圆角矩形"按钮,则在指定的矩形范围(重庆两江新区)内显示一个线性渐变的圆角矩形及文字"重庆两江新区",效果分别如图 291.1 的左图和右图所示。

图 291.1

主要代码如下:

```
public void onClickBtn1(View v) {                    //响应单击"在指定的地理范围添加圆角矩形"
                                                      //按钮

    myBaiduMap.clear();
    //定义圆角矩形的地理范围
    LatLng myPoint1 = new LatLng(29.810734,106.405711);   //左上角纬度经度
    LatLng myPoint2 = new LatLng(29.517988,106.949006);   //右下角纬度经度
    LatLngBounds myBounds = new LatLngBounds.Builder()
                            .include(myPoint1).include(myPoint2).build();
    TextView myTextView = new TextView(getApplicationContext());
    myTextView.setWidth(250);
    myTextView.setHeight(250);
    myTextView.setText("重庆两江新区");
    myTextView.setGravity(Gravity.CENTER);              //设置文字居中对齐
    myTextView.setTextSize(28);                         //设置文字大小
    myTextView.setTextColor(Color.RED);                 //设置文字颜色
```

```java
//在TextView上设置圆角矩形背景
myTextView.setBackground(getDrawable(R.drawable.myshape));
//创建线性布局管理器
LinearLayout myLinearLayout = new LinearLayout(this);
//以垂直方式布局界面控件
myLinearLayout.setOrientation(LinearLayout.VERTICAL);
//在线性布局管理器中添加myTextView控件
myLinearLayout.addView(myTextView);
//将线性布局转换为BitmapDescriptor
BitmapDescriptor myShape = BitmapDescriptorFactory.fromView(myLinearLayout);
//根据地理范围和圆角矩形创建GroundOverlayOptions
GroundOverlayOptions myGroundOverlayOptions = new GroundOverlayOptions()
        .positionFromBounds(myBounds)              //设置地理范围
        .image(myShape)                            //设置覆盖图形
        .transparency(0.7f);                       //设置图像透明度
myBaiduMap.addOverlay(myGroundOverlayOptions);
}
```

上面这段代码在 MyCode\MySampleX36\app\src\main\java\com\bin\luo\mysample\MainActivity.java 文件中。在这段代码中，myTextView.setBackground(getDrawable(R.drawable.myshape))表示使用 XML 文件 myshape 配置的线性渐变的圆角矩形设置 TextView 控件的背景，myshape 文件的主要内容如下：

```xml
<?xml version = "1.0" encoding = "utf-8"?>
<shape xmlns:android = "http://schemas.android.com/apk/res/android"
        android:shape = "rectangle">
    <size android:width = "250dp" android:height = "250dp"/>
    <stroke android:width = "2dp"
            android:color = "#ff0000"
            android:dashGap = "4sp"
            android:dashWidth = "4dp" />
    <gradient android:angle = "270"
            android:endColor = "#0000ff"
            android:startColor = "#ffffff" />
    <corners android:radius = "20dp" />
</shape>
```

上面这段代码在 MyCode\MySampleX36\app\src\main\res\drawable\myshape.xml 文件中。注意：显示百度地图的 com.baidu.mapapi.map.MapView 控件在 MyCode\MySampleX36\app\src\main\res\layout\activity_main.xml 文件中添加。此外，此实例需要引入百度地图组件，即 MyCode\MySampleX36\app\libs 目录下的所有内容，然后在 MyCode\MySampleX36\app\build.gradle 文件中添加 compile files('libs/BaiduLBS_Android.jar') 和 compile 'com.google.android.gms:play-services-appindexing:8.4.0' 依赖项，并在 AndroidManifest.xml 文件中添加开发密钥和操作权限（参考实例 279 或直接看源代码）。百度地图开发说明请参考官方网址：http://lbsyun.baidu.com/sdk。此实例的完整项目在 MyCode\MySampleX36 文件夹中。

292 使用百度 SDK 查询指定地点的热力图

此实例主要通过使用百度地图 SDK，实现查询指定经度和纬度的热力图。当实例运行之后，在"纬度："输入框中输入"29.6876"，在"经度："输入框中输入"106.5979"，然后单击"显示热力图"按钮，则在百度地图中显示重庆主城的热力图，效果如图292.1的左图所示；如果在"纬度："输入框中输入"30.67"，在"经度："输入框中输入"104.06"，然后单击"显示热力图"按钮，则在百度地图中显示成都主城的热力图，效果如图292.1的右图所示。

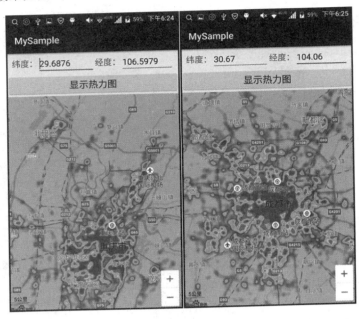

图 292.1

主要代码如下：

```
public void onClickBtn1(View v) {                    //响应单击"显示热力图"按钮
    myBaiduMap = myMapView.getMap();
    LatLng myLocation =
        new LatLng(Double.parseDouble(myEditLatitude.getText().toString()),
        Double.parseDouble(myEditLongitude.getText().toString()));
    MapStatus myMapStatus = new MapStatus.Builder().target(myLocation).build();
    myBaiduMap.setMapStatus(MapStatusUpdateFactory.newMapStatus(myMapStatus));
    myBaiduMap.setBaiduHeatMapEnabled(true);         //允许显示热力图层
}
```

上面这段代码在 MyCode\MySample916\app\src\main\java\com\bin\luo\mysample\MainActivity.java 文件中。注意：显示百度地图的 com.baidu.mapapi.map.MapView 控件在 MyCode\MySample916\app\src\main\res\layout\activity_main.xml 文件中添加。此外，此实例需要引入百度地图组件，即 MyCode\MySample916\app\libs 目录下的所有内容，然后在 MyCode\MySample916\app\build.gradle 文件中添加 compile files('libs/BaiduLBS_Android.jar')和 compile 'com.google.android.gms:play-services-appindexing:8.4.0'依赖项，并在 AndroidManifest.xml 文件中添加开发密钥和操作权限（参考实例279或直接看源代码）。百度地图开发说明请参考官方网址：http://lbsyun.baidu.com/sdk。此实例的完整项目在 MyCode\MySample916 文件夹中。

293 使用百度SDK实现隐藏或显示地名标注信息

此实例主要通过使用百度地图SDK的showMapPoi()方法，实现隐藏或显示百度地图的地名标注等信息。当实例运行之后，单击"显示标注信息"按钮，则在百度地图中显示每个POI的地名标注信息,如图293.1的左图所示。单击"隐藏标注信息"按钮,则将隐藏在百度地图的每个POI的地名标注信息,如图293.1的右图所示。

图 293.1

主要代码如下:

```java
public class MainActivity extends Activity {
    BaiduMap myBaiduMap;
    MapView myMapView;
    @Override
    protected void onCreate(Bundle savedInstanceState) {
        super.onCreate(savedInstanceState);
        SDKInitializer.initialize(getApplicationContext());
        setContentView(R.layout.activity_main);
        myMapView = (MapView) findViewById(R.id.myMapView);
        myBaiduMap = myMapView.getMap();
        myMapView.showZoomControls(true);           //显示右下角百度地图原生的放大和缩小按钮
        //重庆红旗河沟的经纬度(106.53279,29.59129)
        LatLng myPoint = new LatLng(29.59129,106.53279);
        MapStatus myMapStatus = new MapStatus.Builder().target(myPoint).build();
        myBaiduMap.setMapStatus(MapStatusUpdateFactory.newMapStatus(myMapStatus));
        MapStatusUpdate myMapStatusUpdate = MapStatusUpdateFactory.zoomTo(16);
        myBaiduMap.animateMapStatus(myMapStatusUpdate);
    }
    //响应单击"显示标注信息"按钮
    public void onClickBtn1(View v) {
        myBaiduMap.showMapPoi(true);
    }
    //响应单击"隐藏标注信息"按钮
    public void onClickBtn2(View v) {
        myBaiduMap.showMapPoi(false);
    }
}
```

上面这段代码在 MyCode\MySampleX77\app\src\main\java\com\bin\luo\mysample\MainActivity.java 文件中。注意：显示百度地图的 com.baidu.mapapi.map.MapView 控件在 MyCode\MySampleX77\app\src\main\res\layout\activity_main.xml 文件中添加。此外，此实例需要引入百度地图组件，即 MyCode\MySampleX77\app\libs 目录下的所有内容，然后在 MyCode\MySampleX77\app\build.gradle 文件中添加 compile files('libs/BaiduLBS_Android.jar')和 compile 'com.google.android.gms:play-services-appindexing：8.4.0'依赖项，并在 AndroidManifest.xml 文件中添加开发密钥和操作权限（参考实例 279 或直接看源代码）。百度地图开发说明请参考官方网址：http://lbsyun.baidu.com/sdk。此实例的完整项目在 MyCode\MySampleX77 文件夹中。

294　使用百度 SDK 实现以俯视角观察街道三维图

此实例主要通过使用百度 SDK 的 MapStatus，实现在指定位置以 45 度俯视角观察三维百度地图街道。当实例运行之后，将显示以重庆江北观音桥为中心的百度地图，如图 294.1 的左图所示。单击"以 45 度俯视角观察三维百度地图街道"按钮，则该地图上的建筑物将呈现三维效果，如图 294.1 的右图所示。

图　294.1

主要代码如下：

```
public class MainActivity extends Activity {
  BaiduMap myBaiduMap;
  MapView myMapView;
  @Override
  protected void onCreate(Bundle savedInstanceState) {
    super.onCreate(savedInstanceState);
    SDKInitializer.initialize(getApplicationContext());
    setContentView(R.layout.activity_main);
    myMapView = (MapView) findViewById(R.id.myMapView);
    myBaiduMap = myMapView.getMap();
```

```
    myMapView.showZoomControls(true);
    //重庆江北观音桥的经纬度(106.539578,29.579105)
    LatLng myPoint = new LatLng(29.579105,106.539578);
    MapStatus myMapStatus = new MapStatus.Builder().target(myPoint).build();
    myBaiduMap.setMapStatus(MapStatusUpdateFactory.newMapStatus(myMapStatus));
    MapStatusUpdate myMapStatusUpdate = MapStatusUpdateFactory.zoomTo(18);
    myBaiduMap.animateMapStatus(myMapStatusUpdate);
}
//响应单击"以45度俯视角观察三维百度地图街道"按钮
public void onClickBtn1(View v) {
    MapStatus myMapStatus =
            new MapStatus.Builder().zoom(18).overlook(-45).build();
    myBaiduMap.setMapStatus(MapStatusUpdateFactory.newMapStatus(myMapStatus));
} }
```

上面这段代码在 MyCode\MySampleX82\app\src\main\java\com\bin\luo\mysample\MainActivity.java 文件中。注意：显示百度地图的 com.baidu.mapapi.map.MapView 控件在 MyCode\MySampleX82\app\src\main\res\layout\activity_main.xml 文件中添加。此外，此实例需要引入百度地图组件，即 MyCode\MySampleX82\app\libs 目录下的所有内容，然后在 MyCode\MySampleX82\app\build.gradle 文件中添加 compile files('libs/BaiduLBS_Android.jar')和 compile 'com.google.android.gms:play-services-appindexing:8.4.0'依赖项，并在 AndroidManifest.xml 文件中添加开发密钥和操作权限(参考实例 279 或直接看源代码)。百度地图开发说明请参考官方网址：http://lbsyun.baidu.com/sdk。此实例的完整项目在 MyCode\MySampleX82 文件夹中。

295 使用百度 SDK 实现根据经纬度计算两地距离

此实例主要通过使用百度地图 SDK 的 DistanceUtil，实现根据经纬度值计算两地之间的距离。当实例运行之后，在"甲地纬度："输入框中输入重庆的纬度值，如"29.615452"，在"甲地经度："输入框中输入重庆的经度值，如"106.557317"；在"乙地纬度："输入框中输入成都的纬度值，如"30.701717"，在"乙地经度："输入框中输入成都的经度值，如"104.080302"，然后单击按钮"计算甲乙两地的直线距离"，则在弹出的 Toast 中显示重庆和成都之间的直线距离值，如图 295.1 的左图所示。如果在"甲地纬度："输入框中输入重庆的纬度值，如"29.615452"，在"甲地经度："输入框中输入重庆的经度值，如"106.557317"；在"乙地纬度："输入框中输入北京的纬度值，如"39.900835"，在"乙地经度："输入框中输入北京的经度值，如"116.328103"，然后单击按钮"计算甲乙两地的直线距离"，则在弹出的 Toast 中显示重庆和北京之间的直线距离值，如图 295.1 的右图所示。

主要代码如下：

```
public void onClickBtn1(View v) {                //响应单击"计算甲乙两地的直线距离"按钮
    LatLng myPoint1 =
            new LatLng(Double.valueOf(myEditPoint1A.getText().toString()),
                    Double.valueOf(myEditPoint1B.getText().toString()));
    LatLng myPoint2 =
            new LatLng(Double.valueOf(myEditPoint2A.getText().toString()),
                    Double.valueOf(myEditPoint2B.getText().toString()));
    double myDistance = DistanceUtil.getDistance(myPoint1, myPoint2)/1000;
    Toast.makeText(MainActivity.this,"两地的距离是:"
                    + myDistance + "公里", Toast.LENGTH_SHORT).show();
}
```

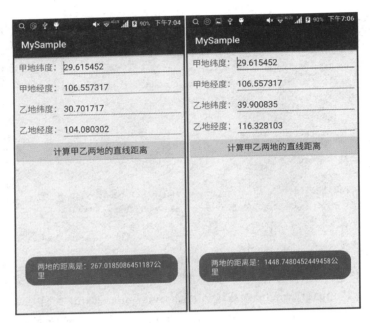

图　295.1

上面这段代码在 MyCode\MySampleX47\app\src\main\java\com\bin\luo\mysample\MainActivity.java 文件中。此外，此实例需要引入百度地图组件，即 MyCode\MySampleX47\app\libs 目录下的所有内容，然后在 MyCode\MySampleX47\app\build.gradle 文件中添加 compile files('libs/BaiduLBS_Android.jar') 和 compile 'com.google.android.gms：play-services-appindexing：8.4.0' 依赖项，并在 AndroidManifest.xml 文件中添加开发密钥和操作权限（参考实例 279 或直接看源代码）。百度地图开发说明请参考官方网址：http://lbsyun.baidu.com/sdk。此实例的完整项目在 MyCode\MySampleX47 文件夹中。

296　使用新浪 SDK 实现跳转到微博主页

此实例主要通过使用新浪微博 SDK，实现从当前应用直接跳转新浪微博的个人账户主页。当实例运行之后，单击按钮"跳转到新浪微博主页"，则将启动新浪微博应用并跳转到个人账户主页，效果分别如图 296.1 的左图和右图所示。

主要代码如下：

```java
public class MainActivity extends Activity {
    AuthInfo myAuthInfo;
    @Override
    protected void onCreate(Bundle savedInstanceState) {
        super.onCreate(savedInstanceState);
        myAuthInfo = new AuthInfo(this, "639620816", "http://www.sina.com", null);
        WbSdk.install(this, myAuthInfo);                    //初始化微博 SDK
        setContentView(R.layout.activity_main);
    }
    public void onClickBtn1(View v) {                       //响应单击"跳转到新浪微博主页"按钮
        WeiboPageUtils.getInstance(MainActivity.this,
            myAuthInfo).startUserMainPage("",false);        //在微博应用中显示用户的个人主页
    }
}
```

图 296.1

上面这段代码在 MyCode\MySample935\app\src\main\java\com\bin\luo\mysample\MainActivity.java 文件中。需要说明的是，此实例需要在 MyCode\MySample935\app\libs 目录下添加库文件 core-4.1.0-openDefaultRelease.aar，然后在 MyCode\MySample935\app 目录下的 build.gradle 文件的根节点下添加如下代码：

```
repositories { flatDir { dirs 'libs' } }
```

并在该文件的 dependencies 节点下添加依赖项：

```
compile 'com.sina.weibo.sdk:core:4.1.0:openDefaultRelease@aar'
```

同时还要在 MyCode\MySample935\app\src\main\AndroidManifest.xml 文件中添加下列权限：

```
<uses-permission android:name="android.permission.INTERNET"/>
<uses-permission android:name="android.permission.ACCESS_WIFI_STATE"/>
<uses-permission android:name="android.permission.ACCESS_NETWORK_STATE"/>
```

此外，使用新浪微博 SDK 需要在微博开放平台上创建应用并获取 AppKey，应用创建步骤如下。

（1）登录新浪微博开放平台（http://open.weibo.com/development/mobile），单击"移动应用"下方的"立即接入"按钮，并在弹出的对话框中单击"继续创建"按钮。

（2）在弹出的页面表单中填入移动应用相关信息，完成后单击"创建"按钮即可完成应用创建操作。

（3）此外，还须完善一部分应用信息，即需要在表单内填入应用包名、应用签名、应用下载地址、Android 下载地址、应用简介、应用介绍等必需信息。

详细情况请登录新浪微博开放平台。此实例的完整项目在 MyCode\MySample935 文件夹中。

297 使用新浪 SDK 获取授权微博账户的简介

此实例主要通过使用新浪微博 SDK,实现在登录微博账户成功之后,从当前应用获取新浪微博的个人账户的简介信息。当实例运行之后,单击"登录微博账户"按钮,则执行登录微博账户的操作,如果登录成功,则在弹出的 Toast 中显示"登录成功!",如图 297.1 的左图所示。只有在登录微博账户成功之后,单击"获取账户信息"按钮,才能在弹出的 Toast 中显示账户的简介信息,如图 297.1 的右图所示。

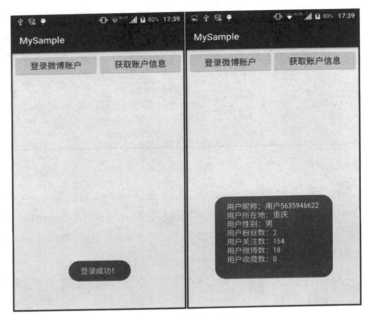

图 297.1

主要代码如下:

```
public class MainActivity extends Activity {
 AuthInfo myAuthInfo;
 SsoHandler myHandler;
 String myToken, myUid;
 @Override
 protected void onCreate(Bundle savedInstanceState) {
  super.onCreate(savedInstanceState);
  myAuthInfo = new AuthInfo(this, "639620816", "http://www.sina.com", null);
  WbSdk.install(this, myAuthInfo);                    //初始化微博 SDK
  setContentView(R.layout.activity_main);
 }
  //响应单击"登录微博账户"按钮
 public void onClickBtn1(View v) {
  myHandler = new SsoHandler(this);
  myHandler.authorizeClientSso(new WbAuthListener() {
   @Override
   public void onSuccess(Oauth2AccessToken token) {
    myToken = token.getToken();
    myUid = token.getUid();                    //获取授权 Token 和 Uid 值
```

```java
            Toast.makeText(MainActivity.this, "登录成功!", Toast.LENGTH_SHORT).show();
        }
        @Override
        public void cancel() { }
        @Override
        public void onFailure(WbConnectErrorMessage message) { }
    }); }
    //响应单击"获取账户信息"按钮
    public void onClickBtn2(View v) {
        JsonObjectRequest myJsonRequest = new JsonObjectRequest(
                "https://api.weibo.com/2/users/show.json?access_token=" + myToken +
                        "&uid=" + myUid, null, new Response.Listener<JSONObject>() {
            @Override
            public void onResponse(JSONObject response) {      //请求当前授权用户账户信息
                try {                                          //解析并显示用户信息
                    String myScreenName = response.getString("screen_name");
                    String myLocation = response.getString("location");
                    String myFollowersCount = response.get("followers_count").toString();
                    String myFriendsCount = response.get("friends_count").toString();
                    String myStatusesCount = response.get("statuses_count").toString();
                    String myFavoritesCount = response.get("favourites_count").toString();
                    String myGender = response.getString("gender").equals("m") ?
                            "男" : response.getString("gender").equals("f") ? "女" : "未知";
                    //通过 StringBuilder 拼接字符串内容
                    StringBuilder myBuilder = new StringBuilder();
                    myBuilder.append("用户昵称:" + myScreenName + "\n");
                    myBuilder.append("用户所在地:" + myLocation + "\n");
                    myBuilder.append("用户性别:" + myGender + "\n");
                    myBuilder.append("用户粉丝数:" + myFollowersCount + "\n");
                    myBuilder.append("用户关注数:" + myFriendsCount + "\n");
                    myBuilder.append("用户微博数:" + myStatusesCount + "\n");
                    myBuilder.append("用户收藏数:" + myFavoritesCount + "\n");
                    Toast.makeText(MainActivity.this,
                            myBuilder.toString(), Toast.LENGTH_SHORT).show();
                } catch (Exception e) { e.printStackTrace(); }
            } }, null);
        Volley.newRequestQueue(this).add(myJsonRequest);
    }
    @Override
    protected void onActivityResult(int requestCode, int resultCode, Intent data) {
        if (myHandler != null)                       //通过 SsoHandler 对象处理授权回调结果
            myHandler.authorizeCallBack(requestCode, resultCode, data);
    } }
```

上面这段代码在 MyCode\MySample937\app\src\main\java\com\bin\luo\mysample\MainActivity.java 文件中。需要说明的是,此实例需要在 MyCode\MySample937\app\libs 目录下添加库文件 core-4.1.0-openDefaultRelease.aar,然后在 MyCode\MySample937\app 目录下的 build.gradle 文件的根节点下添加如下代码:

```
repositories { flatDir { dirs 'libs' } }
```

并在该文件的 dependencies 节点下添加依赖项:

```
compile 'com.sina.weibo.sdk:core:4.1.0:openDefaultRelease@aar'
compile 'eu.the4thfloor.volley:com.android.volley:2015.05.28'
```

同时还要在 MyCode\MySample937\app\src\main\AndroidManifest.xml 文件中添加下列权限：

```
<uses-permission android:name="android.permission.INTERNET" />
<uses-permission android:name="android.permission.ACCESS_WIFI_STATE" />
<uses-permission android:name="android.permission.ACCESS_NETWORK_STATE" />
```

此外，使用新浪微博 SDK 服务需要在微博开放平台上创建应用并获取 AppKey，详细情况请登录新浪微博开放平台（http://open.weibo.com/development/mobile）。此实例的完整项目在 MyCode\MySample937 文件夹中。

298　使用新浪 SDK 将微博账户简介生成二维码

此实例主要通过使用新浪微博 SDK，实现将微博账户的简介信息生成二维码。当实例运行之后，单击"将微博账户的简介信息生成二维码"按钮，则执行登录微博账户的操作，然后将登录成功之后的微博账户的简介信息生成二维码，如图 298.1 的左图所示。使用微信扫描该二维码，则将显示该微博账户的简介信息，如图 298.1 的右图所示。

图　298.1

主要代码如下：

```
public class MainActivity extends Activity {
    AuthInfo myAuthInfo;
    SsoHandler myHandler;
    String myJsonParam = "{\"action_name\":\"QR_LIMIT_SCENE\","
                      + "\"action_info\":" + "{\"scene\":{\"scene_id\":123}}}";
    ImageView myImageView;
```

```java
@Override
protected void onCreate(Bundle savedInstanceState) {
    super.onCreate(savedInstanceState);
    myAuthInfo = new AuthInfo(this, "639620816", "http://www.sina.com", null);
    WbSdk.install(this, myAuthInfo);                    //初始化微博 SDK
    setContentView(R.layout.activity_main);
    myImageView = (ImageView)findViewById(R.id.myImageView);
}
//响应单击"将微博账户的简介信息生成二维码"按钮
public void onClickBtn1(View v) {
    myHandler = new SsoHandler(this);
    //设置授权登录回调监听器
    myHandler.authorizeClientSso(new WbAuthListener(){
        @Override
        //登录成功时的回调函数
        public void onSuccess(Oauth2AccessToken token){
            try{
                AccessTokenKeeper.writeAccessToken(MainActivity.this,token);
                JsonObjectRequest myTicketRequest =
                        new JsonObjectRequest(Request.Method.POST,
                        "https://api.weibo.com/2/eps/qrcode/create?access_token = " +
                        token.getToken(),new JSONObject(myJsonParam),
                        new Response.Listener<JSONObject>(){
                    @Override
                    public void onResponse(JSONObject response){
                        try{
                            ImageRequest myRequest = new ImageRequest(
                                "https://api.weibo.com/2/eps/qrcode/show?ticket = " +
                                response.get("ticket").toString(),new Response.Listener<Bitmap>(){
                                @Override
                                public void onResponse(Bitmap bitmap){
                                    myImageView.setImageBitmap(bitmap);}
                                },0,0,Bitmap.Config.RGB_565,null);
                            Volley.newRequestQueue(MainActivity.this).add(myRequest);
                        }
                        catch(JSONException e){e.printStackTrace();}
                    } },null);
                Volley.newRequestQueue(MainActivity.this).add(myTicketRequest);
            }
            catch(Exception e){e.printStackTrace();}
        }
        @Override
        public void cancel(){}
        @Override
        public void onFailure(WbConnectErrorMessage wbConnectErrorMessage){}
    }); }
@Override
protected void onActivityResult(int requestCode,int resultCode,Intent data){
    super.onActivityResult(requestCode,resultCode,data);
    //通过 SsoHandler 对象处理回调结果
    if(myHandler!= null) myHandler.authorizeCallBack(requestCode,resultCode,data);
} }
```

上面这段代码在 MyCode \ MySample938 \ app \ src \ main \ java \ com \ bin \ luo \ mysample \ MainActivity.java 文件中。有关权限问题请参考实例 296 进行修改。此实例的完整项目在 MyCode\ MySample938 文件夹中。

299　使用新浪 SDK 实现搜索指定关键字的微博

此实例主要通过使用新浪微博 SDK，实现根据搜索关键字在新浪微博中搜索指定的内容。当实例运行之后，在"关键字："输入框中输入关键字，如"工商约谈"，如图 299.1 的左图所示；然后单击"搜索微博"按钮，则该关键字的搜索结果将出现在微博列表中，如图 299.1 的右图所示。

图　299.1

主要代码如下：

```java
public class MainActivity extends Activity {
    AuthInfo myAuthInfo;
    EditText myEditText;
    @Override
    protected void onCreate(Bundle savedInstanceState) {
        super.onCreate(savedInstanceState);
        myAuthInfo = new AuthInfo(this,"639620816","http://www.sina.com",null);
        WbSdk.install(this, myAuthInfo);              //初始化微博 SDK
        setContentView(R.layout.activity_main);
        myEditText = (EditText) findViewById(R.id.myEditText);
    }
    //响应单击"搜索微博"按钮
    public void onClickBtn1(View v) {
        //获取微博工具类实例,并通过该实例对象根据指定关键字搜索微博
        WeiboPageUtils.getInstance(MainActivity.this,myAuthInfo).
                openWeiboSearchPage(myEditText.getText().toString(),false);
    }
}
```

上面这段代码在 MyCode\MySample930\app\src\main\java\com\bin\luo\mysample\MainActivity.java 文件中。有关权限问题请参考实例 296 进行修改。此实例的完整项目在 MyCode\MySample930 文件夹中。

300 使用新浪 SDK 实现发布图像至微博

此实例主要通过使用新浪微博 SDK,实现将图像发布到新浪微博。当实例运行之后,单击"选择图像"按钮,然后在弹出的窗口中选择一幅图像,则该图像将显示在下面的 ImageView 控件中,如图 300.1 的左图所示。单击"分享至新浪微博"按钮,则在 ImageView 控件中显示的图像将被分享到新浪微博中,如图 300.1 的右图所示。

图 300.1

主要代码如下:

```java
public class MainActivity extends Activity {
    AuthInfo myAuthInfo;
    ImageView myImageView;
    WbShareHandler myHandler;
    @Override
    protected void onCreate(Bundle savedInstanceState) {
        super.onCreate(savedInstanceState);
        myAuthInfo = new AuthInfo(this, "639620816", "http://www.sina.com", null);
        WbSdk.install(this, myAuthInfo);                    //初始化微博 SDK
        myHandler = new WbShareHandler(this);
        myHandler.registerApp();                            //注册应用
        setContentView(R.layout.activity_main);
        myImageView = (ImageView) findViewById(R.id.myImageView);
    }
    public void onClickBtn1(View v) {                       //响应单击"选择图像"按钮
        Intent myIntent = new Intent(Intent.ACTION_PICK);
```

```java
    myIntent.setType("image/*");
    startActivityForResult(myIntent,0);
}
@Override
protected void onActivityResult(int requestCode, int resultCode, Intent data){
    myImageView.setImageURI(data.getData());        //通过 ImageView 控件预览所选图像
}
public void onClickBtn2(View v) {                   //响应单击"分享至新浪微博"按钮
    ImageObject myImageObject = new ImageObject();
    myImageObject.setImageObject(((BitmapDrawable)
        myImageView.getDrawable()).getBitmap());    //通过 ImageObject 存储图像内容
    WeiboMultiMessage myMessage = new WeiboMultiMessage();
    myMessage.imageObject = myImageObject;
    myHandler.shareMessage(myMessage,false);        //分享指定图像至新浪微博
} }
```

上面这段代码在 MyCode\MySample933\app\src\main\java\com\bin\luo\mysample\MainActivity.java 文件中。有关权限问题请参考实例296进行修改。此实例的完整项目在 MyCode\MySample933 文件夹中。